Concise Handbook of Fluorocarbon Gases

Scrivener Publishing
100 Cummings Center, Suite 541J
Beverly, MA 01915-6106

Publishers at Scrivener
Martin Scrivener (martin@scrivenerpublishing.com)
Phillip Carmical (pcarmical@scrivenerpublishing.com)

Concise Handbook of Fluorocarbon Gases

Applications in Refrigeration and Other Industries

Sina Ebnesajjad, PhD
President, FluoroConsultants Group, LLC
Chadds Ford, Pennsylvania, USA

Scrivener Publishing

WILEY

This edition first published 2021 by John Wiley & Sons, Inc., 111 River Street, Hoboken, NJ 07030, USA and Scrivener Publishing LLC, 100 Cummings Center, Suite 541J, Beverly, MA 01915, USA

For more information about Scrivener publications please visit www.scrivenerpublishing.com.

Wiley Global Headquarters
111 River Street, Hoboken, NJ 07030, USA

For details of our global editorial offices, customer services, and more information about Wiley products visit us at www.wiley.com.

Library of Congress Cataloging-in-Publication Data

ISBN 978-1-119-32297-9

Cover image: Pixabay.Com
Cover design by Russell Richardson

Set in size of 11pt and Minion Pro by Manila Typesetting Company, Makati, Philippines

MIX
Paper from responsible sources
FSC www.fsc.org FSC® C013604

Contents

Preface

I have worked in the fluoropolymers industries for the better part of four decades. During those years I have encountered fluorocarbon gases as monomers of the fluoropolymers. In time I learned about the detrimental impact of some of those gases on the ozone layer and contribution to global warming. Curiosity led me to attempt educating myself about those impacts. The facts relied on the well-known basic chemical reactions of chlorine free radicals with ozone and as such they were undeniable. Interest in the environmental impact of chlorofluorocarbons gases ignited my long apprenticeship in this field. I set a goal to learn about the topics related to commercial applications of fluorocarbons.

Today, one often hears about the ozone layer, global warming, climate change, recycling and sustainability. Some of these terms had not yet been coined at the onset. In the1970s there was little if any reported news about the ozone monitoring that had been started by the scientists from the British Antarctic Survey who began monitoring ozone in 1957. Paul Crutzen, Mario Molina and Sherwood Rowland raised concerns about the effect of man-made chemicals, especially chlorofluorocarbons, on the Ozone Layer as early as in the 1970's, to little effect. Exception is the unilateral move by the US Environmental Protection Agency banning the production of chlorofluorocarbons in 1978.

Then came the momentous article written by Joe Farman, Brian Gardiner and Jonathan Shanklin, in the prestigious journal *Nature* in May 1985. The authors reported the discovery of the annual depletion of ozone above the Antarctica for the first time (Figure). The information withstood reanalysis by NASA scientists who confirmed the findings. Loss of atmospheric ozone would weaken/remove the filter for harmful ultraviolet light thus subject the entire earth and the living organisms to excess UV radiation, expected to lead to known and unknowns damages. The subject caught the attention of the public. Something had to be done.

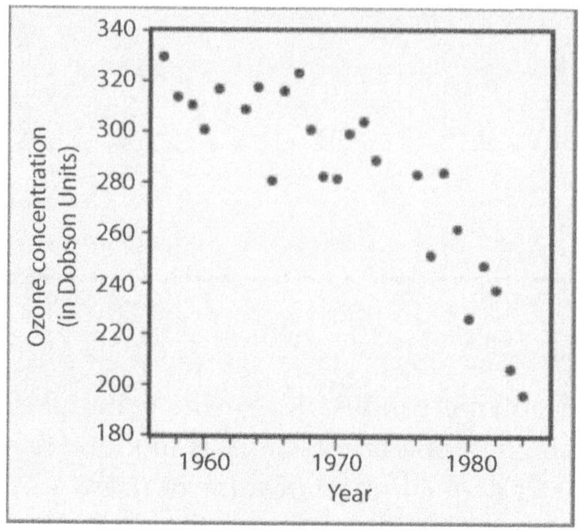

Source: https://undsci.berkeley.edu/article/0_0_0/ozone_depletion_09

[The international community was uncharacteristically quick to act, perhaps because the seemingly sudden appearance of a "hole" in the atmosphere made for such a compelling and easily understandable story. Within two years, in direct response to the *Nature* article and corroborating studies, 46 nations signed the Montreal Protocol, pledging to phase out substances known to cause ozone depletion. All 197 members of the United Nations would eventually ratify the treaty…], as described by the www.History.com. Today, over 98% of the 100 ozone depleting chemicals have been phased out. A great deal of change has taken place in the decades since 1987 to not only eliminate/minimize ozone depleting substances but reduce the emissions of global warming gases.

Many treaties have followed the original Montreal Protocol and colossal numbers of scientific articles have been published. Learning about the technological changes and the state of the ozone layer is a bewildering task. An in depth understanding requires the knowledge of such a large volume of literature that it is impractical even for the practitioners in the areas of fluorocarbon gases. That was the dilemma the author faced in his quest to learn about the topics surrounding the ozone layer. There was no single book to read or refer to for learning. Not finding a source, in 2016 the author set out to write the book he wished he could buy and read: in-depth coverage of important topics yet concise to fit practical time limits. The focus of this book is placed on the commercial and practical subjects as opposed to the theoretical ones.

The outcome of the author's efforts during nearly five years is the book before you. The contents begin at the ground level by providing an introduction to fluorocarbons, that is fluorinated small molecules and polymeric materials. The rest of the book is focused on industrial fluorocarbon, gases and liquids. The issue of ozone depletion is discussed in conjunction with the subject of most chapters to ensure clarity of the relationship of the compounds with the current rules and regulations. In the early chapters some basic topics have been covered including fluorine sources, basic reactions for fluorine entry into organic compounds and applications of fluorocarbon gases and liquids.

Emphasis has been placed on refrigeration though other applications have also been discussed. The preparation methods and properties of fluorocarbons occupy Chapters 6 and 7. Chapter 8 delves into the topics of ozone depletion and global warming by discussing the various treaties and changes since 1987. A reader should be able to gain a complete understanding of the current state of technology and the future directions as foreseen by experts by reading the Chapters 8 and 11. Chapter 9 contains a great deal of data about the industrial fluorocarbons blends. Chapter 10 focuses on fluorocarbon substitutes, including non-fluorinated substances, in current use in the industry.

I would like to thank all authors, companies, government agencies, international organizations and others who have contributed data and information to this book. The author has made every effort to acknowledge the contributors in the references section of each chapter and throughout the narrative.

Last but not least I would like to offer my sincere apologies for any errors the readers may find in this book. The errors are the responsibility of the author. A note to the publisher would be most appreciated so that corrections could be made in the future editions.

I would like to thank Scrivener Publishing for the contracting and production of the book and John Wiley & Sons for its distribution and marketing.

Sina Ebnesajjad, PhD
www.FluoroConsultants.com
Chadds Ford, Pennsylvania
January 2, 2021

1
Introduction

Dictionaries define *fluorocarbons* as any of various chemically inert compounds containing mainly carbon and fluorine. They are used to produce refrigerants, fire extinguishing and foaming agents, aerosol propellants, polymers, nonstick coatings and lubricants. In the industry parlance the term fluorochemicals is often used instead of fluorocarbon. Rather *fluorocarbons* usually refer to small molecule fluorinated gases and liquids. The magic in these compounds of all sizes is in the carbon fluorine bond (C-F). It has a dissociation energy of 536 kJ/mol ranking as the single strongest carbon bond [1–4]. The replacement of C-H bond in hydrocarbons with C-F is the single most important factor giving rise to the unique properties of fluorocarbons.

1.1 Terminology

In this book the terms *fluorochemicals* and *fluorocarbons* refer to all aliphatic organic compounds containing fluorine and carbon, i.e. fluorinated chemicals. Aliphatic is defined as straight chain, branched chain, or cyclic, saturated, as in the paraffin; or unsaturated, as in the olefins and alkynes. Cyclic fluorocarbons are not discussed in this book because their limited commercial applications.

More specifically the term fluorocarbon refers to small molecule of fluorinated alkane and olefin compounds. Fluorinated polymers and fluoropolymers are used to refer to, mostly, olefinic macromolecules of fluorinated thermoplastics and elastomers. *Partially fluorinated* refers to a material that contains residual C-H or other bonds. *Perfluorinated* refers to the absence of hydrogen in a chemical due to complete fluorine substitution. Other atoms such as oxygen, sulfur and nitrogen may be present in the structure of the perfluorinated chemicals, often in small quantities.

Commercial is the guiding beacon for the selection of the contents of the present book. The focus is on fluorochemicals produced and consumed in commercial applications. An important aspect of the commercial consumption of fluorocarbons is their impact on the ozone layer as a result of emissions of those chemicals. The issue came to surface when chlorofluorocarbons were discovered to deplete atmospheric ozone layer. That discovery has resulted in decades of development of replacement compounds and elimination of all ozone depleting fluorocarbons.

The three major groups of commercial products consumed in large quantities are listed in Table 1.1. Of the major fluorinated organic products, only fluorocarbons have small molecules and interact with the ozone layer. Some of the precursors of fluoropolymers

Sina Ebnesajjad, PhD. Concise Handbook of Fluorocarbon Gases: Applications in Refrigeration and Other Industries, (1–12) © 2021 Scrivener Publishing LLC

Table 1.1 Examples of commercial fluorochemicals.

Product	Size of molecule	Example of compound	Example of application
Fluorocarbons	Small	HFC-134a (1,1,1,2-tetrafluoroethane)	Refrigerant, Propellant for aerosols, Blowing agent for foams
Thermoplastic Fluoropolymers (TPF) and Fluoroelastomers (FE)	Large macromolecules	Polyvinylidene fluoride (TPF), copolymer of vinylidene fluoride/ tetrafluoroethylene	Tubes, pump bodies (TPF), seals and gaskets (FE)
Fluorinated Coating - Dispersion (D), powder (P)	Large macromolecules	Polytetrafluoroethylene coatings, Tetrafluoroethylene/ perfluoroalkylvinylether copolymer powder	Interior surface of oil pipelines (D), Corrosion resistant (P)

and fluoroelastomers impact the ozone layer thus must be contained while being converted to polymers.

The terms *ozone depletion* refer to the destruction of atmospheric ozone by the decomposition products of certain gases. Ozone by its action filters out low wavelength ultraviolet light waves known as "UV B" which are hazardous to humans, and some plants and wildlife. Destruction of the atmospheric ozone reduces its filtration action thus allowing an increase in the UV B light waves that reach the earth surface. The First generation of fluorocarbon gases were discovered to be strong causes of ozone depletion. A great deal of global regulatory changes, research, publication, development and replacement have taken place since the 1980s. These issues are discussed in detail in the rest of this book.

Most public and technical discussions about ozone layer and depletion are not quantitative. Nevertheless, atmospheric ozone has been measured to allow factual analysis of changes in the ozone layer. The Dobson Unit (DU) is the unit of measure for total ozone ranging from 100 to 500. If one were to take all the ozone in a column of air stretching from the surface of the earth to space (Figure 1.1) and bring all that ozone to 0°C and pressure of one atmosphere, the column would be about 0.3 centimeters thick. Thus, the total ozone would be 0.3 atm-cm. To make the units easier to work with, the "Dobson Unit" is defined to be 0.001 atm-cm, 0.3 atm-cm would be 300 DU [5].

Incidentally, Dobson Unit is named after Gordon Miller Bourne Dobson (1889-1976) who was a British physicist and meteorologist. Professor Dobson was an ingenious experimentalist who devoted much of his life to the observation and study of atmospheric ozone. He made measurements at a number of locations in Europe to study the relation between ozone distribution and synoptic meteorological variables. The results of Dobson's studies turned out to be of great importance. They lead to an understanding of the structure and circulation of the stratosphere including the ozone layer [7].

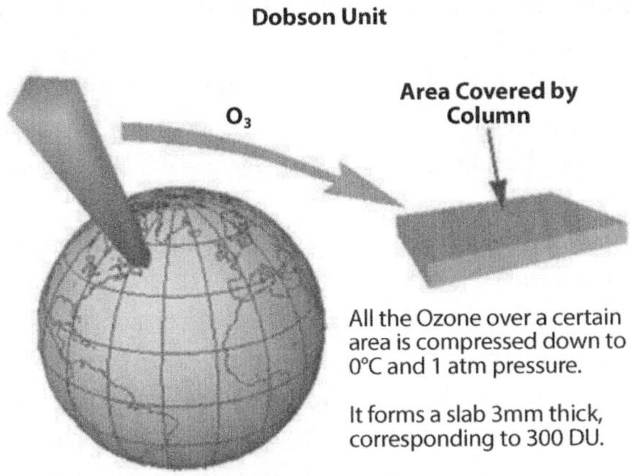

Figure 1.1 Depiction of Dobson Unit for ozone quantitative measurement [6].

1.2 Production and Consumption Statistics of Fluorocarbons

Statistics on the production and consumption of fluorocarbons illustrate the continued growth of those products. A major study published by the US Government in February 2020 sheds light on the global development, trade and consumption of fluorocarbon gases and their alternatives. This study is the result of collaboration of Department of Energy's *National Renewable Energy Laboratory* and *Oak Ridge National Laboratory* [8] Some of the highlights of this study are described in the next Section (1.2.1).

1.2.1 Refrigerants: Market Trends and Supply Chain Assessment

This section has been adopted with some modifications from a Department of energy study published in May 2020 [8]. The conclusions of the reported study are listed in this section. The term refrigerants has a broad meaning in this report because most of the *refrigeration* gases are also used in *non-refrigeration* applications.

The global refrigerants market is large and is projected to grow rapidly as developing countries in warmer areas of the globe continue to grow, become more affluent, and consume more and more air conditioning, refrigeration, foam, and aerosol products and services. Innovations in the global refrigerants market is often led by major U.S. companies; however, the markets for their products are global. Understanding this global market landscape is a critical component for maintaining U.S. leadership in innovation and manufacturing in a strategically important industry.

Key Findings from the National Renewable Energy Laboratory (NREL)/Oak Ridge National Laboratory (ORNL) include:

- Refrigerant markets are global and growing rapidly. 2010–2050: 4.5x increase in air conditioning for non-Organization of Economic Coordination and Development (OECD) countries (developing economies) and 1.3x increase for OECD countries (developed economies).
- Regional, national, and international commitments will create large market opportunities for innovative refrigerants and products that use them.

- U.S.-based companies are leaders in intellectual property and production of advanced refrigerants (Figure 1.2).
- China has aggressively expanded production of conventional refrigerants for domestic use as well as export.
- Refrigerants are used in large quantities for more than just cooling (Figure 1.3).
- Foam production, aerosols, fire suppression, and chemical production are important end uses for these materials.
- Vapor compression systems primarily use fluorocarbon refrigerants. This is the most difficult and impactful area for refrigerant innovation (Figure 1.4).
- Alternative refrigerants to fluorocarbons are well established.
- The refrigerant market is defined as all materials used in applications where fluorocarbons are used: vapor compression, foam blowing, aerosols, chemical feedstock, fire suppression, and solvents.
- Alternative refrigerants comprise more than 50% of the total market as it exists today (Figure 1.5).
- Common natural/hydrocarbon refrigerants include ammonia, pentane, carbon dioxide, propane, and butane. They have substantially lower global warming potential than most fluorocarbons.

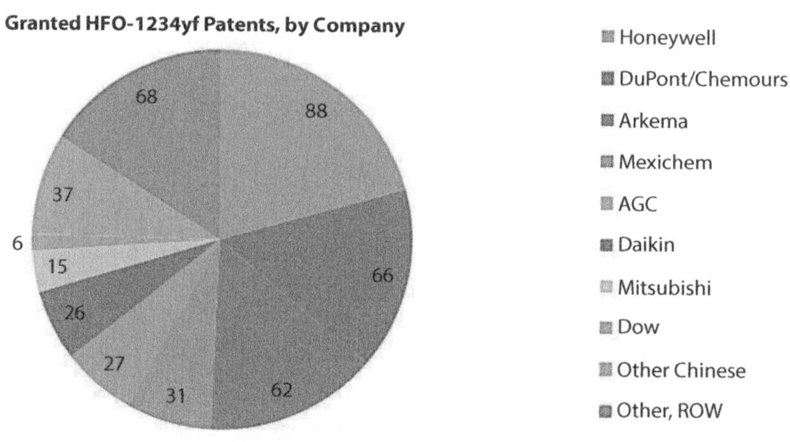

Figure 1.2 Granted patents related to HFO-1234yf [8].

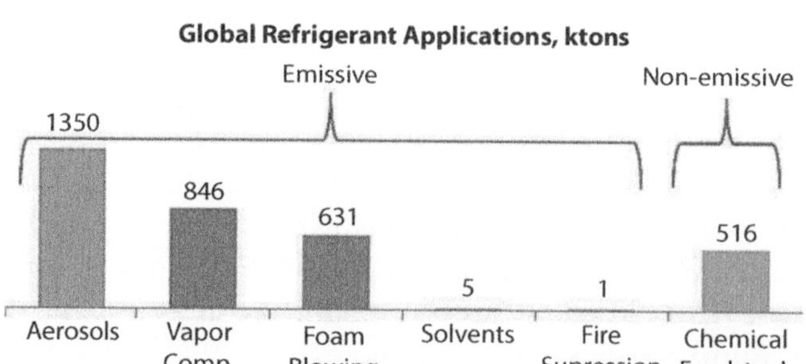

Figure 1.3 Global refrigerant usage, by application [8].

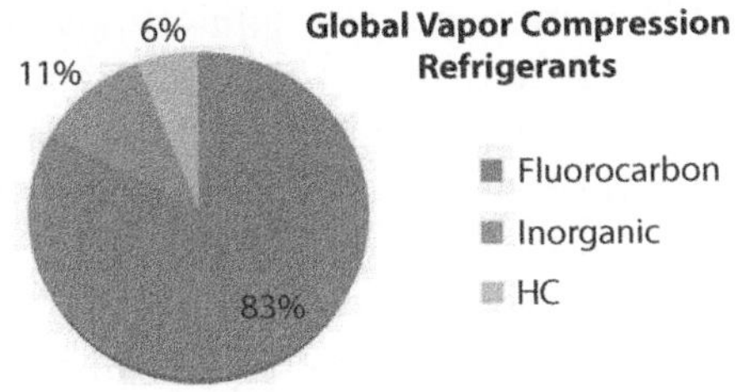

Figure 1.4 Global vapor compression refrigerant types [8].

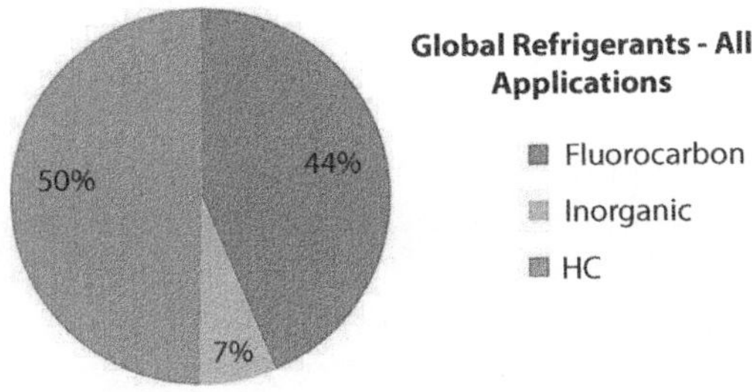

Figure 1.5 Global refrigerants for all applications [8].

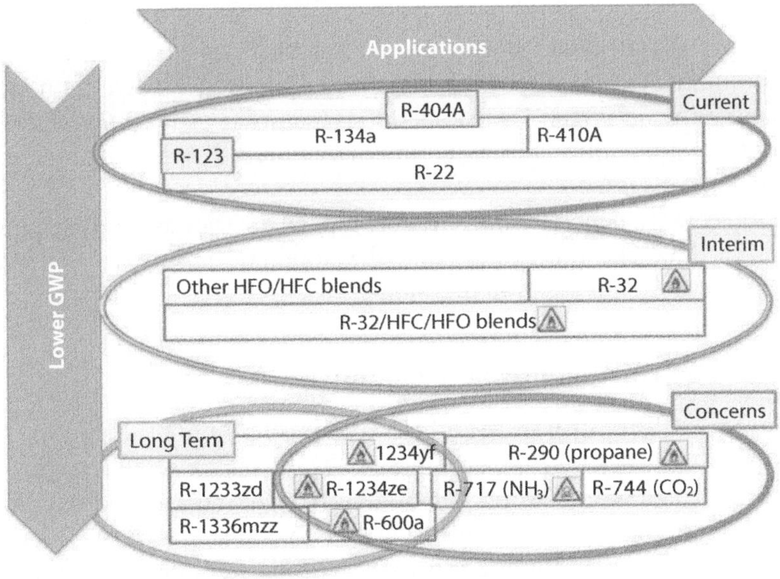

Figure 1.6 Groupings of refrigerants by GWP [8].

- Nonfluorinated refrigerants may provide comparable or superior performance to fluorocarbons in some end uses.
- Advanced fluorocarbon refrigerants are commercially available that reduce environmental impact while maintaining or potentially improving performance (Figure 1.6).

- One size doesn't fit all—some common applications are more difficult to solve than others; this necessitates ongoing research and development.
- U.S. companies are currently at the forefront of innovations.
- The hydrochlorofluorocarbon phaseout is nearing completion in developed countries; attention has formally turned to the phasedown of high GWP hydrofluorocarbons with the passage of the Kigali Amendment in 2016. Significant global adoption of advanced fluorocarbon refrigerants and hydrocarbon alternatives will be instrumental to the success of this imminent HFC phasedown.
- Refrigerant manufacturing locations are primarily guided by:
    - Proximity to fluorspar, hydrofluoric acid, or other chemical feedstock
    - Existing refrigerant manufacturing capital and experienced labor force
    - Availability of cheap energy and labor
    - Financial incentives from local governments or development authorities.
- The United States is positioned to be a major production center for advanced refrigerants, including hydrofluoroolefins (HFOs) and their blends.
- Market share of production is likely to be larger if there is a substantial U.S. market for advanced refrigerants.
- Fluorspar will remain in demand as manufacturers transition to producing HFOs, and the U.S. fluorspar supply is stable. Demand is now supplied mainly by Mexico, whereas historically, China had been the leading supplier.
- U.S. companies, such as Honeywell and Chemours, own much of the intellectual property associated with the production and usage of HFOs.
- Antidumping lawsuits regarding Chinese imports have played a major role in shaping this industry. Decisions on HFC imports have generally been in favor of U.S. companies, setting an important precedent for any future HFO-related trade disputes.
- Recently constructed HFO production capacity serves as an example of the effect that financial incentives from development authorities can have on manufacturers' plant location decisions.

1.2.2 Fluorocarbon Consumption Demand

Global demand for refrigerants is expected to exceed $30 Billion by 2025 at a compounded annual growth rate (CAGR) of 5.3% from 2018 to 2025. The growth of refrigerants is driven my various factors such as increasing demand of refrigerants in Asia-pacific, increasing demand of cooling products, and growth in global cold chain market. The growing end-user industries such as construction, pharmaceutical, automobile, oil & gas, and food are also driving the refrigerant marke [9].

The fluorocarbon segment is witnessing a slow growth due to stringent environmental regulations such as F-Gas regulation [26, 27], and Montreal Protocol, which demands the phasing out of such harmful gases. Countries such as North America have already started the phasing out of HCFCs and HFCs, due to their harmful effect on environment. These strict regulations provide a wide opportunity to refrigerants such as inorganics and hydrocarbons. Inorganics chemicals have the second largest market share and are also expected to have the highest growth rate between 2015 and 2020.

The Asia-Pacific region remains the largest market for refrigerants, followed by Americas, and Europe, which are comparatively very small market as compared to Asia-Pacific. Asia-Pacific refrigerant market is also expected to witness the highest CAGR of 7.9% from 2015 to 2020. The refrigerants are widely used in various end-user industries such as automobile, construction, chemicals, and oil & gas and these industries are booming in the Asia-Pacific region. Asia Pacific is home to both, the largest and the fastest-growing car markets. Construction is also a key emerging market in Asia-Pacific, which is witnessing a significant growth because of the presence of major economies such as China, India, and Japan [10].

Europe accounted for over 10% of the total demand in 2014 and is anticipated to grow at below average growth rates of over 3% from 2015 to 2022. High usage in applications such as pharmaceutical and household surfactants is expected to drive the demand. The regulations such as waste proposals by EU and adoption of Montreal Protocol are anticipated to slowdown its use in Europe [11].

America's fluorocarbon *refrigerants* market was valued at $2.5 billion in 2013 and is estimated to reach $3.1 billion by 2019, at a CAGR of 3.6%, for the given period. The growing demand for chilled and frozen food and beverages has fuelled demand for commercial refrigeration, consequently driving demand for fluorocarbon refrigerants in the region [12]. The global refrigerants market size was US$22.9 billion in 2018 and is projected to reach USD 31.0 billion by 2023, at a CAGR of 6.2% between 2018 and 2023 [13].

Blowing agent application accounted for over 9.5% of the total volume in 2014 and is likely to witness gains at CAGR of over 3% from 2015 to 2024. Fluorochemicals produce hydrogen fluoride, which is broadly used in production of fluorinated hydrocarbons, which serves as foam blowing agents. This is used in wide range of plastics like polystyrene and polyurethane for improving their insulation properties in industrial and domestic appliances such as cooling plants, food processing equipment, refrigerators, cold storage rooms and packaging materials [11].

The largest refrigerant manufacturing countries are China (689,000 metric tons), the USA (604,000 metric tons), Japan (81,000 t metric tons), Germany (32,000 metric tons), India (30,000 metric tons) and the Netherlands (30,000 M tons). The refrigerant quantities produced are not just used in the ventilation, air conditioning and refrigeration technology sector however, but are also used for other industrial processes such as those carried out for aerosols, foams and insulation [14].

The world's most important refrigerant manufacturers include Arkema, Chemours, Honeywell, Solvay, Daikin, Showa Denko and Mexichem, as well as numerous Chinese manufacturers, some of which are Sinochem Lantian, Shandong Dongya and 3F Fluorochemical Industry, a joint venture between DuPont and Zhonghao New Materials [14].

1.3 Production and Consumption Statistics of Fluoropolymers

Extreme properties of fluoropolymers have made them among the most useful modern materials. They provide: nonstick surfaces for cookware and industrial products, waterproofing surface treatments for clothing and other substrates, stain barriers for textiles, high-purity fluid handling in the form of tubes, pipes, gaskets, seals and thread sealant tape, medical devices, wire and cable insulation jackets, high-performance coatings, back sheet for photovoltaic panels, films and membranes for technical, waterproof clothing, and

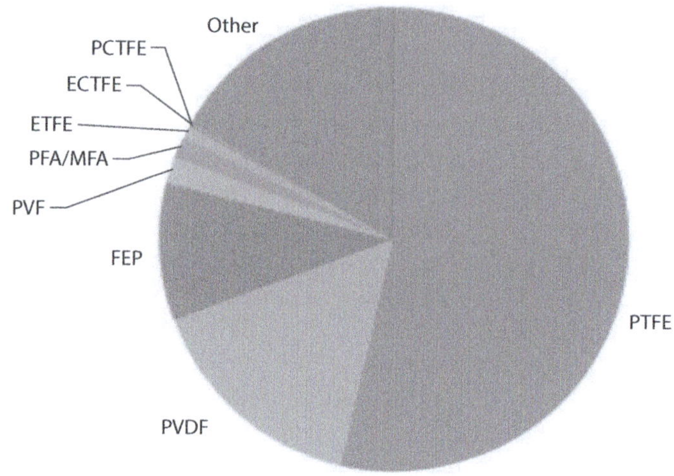

Figure 1.7 Breakdown (2018) of worldwide consumption of fluoropolymers by Type [15].

industrial applications. Figure 1.7 shows the breakdown of consumption of fluoropolymers by type.

PTFE at 58% (in 2015) has the highest consumption volume among fluoropolymers. Significantly, PTFE is the oldest product of all fluoropolymers its commercialization dating back to 1946. Polyvinylidene fluoride (PVDF) has the second highest volume (under 20%) followed by perfluorinated ethylene propylene copolymers (FEP) at about 10%. PVDF and FEP are fastest growing fluoropolymers.

A number of fluoropolymers are consumed at lower volumes including: polychlorotrifluoroethylene (PCTFE), ethylene tetrafluoroethylene copolymer (ETFE), pefluoropropyl vinyl ether copolymer (PFA), pefluoromethyl vinyl ether copolymer (MFA), polyvinyl fluoride (PVF), ethylene chlorotrifluoroethylene copolymer (ECTFE), tetrafluoroethylene-hexafluoropropylene-vinylidene fluoride terpolymer (THV), chlorotrifluoroethylene vinylidene fluoride copolymer, CTFE-VDF (<15% CTFE), and amorphous fluoropolymers. "Other" in Figure 1.1 includes fluorosilicones, fluorinated polyimides and polyamides, fluorinated acrylic polymers, polyurethanes and amorphous fluoropolymers and others.

Consumption of fluoropolymers, fluorocarbons, inorganics and specialties expected to reach 5.49 million tons by 2024, growing at a CAGR of 4.6% from 2015 to 2024 [11]. Global fluoropolymer market in terms of revenue was at $7.2 billion in 2012 and is expected to reach $9.8 billion by 2018 and to US$11.5 million by 2025. Projections for future growth rate of fluoropolymer consumption are in the annual range of 5.2-5.7%. Global consumption volume is projected to grow from 295 kT (1,000 metric tons) in 2014 to over 380 kT by 2019 [16].

China is the by far the most dominant consumer of PTFE, accounting for about 44% of world consumption in 2015. The United States and China each use about 30% of other fluoropolymers [17]. The historical growth rate of PTFE in China has been around 7.5% per year while the rest of fluoropolymer have grown at a rate of 15%.

Fluoropolymer consumption is relatively small in India. The entire value in 2014 was \$100 million, which is expected to \$250 by 2022 [18].

1.4 Production and Consumption Statistics of Fluoroelastomers

Fluoroelastomers are divided into perfluorinated and partially fluorinated polymers. They possess a combination of good chemical resistance, excellent electrical insulation properties and favorable mechanical properties. Fluoroelastomers are highly weather resistant and can function under extreme temperature and pressure.

Applications of fluoroelastomers are increasingly because of the performance advantages these materials offer. They have a combination of physical and chemical properties, comprising excellent chemical resistance, high temperature stability, and good dielectric properties. Fluoroelastomers are extensively used in industrial applications involving severe process conditions. The special properties and relatively low production volume, compared with engineering resins, have kept fluoroelastomer prices are fairly high [19].

Based on volume (base raw material), China accounted for 33% of world consumption in 2018, with Europe and the United States accounting for 19% each, and Japan for 9%. These four regions represented about 80% of the world consumption of fluoroelastomers in 2018 [20].

Fluoroelastomer market size was valued at US\$1.15 billion in 2016, US\$1.21 billion in 2017 and is projected to reach US\$1.64 billion by 2022, representing a CAGR of 6%. Asia-Pacific, with its thriving economies and rapidly expanding populations experienced the fastest growth rate between 2014 and 2019 [19]. Fluoroelastomers consumption is estimated at 24.0 kilo tons (kT) in 2014 reaching 30.4 kT by 2019.

China accounted for 33% of world consumption in 2018 mostly because of growth in its automobile production. Europe and the US account for 19% each, and Japan for 9%. These four regions represented about 80% of the world consumption of fluoroelastomers in 2018. Fluorocarbon elastomers (FKMs) account for about 92% of the world market for fluoroelastomers. Fluorosilicones account for almost 8% of the world market for fluoroelastomers.

1.5 Production and Consumption Statistics of Fluorinated Coatings

Fluoropolymers produce high-performance coatings for harsh industrial environments; mar-free coatings for touch screen electronic devices, architectural and marine coatings and cookware coatings. The coatings are applied as powder or liquid dispersions. Major examples of coating fluoropolymers include PTFE, PFA, FEP, PVF and PVDF. China and the U.S. are the largest consumers of fluoropolymer coating, while Italy, Japan, and the Germany are other key markets for fluoropolymer coating consumption.

The fluoropolymer coatings market is estimated to reach \$2 billion in 2025 indicating annualized growth of 6% between 2019 and 2025. PVDF based coating systems are highly durable coating systems extensively used for coating architectural aluminum surfaces as well as post-formed steel sheets (metal coils). North America fluoropolymer coating market is expected to be the largest market, significantly growing over a CAGR of 5.0% [21].

1.6 Specialty Fluorochemicals

There are many types of specialty fluorochemicals with important applications. They are usually quite expensive and are used in small quantities. Here are some examples. Fluorinated surfactants, significant as they are, are consumed in relatively small quantities. They have not been included in this book because there are already competent books about them [22].

Another group of "fluorinated" compounds are drugs and pharmaceuticals, nearly 20%, containing one or more atoms of fluorine in their structures. Examples of common medicine containing fluorine include Prozac, Cipro and Paxil. In spite of minor presence in the structure of drugs fluorine atom affects the characteristics of pharmaceutical substantially. They are not covered in this book because a separate volume would be needed for their coverage. Journal articles and other books continue to publish information about research and development of new fluorinated chemicals synthesized [23–25].

Another important group of fluorocarbons are used in surgeries as inhalation anesthesia. The benefits over the old compounds include non-toxicity, non-flammability, general metabolically stability, ease of detoxification in the liver, rapid recovery and very few side effects. Commercial examples of those fluorocarbons include methoxyflurane, isoflurane, sevoflurane and desflurane. A popular compound was enflurane which is no longer commonly used but it is effective in relaxing the uterus in pregnant women.

References

1. Cottrell, T.L., *The Strengths of Chemical Bonds*, 2d ed., Butterworth, London, 1958.
2. Darwent, B. deB., National Standard Reference Data Series, National Bureau of Standards, no. 31, Washington, 1970.
3. Benson, S.W., *J. Chem. Educ.*, 42, 502, 1965.
4. Kerr, J.A., *Chem. Rev.*, 66, 465, 1966.
5. National Aeronautics and Space Administration (NASA), https://ozonewatch.gsfc.nasa.gov/, May 2020.
6. The Ozone Hole, www.theozonehole.com/dobsonunit.htm, May 2020.
7. Dobson, G.M.B., Dept of Physics, Oxford University, May 2020.
8. Booten, C., Nicholson, S., Mann, M., Abdelaziz, O., *Refrigerants: Market Trends and Supply Chain Assessment*, Clean Energy Manufacturing Analysis Center, Department of Energy, NREL/TP-5500-70207, February 2020.
9. *Refrigerant Market Size Worth $30.37 Billion By 2025 at CAGR: 5.3%, Refrigerant Market Size & Share Report By Product (Fluorocarbon, Hydrocarbon, Inorganic)*, Grand View Research, www.grandviewresearch.com, May 2020.
10. *Refrigerant Market by Type, and by Application - Global Trends & Forecasts to 2020*, MarketsandMarkets, www.marketsandmarkets.com.
11. Global Market Insights, www.gminsights.com, Fluorochemicals Market Size by Product, by Application, Industry Outlook Report, 2015–2024.
12. Micromarket Monitor, www.micromarketmonitor.com, Nov 2019.

13. *Global $31 Billion Refrigerants Market Forecast to 2023*, published by CISION PR, www.prnewswire.com, March 2019.
14. CHILLVENTA Insights - The Market Analysis, www.chillventa.de, 2019.
15. *Chemical Economics Handbook*, HIS Markit, https://ihsmarkit.com, 2019.
16. Fluoropolymers Market - Global Industry Trends & Forecasts to 2019 www.marketsandmarkets.com, 2019.
17. *Fluoropolymers - Chemical Economics Handbook (CEH) |HS*, www.ihs.com, 2019.
18. India Fluoropolymer & Fluorosurfactant Market Report, www.grandviewresearch.com, 2019.
19. Fluoroelastomer Market - Global Trends & Forecast to 2019, www.marketsandmarkets.com, May 2019.
20. *Fluoroelastomers, Chemical Economics Handbook*, IHS Markit, https://ihsmarkit.com/products/fluoroelastomers-chemical-economics-handbook.html, April 2019.
21. Fluoropolymer Coating Market Report - Global Forecast till 2025, www.marketresearchfuture.com/reports/flouropolymer-coating-market-4844, ID: MRFR/CnM/3416-HCR, May 2020.
22. Kissa, E., *Fluorinated Surfactants: Synthesis, Properties, Applications*, 2nd ed, CRC Press, 2001.
23. Gakh, A. and Kirk, K.L., *Fluorinated Heterocycles*, 1st ed, American Chemical Society, 2009.
24. Knepper, T.P. and Lange, F.T., Polyfluorinated Chemicals and Transformation Products, in: *The Handbook of Environmental Chemistry*, 2012 Ed, Springer, 2011.
25. Ameduri, B. and Boutevin, B., *Well-Architectured Fluoropolymers: Synthesis, Properties and Applications*, 1st ed, Elsevier Science, 2004.
26. Regulation (EC) No 842/2006 of The European Parliament and of the Council of 17 May 2006 on certain fluorinated greenhouse gases. *Off. J. Eur. Union*, 2006.
27. Directive 2006/40/EC of The European Parliament and of the Council of 17 May 2006 relating to emissions from air-conditioning systems in motor vehicles and amending Council Directive 70/156/EC. *Off. J. Eur. Union*, 2006.

2

Classification and Description of Commercial Fluorinated Compounds

Fluorine is present in the air we breathe, the water we drink and the rocks beneath us, as well as in numerous synthetic polymers, pharmaceuticals, agrochemicals and other formulations. Fluorine in the atmosphere, oceans, biosphere and earth's crust might be grouped together as the *fluorosphere* to understand the effects of human activities on the environment [1].

This chapter begins with a short discussion of sources of fluorinated chemicals. Most existing fluorochemicals are organic and virtually all of them are synthetic, that is manufactured. Fluorochemicals as a class of materials are fairly new; all of them have been produced since 1930's. The rest of this chapter primarily covers fluorocarbon gases and liquids. A short explanation of commercial fluoropolymers rounds off the Chapter.

2.1 Fluorine and Fluorochemicals

Organic fluorine compounds rarely occur naturally in contrast to bromine and chlorine products. The most famous naturally existing organic fluorine-containing compound is probably monofluoroacetic acid (FCH_2COOH). This compound is found in a South African plant called "Gifblaar," which is known to be so poisonous that ingesting only a half of its leaf is enough to kill a cow. In 2003, a Chinese research group discovered fluorine-containing compounds from the extract of a marine sponge called *Phakellia fusca*. Another example of a naturally occurring organic fluorine containing compound is called nucleocidin. Even the presence of elemental fluorine has been reported in fluorspar (CaF_2) containing radioactive uranium [2].

None of the rare natural organic compounds are useful as a source for C-F bonds for commercial production of chemicals. The easiest route to produce carbon fluorine (C-F) bond is via hydrofluoric acid (HF). Acid grade fluorspar (CaF_2) is combined with sulfuric acid to produce HF and calcium sulfate. Chloroform is produced by and chlorination of methane. Catalytic reaction of HF and $CHCl_3$ produces $CHClF_2$, which is the starting point for producing many fluorocarbons.

2.2 Fluorocarbons

Development of fluorine chemistry allowed economically viable fluorinated chemicals and polymers to be produced during the first half of 20th century. The first commercial fluorine products of any kind go back to 1930's, developed to meet the evolving industrial needs for compounds such as new refrigerants. The traditional agents used in refrigeration included

Sina Ebnesajjad, PhD. Concise Handbook of Fluorocarbon Gases: Applications in Refrigeration and Other Industries, (13–28) © 2021 Scrivener Publishing LLC

ammonia, carbon dioxide, sulfur dioxide, hydrocarbons and methyl chloride. These refrigerants suffered from shortcomings including toxicity, flammability, instability and poor efficiency in the refrigeration cycle. Consequently, the challenges in the development of new agents consisted of the following characteristics:

1. Non-flammable
2. Good thermal stability
3. Low toxicity
4. Atmospheric boiling point of -40°C to 0°C
5. Refrigeration cycle efficiency

Some of the early patents described methods using rather exotic compounds to produce fluorocarbons [3, 4]. Simpler preparation methods have been described for the reaction of hydrofluoric acid with halocarbons (chlorine and bromine) using a catalyst [5].

Fluorocarbon in this book refers an acyclic alkane hydrocarbon in which all or some of hydrogen atoms have been replaced with fluorine (Table 2.1). An acyclic alkyl has the general formula C_nH_{2n+1}. Some chlorine or bromine atoms may also be present in the molecule. Depending on the structure of chemical compound, fluorocarbon consists of chlorofluorocarbon (CFC), hydrochlorofluorocarbon (HCFC), hydrofluorocarbon (HFC) and hydrofluoroolefin (HFO). All except HFO have been the targets for obsolescence under Montreal Protocol and its amendments. Major fluorocarbon applications include refrigerants, blowing, agents (foam), solvents, aerosol propellants and fire extinguishers.

Chlorofluorocarbons contain Carbon and some combination of Fluorine and Chlorine atoms. They were the original fluorocarbons developed in the 1930's. Hydrochlorofluorocarbons contain Hydrogen, Chlorine, Fluorine, and Carbon atoms; they were quickly developed and commercialized after the enactment of the Montreal Protocol. Hydrofluorocarbons contain Hydrogen, Fluorine, and Carbon (no chlorine) known as the *third generation* of fluorocarbons.

Hydrobromofluorocarbons contain Hydrogen, Bromine, Fluorine, and Carbon atoms. Perfluorocarbons contain Fluorine, Carbon, and Bromine atoms, and some contain Chlorine and/or Hydrogen atoms.

Table 2.1 Examples of various acyclic alkane fluorocarbons.

Compound	Chemical formula	Form
Hexane	CH_3-CH_3	Alkane
Perfluorohexane	CF_3-CF_3	Perfluorinated
2-bromo-2-chloro-1,1,1-trifluoroethane (Halothane)	CF_3-CHBrCl	Fluorobromochlorinated
1,2-dichloro-1,1,2-trifluoroethane	$CClF_2$-CHClF	Hydrochlorofluorocarbon
1,1,2-Trifluoroethane	CHF_2-CH_2F	Hydrofluorocarbon

Hydrobromofluorocarbons contain Hydrogen, Bromine, Fluorine, and Carbon atoms. Perfluorocarbons contain Fluorine, Carbon, and Bromine atoms, and some contain Chlorine and/or Hydrogen atoms.

To avoid repeating the rather long chemical names of fluorocarbons a system has been devised to use codes for each of them. Oak Ridge National Laboratory [6] has provided the designations of fluorocarbon codes (Section 2.2.1).

2.3 Designations for Fluorocarbons

Fluorocarbon compounds are designated by a combination of letters and numbers (e.g., CFC-11, HCFC-142b). In the latter example, the lower-case b refers to an isomer. It has no relationship to the chemical formula ($C_2H_3F_2Cl$), rather it designates a particular structural arrangement of the atoms in the molecule. For example, HCFC-142b identifies the isomer in which all three hydrogen atoms are attached to the same carbon atom, and the structural formula is written as CH_3CF_2Cl. By contrast, HCFC-142 (without the b) refers to an arrangement in which one carbon atom is attached to two hydrogen atoms and one chlorine atom, while the other carbon atom is attached to the third hydrogen atom and two fluorine atoms (CH_2ClCHF_2).

To find the number, given the chemical formula: consider the number as consisting of 3 digits: *a*, *b*, and *c*. For 2-digit numbers (e.g., CFC-11) one digit is zero (e.g., CFC-011).

a is the number of carbon atoms minus 1;
b is the number of hydrogen atoms plus 1;
c is the number of fluorine atoms.

For $CFCl_3$: a = the number of carbon atoms (1) minus 1 = 0
b = the number of hydrogen atoms (0) plus 1 = 1
c = the number of fluorine atom = 1

and the compound is CFC-011, or CFC-11.

Similarly:

CCl_2F_2 is CFC-012
$C_2Cl_3F_3$ is CFC-113

To find the chemical formula, given the number: first add 90 to the *abc* number, to obtain a 3-digit definite number (*def*), where:

d is the number of carbon atoms;
e is the number of hydrogen atoms;
f is the number of fluorine atoms; and
$2d + 2 - e - f$ = the number of chlorine atoms.

For CFC-11: 90 + 11 = 101, so the number of carbon atoms is 1, the number of hydrogen atoms is zero, the number of fluorine atoms is 1, and the number of chlorine atoms is (2 + 2 - 0 - 1 = 3). So the chemical formula for CFC-11 is $CFCl_3$. That is, it has 1 carbon atom, no hydrogen, 1 fluorine atom, and 3 chlorine atoms.

Similarly: CFC-12: 12 + 90 = 102; 4-2 = 2 Cl; the chemical formula is: CCl_2F_2
CFC-113: 113 + 90 = 203; 6-3 = 3 Cl; the chemical formula is: $C2Cl_3F_3$
HCFC-142b: 142 + 90 = 232; 6-5 = 1 Cl; the chemical formula is: $C_2H_3F_2Cl$

Halons are fluorocarbons with at least one bromine atom.

Halon (no hydrogen number) nomenclature is relatively simple; each halon has an *abcd* number, where:

a = the number of carbon atoms;
b = the number of fluorine atoms;
c = the number of chlorine atoms, and
d = the number of bromine atoms.

$C_2F_4Br_2$ is Halon 2402. And, an example going the other way Halon 1211 is CF_2ClBr.

It is not possible to speak about fluorocarbons without bringing up the issue of the effect of these compounds on the Earth's Ozone Layer and the Montreal Protocol (described later in this section) that has banned most or all of those fluorocarbons. Examples of the three main commercial classes, based on their interaction with ozone, of fluorocarbons are summarized in Table 2.2. CFC's have been completely phased out and the others, except for HFOs, are in various stages of being phased out.

Hydrofluorocarbons and, to a very limited extent, perfluorocarbons (PFCs), serve as alternatives to ozone depleting substances phased out under the Montreal Protocol. Tables 2.3 and 2.4 summarize HCFC, HFCs and PFCs fluorocarbons. Applications of HFCs are presented in Table 2.5.

The usage of olefinic fluorocarbons required an extension of the designation system by ASHRAE (American Society of Heating, Refrigerating and Air-Conditioning Engineers). The following descriptions have been adopted from an ASHRAE publication [10].

Olefin: an organic (carbon containing) compound characterized by the presence of one or more double bonds between carbon atoms in the molecule. Such a compound can also be described as being unsaturated.

Unsaturated: as used in this standard, an organic (carbon containing) compound in which one or more carbon atom is joined to other carbon atoms by a carbon-carbon double bond. Such a compound can also be described as an olefin.

The identifying numbers assigned to the hydrocarbons and halocarbons of the methane, ethane, ethene, propane, propene, and cyclobutane series are such that the chemical composition of the compounds can be explicitly determined from the refrigerant numbers, and vice versa, without ambiguity. The molecular structure can be similarly determined for the methane, ethane, ethene, and most of the propane and propene series from only the identification number.

The carbon atoms are numbered sequentially, in order of appearance, with the number 1 assigned to the end carbon with the greatest number of hydrogen substituents (i.e., number of halogenated atoms substituted for hydrogen on the alkane end carbon atoms). In the case where both end carbons of a saturated compound contain the same number of (but different) halogen atoms the number 1 shall be assigned to the first end carbon, defined

Table 2.2 Examples of main commercial classes of fluorocarbons [7].

Fluorocarbon type	Description	Applications
CFC	Substance that contains chlorine and has strong characteristics of destroying the ozone layer. Examples: CFC-11, 12, 113 etc.	It has been widely used with refrigerant, blowing, solvent, and aerosol propellant, but in accordance with Montreal Protocol, those have been completely banned production and imports by the end of 1995.
HCFC	It contains chlorine but it also contains hydrogen, therefore propensity to destroy the ozone layer is considerably weakened (low ODP). Examples: HCFC-22, 123, 141b, 142b etc.	It has been widely used with refrigerant, blowing, solvent, and aerosol propellant, but in accordance with Montreal Protocol, those are on the way of total abolition. Depending on types and uses, some have been totally abolished. (Example, HCFC-141b for PU blowing agent)
HFC	It does not contain chlorine and has hydrogen that does not destroy ozone layer. Examples: HFC-23, 32, 125, 134a, 143a, 152a etc. HFC blend refrigerants R-410A, 407C, 404A.	Refrigerant: Refrigerator, various refrigeration equipment, mobile air-conditioning, room air-conditioner, package air-conditioner etc. Blowing agent, solvent, and aerosol propellant.
HFO	It does not contain chlorine and has hydrogen. Does not destroy ozone layer. Plus it has no impact on global warming Examples: R-1234yf ($CF_3CF=CH_2$), R-1234ze ($CF_3CH=CHF$).	Refrigerator, various refrigeration equipment, mobile air-conditioning, room air-conditioner, package air-conditioner etc. Blowing agent, solvent, and aerosol propellant.

as having the largest number of bromine, then chlorine, then fluorine, and then iodine atoms. If the compound is an olefin, then the end carbon nearest to the double bond will be assigned the number "1", as the presence of a double bond in the backbone of the molecule has priority over substituent groups on the molecule.

In the case of isomers of propene series, each has the same number, with the isomers distinguished by two appended lowercase letters. The first appended letter indicates the substitution on the central carbon atom:

- Cl	x
- F	y
- H	z

Table 2.3 List of hydrochlorofluorocarbon (HCFC) refrigerants [8].

Product R-Number	ODP		GWP
22	0,055	Medium	1810
123	0,060	Medium	77
401A	0,033	Medium	1182
401B	0,036	Medium	1288
402A	0,019	Medium	2788
402B	0,030	Medium	2416
408A	0,024	Medium	3152
409A	0,046	Medium	1909

The second letter designates the substitution on the terminal methylene carbon as defined for the methylene carbon of the propane:

$=CCl_2$	a
$=CClF$	b
$=CF_2$	c
$=CHCl$	d
$=CHF$	e
$=CH_2$	f

The following schematic summarizes the nomenclature for hydrofluoroolefins:

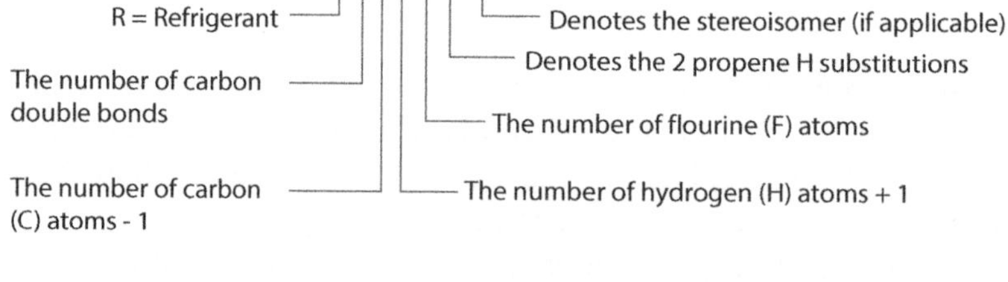

In the cases where stereoisomers exist, the opposed isomer will be identified by the suffix (E) and the same side isomer will be identified by the suffix (Z). The letters "E" or "Z" are appended at the end of the refrigerant number to show the precedence of the atoms or groups, which are attached to the carbon atoms at either end of the double bond. An example of this system is given in Table 2.6. "E" for Entgegen is similar to *trans*, where priority atoms or groups are across the double bond from each other. "Z" for Zusammen is similar to cis, signifying that priority atoms or groups are on the same side of a double bond (Figure 2.1).

Table 2.4 List of Hydrofluorocarbon (HFC) refrigerants [8].

Product R- Number	ODP		GWP
23	0	Zero	14800
32	0	Zero	675
134A	0	Zero	1430
404A	0	Zero	3922
407A	0	Zero	2107
407C	0	Zero	1774
407F	0	Zero	2088
417A	0	Zero	2346
422A	0	Zero	3143
422D	0	Zero	2729
423A	0	Zero	2280
424A	0	Zero	2440
427A	0	Zero	2138
428A	0	Zero	3607
434A	0	Zero	3245
437A	0	Zero	1805
438A	0	Zero	2265
442A	0	Zero	1888
507A	0	Zero	3985
508B	0	Zero	13396

Priority order of atoms connected to either of the unsaturated carbons is determined by standard CIP (Cahn-Ingold-Prelog) rules of organic nomenclature [11]. In essence, attached atoms of higher atomic number have higher priority: I>Br>Cl>F>O>C>H. In case of a priority tie, the next attached atoms or substituents on the next attached carbon atom are considered, until a priority is determined. In the case of refrigerants, it is more exact and less cumbersome to use atomic mass rather than atomic numbers of the atoms. This is because the sum of the atomic numbers of substituents of CHF_2 and CHCl are the same, while the summed atomic masses do differ.

In case of Composition-Designating Prefixes the identifying Number is prefixed by the letter C, for carbon, and preceded by B, C, or F—or their combinations—to signify the

Table 2.5 Main applications for HFC and PFC substitutes[1] for Ozone Depleting Fluorocarbons (ODS) [9].

Chemical	Refrigeration and air conditioning	Fire suppression and explosion protection	Aerosols		Solvent cleaning	Foam blowing	Other Application[2]
			Propellants	Solvents			
HFC-23	X	X					
HFC-32	X						
HFC-125	X	X					
HFC-134a	X	X	X			X	X
HFC-143a	X						
HFC-152a	X		X			X	
HFC-227ea	X	X	X			X	X
HFC-236fa	X	X					
HFC-245fa				X		X	
HFC-365mfe				X	X	X	
HFC-43-10mee				X	X		
PFC-14[3] (CF_4)		X					
PFC-116 (C_2F_6)							X
PFC-218 (C_2F_8)							
PFC-31-10 (C_4F_{10})		X					
PFC-51-14[4] (C_6F_{14})					X		

[1]Several applications use HFCs and PFCs as components of blends. The other components of these blends are sometimes ODSs and/or non-greenhouse gases. Several HFCs, PFCs and blends are sold under various trade names.

[2]Other applications include sterilization equipment, tobacco expansion applications, plasma etching of electronic chips (PFC-116) and as solvents in the manufacture of adhesive coatings and inks.

[3]PFC-14 (chemically CF_4) is used as a minor component of a proprietary blend. Its main use is for semiconductor etching.

[4]PFC-51-14 is an inert material, which has little or nil ability to dissolve soils. It can be used as a carrier for other solvents or to dissolve and deposit disk drive lubricants. PFCs are also used to test that sealed components are hermetically sealed.

Table 2.6 Propene series isomers [10].

Isomer	Chemical formula	Stereoisomer	
		IUPAC	ACS
R-1234yc	$CH_2F\text{-}CF{=}CF_2$		
R-1234zc	$CHF_2\text{-}CH{=}CF_2$		
R-1234ye(E)	$CHF_2\text{-}CF{=}CHF$	Entgegen	Trans
R-1234ye(Z)	$CHF_2\text{-}CF{=}CHF$	Zusammen	Cis
R-1234ze(E)	$CF_3\text{-}CH{=}CHF$	Entgegen	Trans
R-1234ze(Z)	$CF_3\text{-}CH{=}CHF$	Zusammen	Cis
R-1234yf	$CF_3\text{-}CF{=}CH_2$		

IUPAC = International Union of Pure and Applied Chemists ACS = American Chemical Society

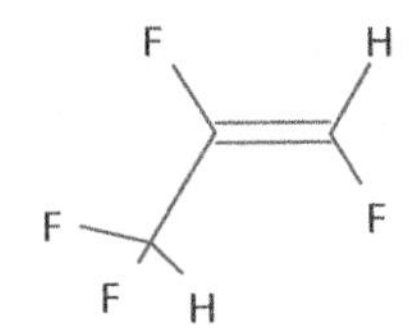

1(E)-1,2,3,3-tetrafluoro-1-propene, or HFO-1234ye(E)

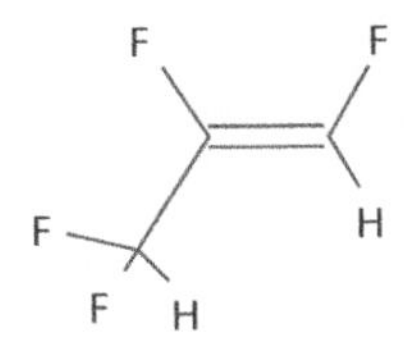

1(Z)-1,2,3,3-tetrafluoro-1-propene, or HFO-1234ye(Z)

Figure 2.1 Two examples of isomeric hydrofluorooelefin [10].

presence of bromine, chlorine, or fluorine, respectively. In compounds that also contain hydrogen the codes are preceded by the letter H, to signify the increased deterioration potential before reaching the stratosphere. The compositional designating prefixes for ether substitute an "E" for "C," such that "HFE," "HCFE," and "CFE" refer to hydrofluoroethers, hydrochlorofluoroethers, and chlorofluoroethers, respectively. The composition designating prefixes for halogenated olefins is "CFC", "HCFC", or "HFC" referring to chlorofluorocarbon, hydrochlorofluorocarbon, or hydrofluorocarbon, respectively. With substitution of an "O" for the carbon "C" as "CFO", "HCFO" and "HFO" refer to chlorofluoro-olefin, hydrochlorofluoro-olefin, or hydrofluoro-olefin, respectively.

Halogenated olefins are a subset of halogenated organic [or carbon-containing] compounds. They have significantly shorter atmospheric lifetimes than their saturated counterparts. Examples include: CFC-11, CFC-12, BCFC-12B1, BFC-13B1, HCFC-22, HC-50, CFC-113, CFC-114, CFC-115, HCFC-123, HCFC-124, HFC-125, HFC-134a, HCFC-141b, HCFC-142b, HFC-143a, HFC-152a, HC-170, and FC-C318, and HFC-1234yf or HFO-1234yf.

2.4 Fluoropolymers and Fluoroelastomers

In this book the term fluoropolymer describes fluorinated polymers and copolymers of a few olefinic monomers that are consumed in significant commercial scale. These monomers include tetrafluoroethylene (CF_2=CF_2), vinylidene fluoride (CF_2=CH_2) and chlorotrifluoroethylene (CFCl=CF_2) and vinyl fluoride (CHF=CH_2). The polymers of the last two are produced at significantly lower volumes than the first two monomers but have been included because of the importance of their applications. Generally, an increase in fluorine content of polymer enhances the desirable properties for which fluorinated polymers are known (Table 2.7).

Fluoroelastomers consist of a number of high performance synthetic rubbers that are partially or fully fluorinated. Fluoroelastomers are made by copolymerizing various combinations of vinylidene fluoride (CH_2=CF_2), hexafluoropropylene (CF_2=$CFCF_3$), chlorotrifluoroethylene (CF_2=CFCl), and tetrafluoroethylene (CF_2=CF_2). These fluorinated elastomers have outstanding resistance to oxygen, ozone, and heat and to swelling by oils, chlorinated solvent, and fuels.

Table 2.7 Effect of increasing fluorine content in polymers.

Property	Impact
Chemical resistance	Up
Melting point	Up
Coefficient of friction	Down
Thermal stability	Up
Dielectric constant	Down
Dissipation factor	Down
Volume and surface resistivity	Up
Mechanical properties	Down
Flame resistance	Up
Resistance to weathering	Up

2.4.1 Fluoropolymers

The inception of fluoropolymers as a group dates back to the serendipitous discovery of polytetrafluoroethylene (PTFE) in a DuPont laboratory by Roy Plunket. His research program was aimed at the discovery and development of new refrigerants. The initial testing of the waxy white powder found in a tetrafluoroethylene gas cylinder revealed something unusual. The powder had a high melting point, was insoluble and was unaffected by any chemical and intractable as a thermoplastic. Manhattan Project's needs for chemical resistance materials led to the first application of PTFE.

Following the successful introduction of PTFE, produced in a pilot plant, into the processing equipment for uranium manufacturing and purification, interest in this plastic began gathering. The DuPont Company in West Virginia, USA in 1946 built the first commercial plant. The story of the development fluoropolymers has been covered, in detail, in other publications [12].

The basic properties of PTFE may be justifiably called *extreme* (Table 2.8). The properties and characteristics of PTFE include high chemical resistance, low and high temperature capability, resistance to weathering, low friction, electrical and thermal insulation, and low friction or slipperiness. One drawback of this polymer is its relative softness, compared to engineering polymers, resulting in cold flow and ease of abrasion both of which are undesirable for many applications. Fillers are incorporated in PTFE to enhance its hardness.

The first fluoropolymer PTFE was quite peculiar in that it would not flow upon melting. It could not be processed by typical melt processing techniques such as extrusion. Metal powder processing techniques were modified and adopted to fabricate parts from PTFE examples of which include modified compression molding, ram extrusion and paste extrusion. Afterward a number of melt processible fluoropolymers were developed in the decades since the Second World War [13, 14].

Table 2.8 Fundamental properties of polytetrafluoroethylene.

High melting point, 342°C
Exceptional thermal stability
Useful mechanical properties at extremely low and high temperatures (-260 to 260°C)
Insolubility
Chemical inertness
Low coefficient of friction
Low dielectric constant/dissipation factor
Low water ab/adsorptivity
Excellent outdoor weatherability
Flame resistance (limiting oxygen index = 95)
Exceptional Purity

A whole family of thermoplastic polymers has been developed based on homopolymers and copolymers of tetrafluoroethylene (TFE). After PTFE the largest volume member is perfluorinated ethylene propylene copolymer (FEP). The second monomer in the FEP is hexafluoropropylene (CF_3-CF=CF_2), which is conveniently co-produced along with TFE in the monomer manufacturing process. Other important products included a copolymer of TFE with ethylene at a molecular ratio of 1:1 and TFE (ETFE) and perfluoroalky vinyl ethers (PFA and MFA).

The third largest volume family of fluoropolymers is based on vinylidene fluoride (VDF) hompolymers and copolymers. These thermoplastics are processed by normal melt processing techniques such as extrusion, film blowing and various molding methods. VDF is copolymerized with other monomers such as chlorotrifluoroethylene (CTFE). Both thermoplastics and elastomers can be prepared depending on the concentration of the CTFE in the copolymer. A new example is Arkema Corporation's VDF copolymers containing a reactive group that allows its ease of bondability [15].

The important polymers based on CTFE are its homopolymer polychlorotrifluoroethylene (PCTFE) and its copolymer with ethylene. The latter chlorotrifluoroethylene ethylene copolymer (ECTFE) is the analog of ETFE. PCTFE is borderline melt processible and is considered somewhat difficult to process. This CTFE homopolymer degrades when it is heated in the air lead to rapid decrease in its molecular weight.

Chemours Corp is the only commercial manufacturer of vinyl fluoride and its homopolymer polyvinyl fluoride (PVF). PVF cannot be processed by the standard melt processing method because it undergoes rapid thermal decomposition when heated above its melting point (195°C). PVF is fabricated into films and coatings using a somewhat unusual process. PVF is first dispersed in a latent (polar) solvent that suppresses its melting point thus permitting its melt processing without decomposition [16].

Fluoropolymer applications traverse across virtually every industrial segment and geographical regions. Examples of industries include automotive, aerospace, chemical and petrochemical processing, electrical, microelectronic, semi-conductor, pharmaceutical, biopharmaceutical, consumer, sports and recreation, food, beverage, laboratory applications, construction and architectural, military and others.

PTFE is the best material for manufacturing parts including gaskets, vessel and pipe linings, seals, spacers, dip tubes and other applications where corrosion resistance against most chemically challenging agents are required. Copolymers of TFE may also be used to manufacture some of those parts and wire and cable insulation. Most acids, bases and organic solvents do not affect these polymers even at elevated temperatures. PVDF applications include wire insulation, tubing and pipes for high purity water transportation, membrane distillation, gas separation, separator for lithium ion battery. PCTFE films are part of the composites for manufacturing bubble packs for pharmaceuticals because of their resistance to moisture vapor permeation.

2.4.2 Fluoroelastomers

Fluorocarbon elastomers are the largest group of fluoroelastomers. Different polymers are fluorinated to variable extents; most of fluorine is bonded to the carbon-carbon backbone of the macromolecule. Commercial fluoroelastomers are based on a number of monomers: tetrafluoroethylene (TFE), hexafluoropropylene (HFP), perfluoromethyl vinyl ether (PMVE), vinylidene fluoride (VDF), chlorotrifluoroethylene (CTFE), 1-hydro-pentafluoropropane (HPFP), ethylene, and propylene. Specific combinations of two or more of these monomers result in amorphous polymers with elastomeric behavior. A more complete list of monomers combined in significant commercial fluoroelastomers has been provided [17].

The first commercially available fluoroelastomer was Kel-F® developed by the M.W. Kellog Co. in the late 1950's. A well-known fluorinated elastomer was a copolymer of TFE and VDF developed by DuPont under the trade name Viton® A. Later, Viton® B was developed that was a terpolymer of TFE/VDF/HFP. Vinylidene fluoride based fluoroelastomers have been the most popular in the group. Fluorocarbon elastomers have grown extensively over time because of the need for high performance products.

Fluoroelastomers may be classified by their fluorine contents, 66%, 68%, & 70% respectively. Fluoroelastomers with higher fluorine content have higher fluids resistance due to higher fluorine content. Peroxide cured fluoroelastomers have inherently improved water, steam, and acid resistance. Fluoroelastomers are, generally, manufactured by emulsion polymerization process. The monomers are charged to a batch reactor and polymerized under elevated temperature and pressure in the presence of surfactants and additives. After polymerization has been completed the latex is discharged from the reactor, the polymer coagulated, washed, dried and prepared for shipping. Chemical formulas of major fluoroelastomers are shown below:

Copolymer fluoroelastomer:

$-(CF_2\text{-}CF)\text{-}(CH_2\text{-}CF_2)\text{-}$

|

CF_3

$-(CH_2\text{-}CF_2)\text{-})\text{-}(CClF_2\text{-}CF_2)\text{-}$

Terpolymer fluoroelastomer:

$-(CF_2\text{-}CF)\text{-}(CH_2\text{-}CF_2)\text{-}(CF_2\text{-}CF_2)\text{-}$ TFE level can be varied for different fluorine contents

|

CF_3

Improved Low Temperature Fluoroelastomer Terpolymer:

$-(CF_2\text{-}CF)\text{-}(CH_2\text{-}CF_2)\text{-}(CF_2\text{-}CF_2)\text{-}$ HFP was replaced with a fluorinated ether

|

$O\text{-}CF_3$

Non-VF_2 Fluoroelastomer Terpolymer:

$-(CF_2\text{-}CF)\text{-}(CH_2\text{-}CH_2)\text{-}(CF_2\text{-}CF_2)\text{-}$ VF2 replaced with ethylene, imparts base/amine resistance

|

$O\text{-}CF_3$

Fluoroelastomers provide high levels of resistance to chemicals, oil and heat, and service life at elevated temperatures, some exceeding 200°C. Fluoroelastomers are fabricated into a variety of shapes like hoses, seals, O-rings, shaft seals, diaphragms, vibration dampeners, expansion joints and electrical connectors and many others for demanding end uses in a number of industries:

Aircraft and aerospace
Automotive
Chemical processing and transportation
Oil and gas exploration and production
Petroleum refining and transportation
Pharmaceutical

2.5 Fluorinated Coatings

Fluoropolymer coatings (also paints or finishes) have also developed based on their stable dispersions obtained by emulsion polymerization. The two coatings with the largest market share are based on polytetrafluoroethylene and polyvinylidene fluoride. One well-known application of these coatings is in non-stick cookware and bake ware. Fluorinated coatings are extensively applied to surfaces in industrial environments to protect them from a variety of corrosive agents [13].

Coatings have the advantage of ease of propagation based on few fluoropolymers by manipulating the formulation. Fluoropolymers can be converted into coatings in two separate ways, as liquids or solids. PTFE must be applied as a liquid coating in contrast to the melt processible fluoropolymers that can also be powder-coated. Fluoropolymer paints are formulated as water-based (aqueous) or solvent-based products. The latter is limited to cases in which aqueous dispersions would not work.

Fluorinated coatings, similar to other paints, are mixtures of a variety of components including liquid carriers, pigments and specific additives in addition to fluoropolymers and additional polymeric binders. Pigments affect the color and appearance of the coatings as well as its functions such as abrasion resistance, thermal conductivity, the extent of porosity and electrical resistance/conductivity. Specific additives are sometimes incorporated to modify the rheology of the paint, wettability of the final coating and facilitate formation of the film at the time of baking. Other polymers are included in the paint to enhance the adhesion of the paint to the substrate or increase its hardness thus overcoming shortcomings of fluoropolymers.

Like other paints those made with fluoropolymers can be applied as one or more "coats" or layers. Flexibility of fluoropolymer content and availability of thickness ranges (2.5-2,000 μm) are two important features of those paints. The base polymers polyvinylidene fluoride and polytetrafluoroethylene provide a broad range of bake temperature ranging from 175 to 440°C. The combination of material properties, coating thickness and process variables determine the performance parameters of the final paint.

Applications of fluorinated paints include chemical process equipment liners, insulating coatings for electronics, non-stick coatings for cookware, surgical patches and glass fiber fabric coatings used for roofing of large structures flue gas heat exchangers, interior of exhaust ductworks and others.

2.6 Summary

HFOs Hydrofluoroolefins (HFOs) are the fourth generation of fluorine-based gases. HFC refrigerants are composed of hydrogen, fluorine and carbon atoms connected by single bonds between the atoms. HFO refrigerants are composed of hydrogen, fluorine and carbon atoms but contain at least one double bond between the carbon atoms. HFO refrigerants have zero ODP and low GWP thus offer a more environmentally friendly alternative to CFCs, HCFCs and HFCs.

DuPont (Chemours) and Honeywell jointly developed HFO 1234yf which is sold under the brand names Opteon™ yf and Soltice® yf. This low GWP fluorocarbon is also a replacement for R134a for use in mobile air conditioning (MAC) systems in the automotive sector.

References

1. Tavener, S.J. and Clark, J.H., Chapter 5: Fluorine: Friend or Foe? A Green Chemist's Perspective, in *Fluorine and The Environment*, Vol. 2, A. Tressaud (ed.), Elsevier, 2006.
2. Sato, K., *Naturally Occurring Organic Fluorine Compounds*, Tokyo Chemical Industry, April 2016, www.tcichemicals.com.
3. US Patent 1,978,840, A. L. Henne, assigned to General Motors Corp, Oct. 30, 1934.
4. US Patent 2,192,143, T. Midgley, Jr., A. L. Henne, assigned to Kinetic Chemicals Co., Feb. 27, 1940.
5. US Patent 2,062,743, H. W. Daudt, M. A. Youker, assigned to Kinetic Chemicals Co., Dec. 1, 1936.
6. Blasing, T.J. and Jones, S., *Environmental Sciences Division*, Oak Ridge National Laboratory, doi: 10.3334/CDIAC/atg.033, February 2012.
7. Daikin Industries, Fluorocarbon, www.daikin.com/chm/products/fluorocarbon, May 2016.
8. *Refrigerants Environmental Data*, The Linde Group, www.linde-gas.com, April 2020.
9. Ashford, P. *et al.*, Chapter 7, Emissions of Fluorinated Substitutes for Ozone Depleting Substance, in: *Processes and Product Use*, vol. 3, IPCC Guidelines for National Greenhouse Gas Inventories, www.ipcc-nggip.iges.or.jp, 2006.
10. ASHRAE Standard, ANSI/ASHRAE Addenda z, ah, ai, and aj to ANSI/ASHRAE Standard 34-2007, www.ashrae.org, Jan 27, 2010.
11. Smith, M.B. and March, J., *March's Advanced Organic Chemistry - reactions, mechanism and Structure*, John Wiley & Sons, 2007.
12. Ebnesajjad, S., *Introduction to fluoropolymers: materials, technology, and applications*, 2nd ed, Elsevier, New York, 2020.
13. Ebnesajjad, S., *Fluoroplastics Vol 1 – Non-melt Processible Fluoropolymers*, 2nd ed, Elsevier, 2015.
14. Ebnesajjad, S., *Fluoroplastics Vol 2 – Melt Processible Fluoropolymers*, 2nd ed, Elsevier, 2015.
15. Kynar® PVDF Fluoropolymer Family, Arkema High Performance Polymers, http://americas.kynar.com, 2020.
16. U.S. Patent 2,810,702, M.F. Bechteld and M. I. Bro, assigned to Du Pont Co, October 22, 1957.
17. Fluoropolymers Market - Global Industry Trends & Forecasts to 2019 www.marketsandmarkets.com, May 2016.

3

Fluorine Sources and Basic Fluorocarbon Reactions

Fluorine represents the most extreme of all elements [1]. It is the most reactive element known to man. It reacts with glass and nearly everything else. Even noble gases such as xenon, krypton and gold are not safe because every one of them reacts with fluorine. Fluorine is quite unique among all other elements because of its properties. It carries its uniqueness into organic substances when fluorine substitutes hydrogen and other elements in their molecules. The first synthesis of fluorine has been attributed to Moissan who exhibited the reactivity of fluorine [2].

The impact of fluorine on other materials, nicknamed *superhalogen*, is more severe than that of other halogens including chlorine. This chapter describes fluorine ores and the basic chemistry of reactions to produce organic fluorine compounds. Narrow aspects of fluorine chemistry related to the preparation of commercial fluorinated alkanes are discussed. An overview of the polymerization of fluorinated oleffinic monomers is also reviewed. The readers should consult the books and articles cited in this chapter for a broader understanding of fluorine chemistry.

3.1 Role of Fluorine in Fluorocarbons

Fluorine is the most electronegative of all elements at electronegativity of 4 in Pauling units. Electronegativity of other elements are 3.4 for oxygen, 3.2 for chlorine, 2.6 for carbon and 2.2 for hydrogen (Table 3.1) [3]. Extreme electronegativity of fluorine renders its covalent bonds highly polarized such as in C-F. Consequently, fluorine gas attacks nearly every substance and chemical because of very high reactivity. It even attacks noble gases like xenon producing XeF_x. It is easy to fluorinate hydrocarbons by fluorine gas, but the intensity of this reaction is too severe to control and causes broad decomposition.

The shortest bond is formed between C and H (0.11 nm) followed by C-F at 0.14 nm. Van der Waals radius (r_w) of fluorine substituent is 0.147 nm, shorter than in any other substituent. Van der Waals radius refers to the radius of an imaginary sphere that an atom occupies. The short bond length and r_w prevent the development of steric strain in perfluorinated compounds contributing to high thermal stability [3].

Carbon and fluorine form one of the strongest covalent bonds with an average bond energy around 480 kJ/mol. It exceeds the strength of carbon bond with other halogens (Table 3.2). C-F strength is one of the important reasons for high thermal and chemical stability of organic fluorochemicals. The F atoms have just the right size to create a protective shield (or sheath) over the carbon backbone when it is attached directly to the chain like in polytetrafluoroethylene (PTFE). If the F atoms were any smaller or larger than they are, the sheath would not form a regular uniform cove. This F shield protects the carbon chain from attack and confers chemical inertness and stability to PTFE. Fluorinated chemical groups play a similar role in hydrocarbons [5].

Sina Ebnesajjad, PhD. Concise Handbook of Fluorocarbon Gases: Applications in Refrigeration and Other Industries, (29–40) © 2021 Scrivener Publishing LLC

Table 3.1 Atomic properties of fluorine and other elements [4].

Element	Van der Waals, radii, nm	Electronegativity, Pauling
F	0.147	3.98
O	0.152	3.44
N	0.155	3.04
C	0.170	2.55
H	0.120	2.20

Table 3.2 Atomic properties of fluorine and other elements [4].

Element	Average bond strength, kJ/mol	Average bond length, nm
C-F	485	0.139
C-C	356	0.153
C-O	336	0.143
C-H	416	0.109

3.2 Fluorine Sources

Fluorine is the 13th most common element on the Earth but it occurs virtually only as inorganic compounds. There are three industrially significant fluorine minerals: fluorite or fluorspar, fluorapatite and cryolite. Fluorspar consists of CaF_2, seen in Figure 3.1 and comes in two grades. The more pure variety is called *acid grade* and is used for HF production. The other fluorspar grade (lower) is used in metallurgy in iron and steel casting, primary aluminum production, glass manufacture, enamels; welding rod coatings, cement production, as a flux in steelmaking and in other applications [6].

Fluorapatite or fluoroapatite is a phosphate with the chemical formula of $Ca_5(PO_4)_3F$ [Figure 3.1(b)]. It is the most abundant fluorine ore but it contains little fluorine. The third fluorine mineral Cryolite [Figure 3.1(c)] has a chemical formula of Na_3AlF_6, used in aluminum production. Cryolite has the highest amount of fluorine but it is a relatively scarce ore.

The dynamics of supply and consumption of fluorspar are quite complex. Fluorspar price and annual consumption/import for the United States is presented in the rest of this section. Figure 3.2 is helpful in clarification of some of the issues relevant to fluorspar. Some of the factors impacting the imported quantities and the prices are listed below:

1. The general economy and business cycle such as recessions and booms.
2. Demand for fluorinated and other materials that requiring HF.
3. United State stopped production of fluorspar after 1996.
4. USA has been importing acid grade fluorspar mostly from China, Mexico and South Africa. These three countries are the major three producers of CaF_2 in the world.

Figure 3.1 Photographs of fluorine minerals: (a) fluorspar, (b) fluorapatite, and (c) cryolite *Sources:* (a) www.themineralgallery.com/elmwoodroom.htm, *Courtesy*: www.mindat.org and the Hudson Institute of Mineralogy (b) www.mindat.org/photo-16871.html, *Courtesy*: David Soler, (c) www.mindat.org/min-1161.html, *Courtesy*: JGW, May 2016.

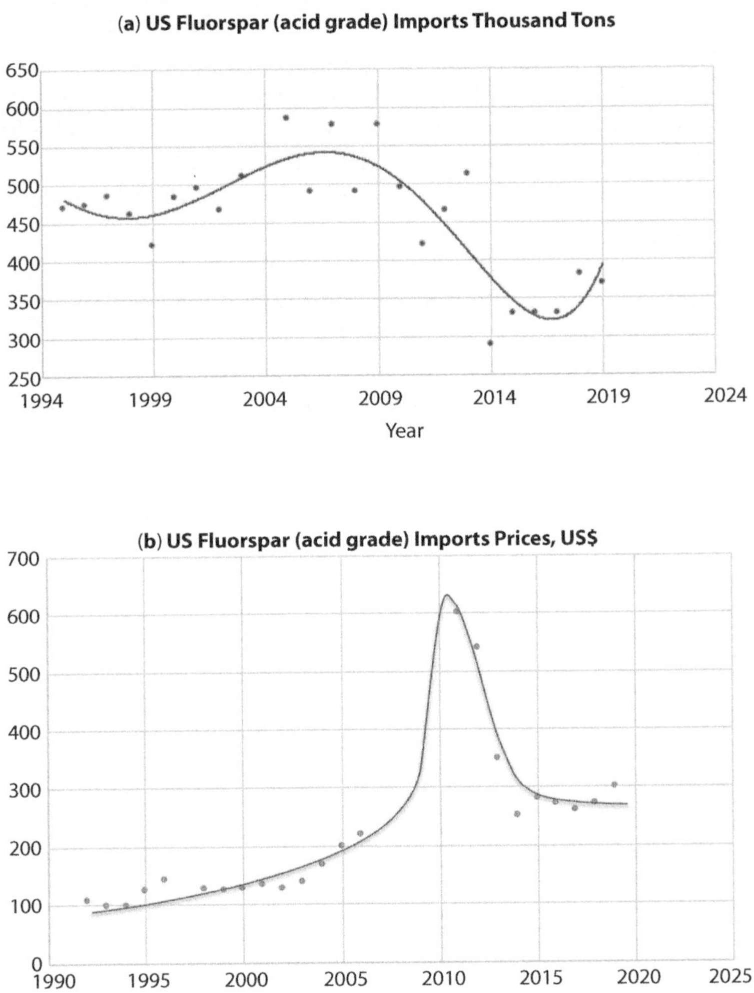

Figure 3.2 Historical US imports and prices of acid grade fluorspar [7].

5. During the early years post stoppage of domestic fluorspar China was the primary supplier to the US. In the recent years Mexico is the top supplier.
6. A supply of fluorspar is maintained by the Defense National Stockpile Center. The Center sells the fluorspar as required.
7. The Montreal Protocol (1987) banned most chlorofluorocarbon (CFC) gases. A two-phase replacement plan was developed: 1. To transition from CFCs to Hydrochlorofluorocarbons (HCFCs) as an intermediate solution; 2. To replace HCFCs with Hydrofluorocarbons (HFCs) and 3. To replace all previous fluorocarbons with low Ozone Depleting and low Global Warming replacements such as hydrofluoroolefins.
8. Most CFCs were phased out by 1996; the rest were removed by 2010; HCFCs replaced CFCs. The replacement required a one-time large block of production to replace all fluids in refrigeration and air conditioning units in addition to other applications of CFCs.
9. European Union countries replaced all HCFCs with HFCs by the end of 2010.
10. In 2010 the Chinese Ministry of land and Resources placed a quota on the annual production rate of fluorspar limiting at 11 million tons.
11. In the recent years many programs have been implemented to capture and recycle fluorocarbons in addition to reducing consumption of these materials.
12. Some of the fluoropolymer manufacturing, specifically polytetrafluoroethylene, has moved away from the US, to China and India.
13. A reason fluorinated chemicals are more expensive than hydrocarbons is the cost of fluorspar.

As shown in Figure 3.3, China mines more acid-grade fluorspar than any other country in the world, at 2,500 ktons annually. Mexico is the second-largest producer of acid-grade fluorspar at around 600 ktons annually, followed by South Africa (150 ktons). Contributions from the rest of the world have declined since the mid-2000s, with the remaining balance of acid-grade fluorspar production at about 300 ktons in 2016 [7].

From the mid-1990s to early 2000s, most of the acid-grade fluorspar mined in China was exported to the major refrigerant producing regions (US, EU, Japan). Starting from around 78% in 1999, the fraction of all mined acid-grade fluorspar in China that was exported has

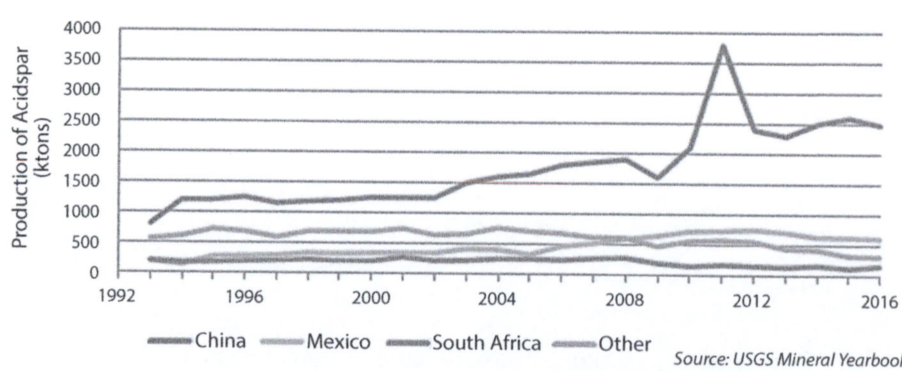

Figure 3.3 Global mined quantities of acid-grade fluorspar during 1993 to 2016 [8].

declined substantially. As a result of the dramatic expansion of the Chinese refrigerant and fluoropolymer manufacturing industry in the 2000s, most acid-grade fluorspar mined in China is now consumed domestically. Only 8% of the acid-grade fluorspar mined in China in 2016 was exported.

In 2008, Mexico overtook China as the leading exporter of acid-grade fluorspar to the US and in 2009 became the largest fluorspar exporter globally. In 2016, Mexico supplied about 65% of the acid-grade fluorspar imported into the United States (see Figure 3.4) [8].

HF itself is also traded globally. it is, however, both toxic and corrosive making it difficult to transport great distances. Exported quantities are thus substantially lower than those of fluorspar, and HF export is mostly limited to nearby countries. For example, China primarily exports HF to Japan and South Korea, whereas nearly all Mexican HF exports go to the US. Likewise, most HF imported into the United States comes from Mexico (see Figure 3.5) [8].

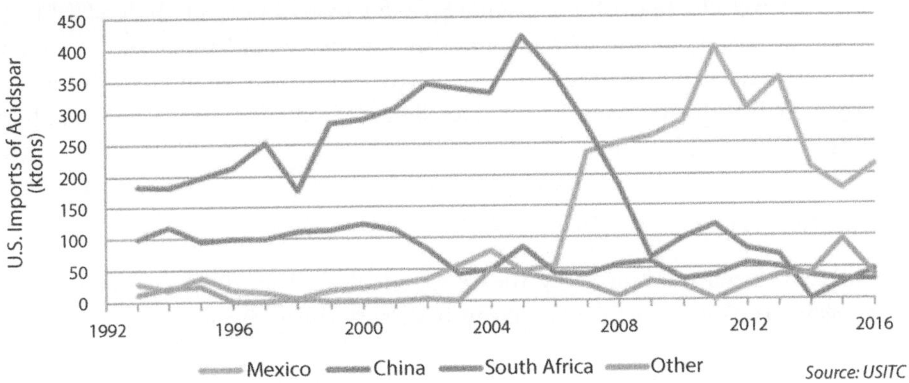

Figure 3.4 Annual U.S. import quantities of acid-grade fluorspar ("acidspar"), 1993–2016. Mexico, China, and South Africa are the major countries of origin for acidspar. Other countries that have exported significant quantities of fluorspar to the United States include Vietnam, Spain, the United Kingdom, and Mongolia [8].

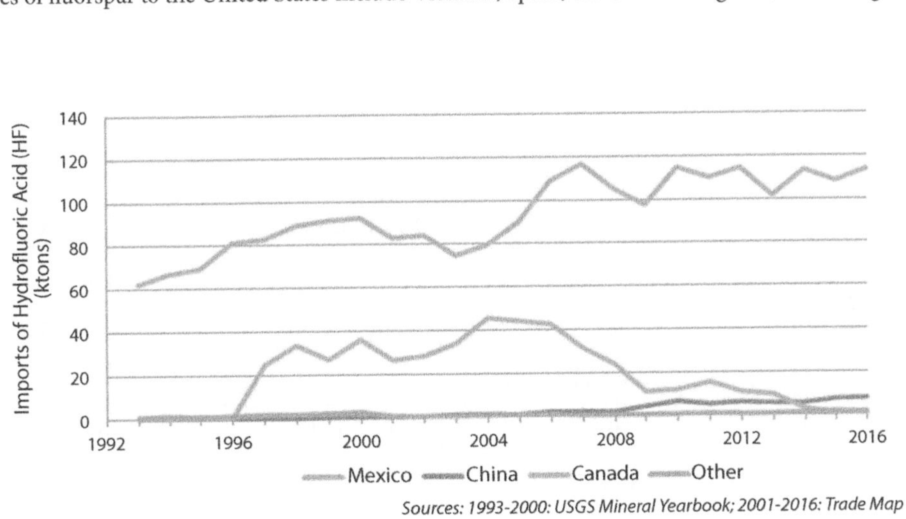

Figure 3.5 U.S. imports of hydrofluoric acid, 1993–2016 [8].

3.3 Fluorocarbon Compounds

There are a number of routes to introduce fluorine into organic compounds [9–11]. Those methods include direct fluorination of carbon; halogen exchange between chlorocarbons and hydrofluoric acid; fluorination of hydrocarbons using electrochemical and catalytic methods; and fluorination techniques using metal fluorides [12]. Chapter 6 describes some of the commercial techniques for fluorocarbon preparation in detail.

3.4 Hydrofluoric Acid

The commercial manufacture of fluorocarbons requires converting fluorine's inorganic ores to a suitable intermediate. That would in turn be used to introduce fluorine into organic compounds. A suitable compound would react with hydrocarbons (though not too reactive), inexpensive and is safe would be ideal. The most frequently used agent, commercially speaking, has proven to be hydrofluoric acid (HF), far from a perfect choice (Table 3.3).

HF when combined with water forms a highly corrosive acid that can even etch glass. Skin contact, inhalation, ingestion and contact with eyes must be avoided because of the extreme danger hydrofluoric poses. Safety data sheet (SDS) of HF must be consulted prior to its handling.

3.4.1 Manufacturing Hydrofluoric Acid

Hydrofluoric acid is manufactured by the reaction of acid-grade fluorspar (≥97% CaF_2) with sulfuric acid (H_2SO_4). The basic reaction is shown in Eq. (3.1).

$$CaF_2 + H_2SO_4 \rightarrow CaSO_4 + 2HF \tag{3.1}$$

Figure 3.6 displays a process diagram for commercial production of HF [14]. In the first step fluorspar is dried for 30-60 minutes in a horizontal rotary kiln that is heated to 200-250°C.

Table 3.3 Typical physical properties of hydrogen fluoride and hydrofluoric acid [13].

Concentration	49% HF	70% HF	100% HF (AHF)
Freezing Point	-33°F (-36°C)	-95°F (-71°C)	-118°F (-84°C)
Boiling Point	223°F (106°C)	146°F (63°C)	67.1°F (19.5°C)
Density (68°F)	9.6 lbs/gal	10.1 lbs/gal	8.3 lbs/gal
pH	<3.4	<3.0	<1.0 (10% solution)
Flash Point	Not Flammable		
Vapor Pressure (68°F)	23 mm Hg	132 mm Hg	771 mm Hg
Vapor Density	2.4 (air = 1.0)		

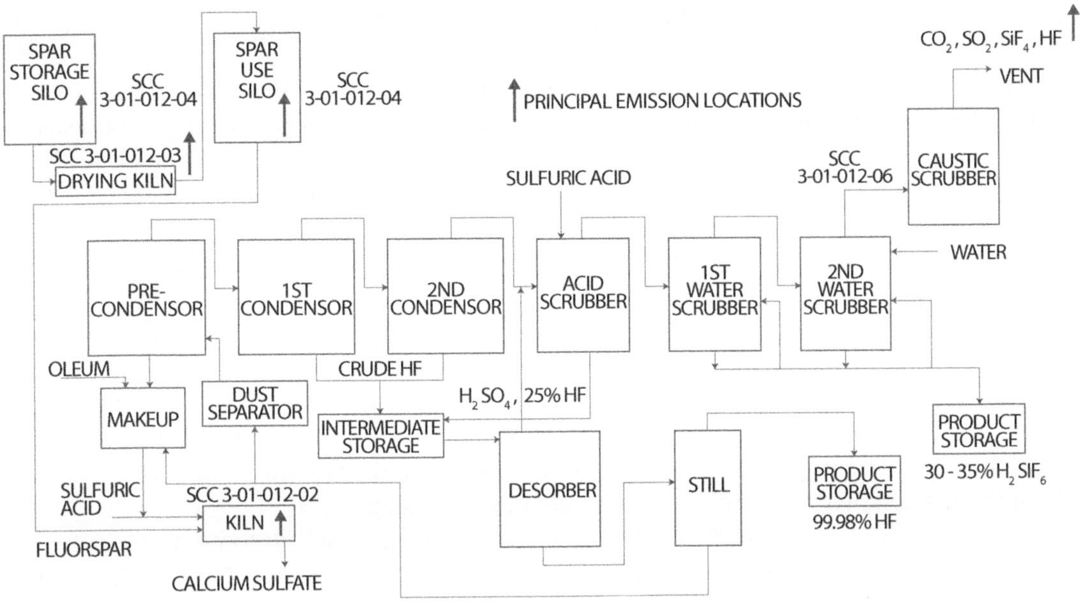

Figure 3.6 Hydrofluoric acid manufacturing process flow diagram [14].

Dry fluorspar and a small excess of sulfuric acid are fed continuously to the front end of a stationary pre-reactor or directly to the kiln by a screw conveyor. The pre-reactor mixes the components prior to charging to the rotary kiln. Calcium sulfate ($CaSO_4$) is removed through an air lock at the opposite end of the kiln. The gaseous reaction mixture - HF and the excess H_2SO_4 from the primary reaction are removed at the front end of the kiln along with entrained particulates. Silicon tetrafluoride (SiF_4), sulfur dioxide (SO_2), carbon dioxide (CO_2), and water produced in secondary reactions are also removed along with the HF.

The particulates are separated from the gas stream using a dust separator and returned to the kiln. Sulfuric acid and water are removed using a pre-condenser. HF vapors are next condensed in refrigerated condensers forming *crude* HF (impure) sent to storage tanks. The remaining gas stream passes through a sulfuric acid absorption tower or acid scrubber, to take out most of the residual HF and H_2SO_4. The gases exiting the scrubber are passed through water scrubbers to recover SiF_4 and HF as fluosilicic acid (H_2SiF_6). The water scrubber gases are passed through a caustic scrubber before release to the atmosphere. Stored HF and H_2SO_4 are distilled to obtain HF at 99.98% purity. Lower concentration HF are prepared by water dilution [14].

3.5 Aliphatic Fluorinated Organic Compounds

Commercial fluorocarbons are classified as *aliphatic* compounds which means they have saturated or unsaturated linear chemical structures. *Cyclic* fluorocarbons are considered part of the aliphatic group but they are not used in any significant quantity in the applications of the rest of the aliphatic fluorocarbons. Consequently, cyclic compounds are not included in this chapter.

Carbon forms its strongest bond with fluorine. The credit for demonstrating the stability of the C–F bond goes to the French chemists Dumas and Peligot, who heated dimethyl sulfate with potassium fluoride and obtained methyl fluoride [see Eq. (3.2)] [15].

$$(CH_3O)_2SO_2 + 2KF \rightarrow 2CH_3F + K_2S_2O_4 \quad (3.2)$$

The first nucleophilic replacement of another halogen by fluorine was attributed to a genius, the Russian musician and gifted chemist Alexander Borodin [16]. He synthesized benzoyl fluoride by replacement of chlorine in benzoyl chloride using Fremy's Salt (KF+HF) [17]. The reaction, called *halex* (abbreviation for halogen exchange) has continued to be the most significant way to produce C–F bonds on a commercial scale [12].

The pioneering work of Belgian chemist Frederic Swarts breathed a new life into the lagging chemistry of aliphatic fluorine compounds. Swartz conducted halogen exchange on polychlorides and polybromides through the use of combined antimony trifluoride and bromine ($SbF_3 + Br_2$). He elucidated dehalogenation reaction using Zn and dehydrohalogenation using K_2CO_3 could selectively remove halogens other than fluorine leading to the formation of fluorinated olefins. Swarts has been credited with the first synthesis of CCl_2F_2 by Midgley and Henne of the Frigidaire Co. (part of General Motors), who pioneered the use of fluorinated hydrocarbons in the refrigeration industry [18, 19].

In the 1950s and 1960s, study of the fluorocarbons began leading to developments for biological activity. Fluorocarbons, for instance, such as Fluroxene® ($CF_3CH_2OH{=}CH_2$) started a massive change in the types of inhalation anesthetics drugs. In the 1970s fluorocarbons became the agent of choice for inhalation anesthesiology. Other fluorocarbon related developments include artificial blood and respiratory fluids [18].

3.6 Synthesis of Fluorocarbons

Hydrogen fluoride has been called the *lifeblood* of the fluorochemical industry [20]. In a more general way, aliphatic fluorocarbons can be made using halogen-exchange by the reaction of HF with alkanes and chlorinated alkanes catalytically. This occurs because fluorine can displace any halogen in a substitution reaction. Commercially speaking the halogen choices consist mostly of chlorine and bromine.

One of the first reactions to prepare a chlorofluorocarbon was reported, by the Belgian chemist Swarts. Anhydrous HF and carbon tetrachloride were reacted using antimony pentachloride catalyst. The reaction, further developed by Midgley and Henner, produces fluorochloromethanes in which the ratio of products is determined by the reaction conditions [21].

$$CCl_4 + 2HF \xrightarrow{SbCl_5} \underset{\text{R-12}}{CF_2Cl_2} + 2HCl \quad (3.3)$$

Another reaction uses antimony trichloride and chlorine gas along with anhydrous HF; Eq (3.4) is an example of this reaction. In both reactions, antimony fluoride is produced in situ. Temperature and pressure of the reaction are 110°C and 30.4 bars [22]. A number of antimony halides are used as catalyst in chlorofluorocarbon synthesis. Reactivity of these salts is in the following order: $SbF_5 > SbF_3Cl_2 > SbF_3/SbCl_5$ mixture > SbF_3. Indeed, carbon tetrachloride used to be the feedstock for the production of R-11 and R-12 before the banning of both gases by the Montreal Protocol.

$$\underset{}{CCl_4} + 2HF \xrightarrow{Cl_2+SbCl_3} \underset{\substack{\text{R-11}\\(9\%)}}{CFCl_3} + \underset{\substack{\text{R-12}\\(90\%)}}{CF_2Cl_2} + \underset{\substack{\text{R-13}\\(0.5\%)}}{CF_3Cl} \tag{3.4}$$

An important consideration in the halogen exchange of chlorine with fluorine, in chloroalkanes, is the ease of substitution at the onset of reactions like Eq. (3.4). As the number of fluorine atoms to chlorine-bearing atoms increases, it becomes increasingly more difficult to make further substitutions. The reason is the steric crowding in a compound such as CCl_4 encourages removal of chlorine ions. But chlorine becomes a poor donor as the electron withdrawing power of the attached groups increases. Consequently, fluorination of CCl_3-CCl_3 (hexa-chloroethane) results in little of the asymmetric chlorofluoroethanes [23]. The disadvantage of CFCs, also called *second generation* refrigerants, is they have both high Global Warming and high Ozon Depletion potentials.

Third generation of refrigerants are based on hydrochlororo fluorocarbons (HCFC). One of the HCFC replacement gases for CFC-11 (R-11) and CFC-12 (R-12) was difluorochloromethane, R-22 (HCFC-22) per Eq. (3.5). It is also an intermediate in the preparation of tetrafluoroethylene, as seen in Eq. (3.6), an important monomer. The former reaction was conducted with $SbCl_3$ catalyst [24].

$$HCCl_3 + 2\ HF \xrightarrow{SbCl_3} \underset{\text{R-22}}{HCF_2Cl} + 2\ HCl \tag{3.5}$$

$$2\ CHClF_2 \rightarrow C_2F_4 + 2\ HCl \tag{3.6}$$

HCFC's like R-22 have ozone-depleting effect and had to be replaced. Hydrochlorofluorocarbons (HCFCs) were replaced with the third generation hydrofluorocarbons (HFCs) that have no impact on the ozone layer though they have large Global Warming Potential. Much of the chemistry developed for the manufacture of the CFCs was used for the production of HFCs [25–29]. Figure 3.7 shows two important industrial routes to HFC-134a in which chromium(III) catalysts are used in conjunction with HF for the halogen exchange steps [29].

While HFC-134a has been one of the global standard automotive air-conditioning refrigerants, it will not meet the European Union F-gas regulation. HFC-134a is being replaced by hydrofluoroolefin refrigerants (fourth generation) such as HFO-1234yf ($CF_3CF{=}CH_2$). HFC refrigerants are comprised of hydrogen, fluorine and carbon atoms connected by single bonds between the atoms. HFO refrigerants are composed of hydrogen, fluorine and carbon atoms, but contain at least one double bond between the carbon atoms. One benefit of HFO-1234yf is it has the potential to be used in current HFC-134a

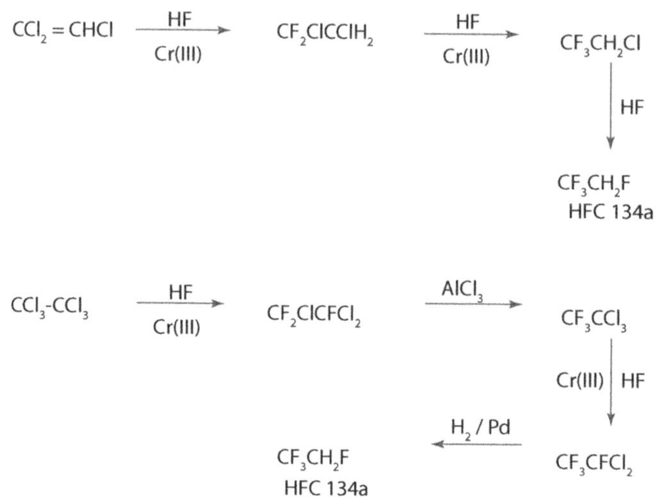

Figure 3.7 Two reactions to prepare 1,1,1 trifluoro-2, fluorodichloroethane [23].

systems with minimal system modifications. The important characteristics of HFO-1234yf are listed below:

Low toxicity
Low GWP; GWP = 4
Zero ozone-depletion potential
Low total contribution to climate change
Same operating pressures as current HFC-134a system

One method for preparation of HFO-1234yf involves feeding a mixture of chlorotrifluoroethylene (CTFE) with a methyl halide into a reactor at temperatures in the range 650-750°C. The mixture forms an intermediate stream producing HFO-1234yf precursors such as CF_2=$CFCH_2Cl$. It is then fed into a second reactor along with HF, which converts the precursors into HFO-1234yf. The second reactor contains fluorinated chromium oxide catalyst heated to 280-550°C [30].

This chapter provides a brief discussion of the basic chemistry of fluorocarbons. Additional descriptions of the chemistry of synthesis of commercial fluorocarbons are presented in the Chapter 6 of this book.

References

1. John's, K. and Stead, G., Fluoroproducts-the extremophiles, *J. Fluorine Chem.*, 104, 5, 2000.
2. Moissan, H. and Dewar, J., *J. Chem. Soc.*, 13, 175, 1897.
3. Kirsch, P., *Modern fluoroorganic chemistry: synthesis, reactivity, applications*, Wiley-VCH, 2004.
4. Zhang, W., *Magic Fluorine Chemistry for Medicinal Chemistry Appl, NEACT 72nd, Annual Summer Conf, St. Joseph's College, Maine, August 1-4*, 2011, www.neact.org.
5. Gangal, S.V., *Perfluorinated Polymers, Polytetrafluoroethylene, in on-line edition of Encyc of chemical technology*, John Wiley & Sons, 2010.

6. *Mineral Commodity Summaries 2016*, U.S. Geological Survey, Department of the Interior, http://dx.doi.org/10.3133/70140094, May 2016.
7. Data collected from various mineral reports publications of the United States Geological Survey, https://www.usgs.gov, August 2019.
8. Booten, C., Nicholson, S., Mann, M., Abdelaziz, O., Refrigerants: Market Trends and Supply Chain Assessment, Clean Energy Manufacturing Analysis Center, Department of Energy, NREL/TP-5500-70207, February 2020.
9. Kitazume T. and Yamazaki, T., *Experimental Methods in Organic Fluorine Chemistry*, 1st ed, CRC Press, 1999.
10. Kirsch, P., *Modern Fluoroorganic Chemistry: Synthesis, Reactivity, Applications*, 2nd ed, Wiley-VCH, 2013.
11. Hudlicky, M., *Fluorine Chemistry for Organic Chemists*, 1st ed, Oxford University Press, 2000.
12. Banks, R.E., Smart, B.E., Tatlow, J.C., *Organofluorine Chemistry—Principles and Commercial Applications*, Plenum Press, 1994.
13. *Product Safety Summary Hydrogen Fluoride and Hydrofluoric Acid, CAS No. 7664-39-3*, Solvay America, Inc, www.solvay.us, 2012-2013.
14. Hydrofluoric Acid, AP-42 Background Rep., AP-42 SECTION 5.8, Prep. For U.S. EPA, OAQPS/TSD/EIB, 1/1996.
15. Kirsch, P., *Modern Fluoroorganic Chemistry: Synthesis, Reactivity, Applications*, John Wiley & Sons, New York, 2004.
16. Festa, R.R., Alexander Borodin: Full-time chemist, part time musician. *J. Chem. Educ.*, 64, 4, 326, 1987 (April).
17. Räsänen, M. *et al.*, HKrF in solid krypton. *J. Chem. Phys.*, 116, 2508–15, 2002.
18. Ebnesajjad, S., *Introduction to Fluoropolymers*, Elsevier, 2013.
19. Banks, R.E., Smart, B.E., Tatlow, J.C., *Organofluorine Chemistry—Principles and Commercial Applications*, Springer, 1994.
20. Brbour, A.K., *Organofluorine Chemicals and Their Industrial Applications*, R.E. Banks (Ed.), Horwood, Chichester, UK, 1979.
21 Midgley, T. Jr., and Henne, A.L., *Ind. Eng. Chem.*, 22, 542, 1930.
22. Hudlickey, M., *Chemistry of Organic Fluorine Compounds*, Pergamon, London, 1961.
23. Chambers, R.D., *Fluorine in Organic Chemistry*, WileyBlackwell Publishing, 2004.
24. Revell, K.S., Brit. Patent 983,222, 1963.
25. Barbour, A.K., Belf, L.J., Buxton, M.W., *Adv. Fluorine Chem.*, 3, 181, 1963.
26. Hamilton, J.M., *Adv. Fluorine Chem.*, 3, 117, 1963.
27. Smart, B.E. and Fernandez, R.E., *Kirk-Othmer Encyclopedia of Chemical Technology*, 4th ed, Vol. 11, M. Howe-Grant (Ed.), p. 499, John Wiley and Sons, New York, 1994.
28. Elliot, A.J., *Organofluorine Chemistry. Principles and Commercial Applications*, R.E. Banks, B.E. Smart and J.C. Tatlow (Eds.), p. 145, Plenum Press, New York, 1994.
29. Rao, V.N.M., *Organofluorine Chemistry. Principles and Commercial Applications*, R.E. Banks, B.E. Smart and J.C. Tatlow (Eds.), p. 159, Plenum Press, New York, 1994.
30. US Patent 8,071,826, M. Van Der Puy, assigned to Honeywell International, Dec. 6, 2011.

4

Applications of Fluorocarbon Gases and Liquids

Fluorocarbon gases have numerous applications (Tables 4.1-4.4) some of which are critical to the health of individuals. The bans promulgated by the Montreal and Kyoto Protocols, phase-outs and emission reduction requirements, have impacted the pattern of fluorocarbon consumption. The use of individual materials and total consumption of fluorocarbons has been reduced. The estimated 2014 consumption profile of fluorocarbons in the United States, Canada and Mexico is summarized in Table 4.5. Nearly 80% of fluorocarbons are consumed by refrigeration and as intermediates with the former comprising the largest quantity. The next largest segments include blowing agents, propellants, solvents and others.

Refrigerant consumption in air conditioning is expected to grow globally and rapidly over the four decades beginning in 2010 to 2050. A study published in 2020 by National Renewable Energy and Oak Ridge National Laboratories provides quantitative estimates of the anticipated growth. The growth for advanced economies, that is members of Organization of Economic Coordination and Development.

(OECD), is estimated to be 1.3 times. The estimate for developing economies, that non-OECD members, is pegged at a whopping 4.5 times [1].

4.1 Refrigeration and Air Conditioning

Fluorocarbons have been used extensively as refrigeration agents since the middle of 20th century (Figure 4.1). Fluorocarbon refrigerants function by removing heat at low temperatures and pressures by evaporation, and then being compressed and condensed to reject that heat at higher temperatures and pressures. Among the salient characteristics of fluorocarbons are chemical stability, excellent thermodynamic properties, relative low non-toxicity, non-flammability, and inert chemical nature. Most people encounter these gases in household refrigeration, home conditioning units and in automotive air conditioners on a daily basis. Because of flammability resistance fluorocarbons had been classified as Class 1 refrigerants by the *American Society of Heating, Refrigerating and Air Conditioning Engineers*. Table 4.6 shows the classification of other refrigerants. The new generations of refrigerants hydrofluoroolefins have lower flammability rating.

Class 3 refrigerants are subject to stringent design and operational standards because of their high flammability. Selection of optimum refrigerant in a given application depends on many factors, including the temperatures at which the heat will be removed and rejected and the amount of heat that will be removed per unit time. Other variables include the nature of the service (i.e. continuous and steady or variable), and size or other mechanical design constraints on the refrigeration system. These factors influence the physical size of

Sina Ebnesajjad, PhD. Concise Handbook of Fluorocarbon Gases: Applications in Refrigeration and Other Industries, (41–86)

Table 4.1 Applications of commercial Chlorofluorocarbons (CFC).

ASHRAE code	CFC name	Chemical formula	Applications
11	Trichlorofluoromethane	CCl_3F	Refrigerant (centrifugal compressors), blowing agent, fluoropolymers, cleaning solvent, propellant and sterilization gas
12	Dichlorodifluoromethane	CCl_2F_2	Refrigerant, blowing agent, fluoropolymers, cleaning solvent and sterilization gas
13	Chlorotrifluoromethane	$CClF_3$	Specialty low temperature refrigerants, blowing agent, fluoropolymers, cleaning solvent and sterilization gas
113	1,1,2 Chloro,trifluoroethane	$CClF_3$	Refrigerant, blowing agent, cleaning solvent, fluoropolymers and propellant
114	1,2-Dichloro-1,1,2,2-tetrafluoro-ethane	CF_4	Refrigerant in centrifugal compressors, blowing agent, fluoropolymers, propellant, cleaning solvent and sterilization
115	1-Chloro-1,1,2,2,2-pentafluoroethane	C_2ClF_5	Refrigerant, blowing agent, fluoropolymers, cleaning solvent, propellant, sterilization

the components, the type of compressor and lubricants used, the inventory of refrigerants required, and the service and maintenance requirements of the system.

Blends of different fluorocarbons and occasionally non-fluorinated compounds are used to replace the phased out refrigerants. The interaction and number of factors limit the designers' discretion to substitute refrigerants without total redesign of the refrigeration system. The breakdown of fluorocarbon consumption in different refrigeration systems in the past and possible substitutes, prior to the development of hydrofluoroolefins HFOs, are given in Table 4.7. The types of blends and refrigeration applications are discussed in Chapter 9.

Ammonia-based refrigeration systems currently account for a large share of refrigerants in cold storage facilities. These systems are also heavily utilized in industrial systems for process cooling and for chillers. They were in use in household refrigeration systems as recently as 70 years ago before being displaced by fluorocarbons. While expanded use of ammonia in industrial and commercial applications is possible, the reintroduction in household appliances does not seem likely because of safety concerns. National and local building codes may preclude this use in residential and some commercial applications unless specific, costly measures are taken to isolate the material [2].

Table 4.2 Highest consumption Hydrochlorofluorocarbons (HCFC).

ASHRAE code	HCFC name	Chemical formula	Applications
123	2,2-Dichloro-1,1,1-trifluoroethane	CF_3CHCl_2	Refrigerant (replacement for CFC-11), blowing agent, cleaning solvent and others
124	1-Chloro-1,2,2,2-tetrafluoroethane	$CHClFCF_3$	Refrigerant (blends for replacement for CFC-12 and CFC-114), blowing agent, cleaning solvent, fire extinguishant and others
141b	1,1-Dichloro-1-fluoroethane	CCl_2FCH_3	Blowing agent (replacement for CFC-11), solvent cleaning (replacement for CFC-113) and Refrigerant (>180°C)
142b	1-Chloro-1,1-difluoroethane	$CClF_2CH_3$	Refrigerant at high temperatures and Blowing agent
21	Dichlorofluoromethane	CCl_2FH	Refrigerant, blowing agent, cleaning solvent, fluoropolymers, propellant and sterilization
22	Difluorochloromethane	$CClF_2H$	Refrigerant and cleaning solvent
225ca	1,1-Dichloro-2,2,3,3,3-pentafluoropropane	$CCl_2HCF_2CF_3$	Cleaning solvent (replacement for CFC-113) and aerosol solvent
225cb	1,3-Dichloro-1,1,2,2,3-pentafluoropropane	$CClF_2CF_2CHClF$	Cleaning solvent and aerosol solvent

Household refrigeration systems based on hydrocarbons have been broadly introduced in Europe, as well as in Japan, China, and elsewhere, and could be one of the likely substitutes for hydrofluorocarbon (HFC) gases. Automakers in the US are required to charge all mobile air conditioning units with an alternative to HFC-134a as of model year 2021. In Europe, the transition away from HFC-134a was to be fully implemented by 2017. Regional regulations that restrict the use of refrigerants with high global warming potential have led to the use of low global warming potential alternatives that is the hydrofluoroolefins (HFO) such as HFO-1234yf (tetrafluoro-propene) for automotive refrigeration. Other low global warming potential alternatives have been evaluated or are under evaluation, including carbon dioxide.

Table 4.3 Highest consumption Hydrofluorocarbons (HFC).

ASHRAE code	HFC name	Chemical formula	Applications
23	Trifluoromethane	CHF_3	Refrigerant, blowing agent, cleaning solvent, fluoropolymers, propellant, fire extinguishant and sterilization
32	Difluoromethane	CH_2F_2	Refrigerant (blend component)
HFC-43-10mee	1,1,1,2,2,3,4,5,5,5-decafluoropentane	CH_3F	Solvent
125	Pentafluoroethane	CF_3CHF_2	Refrigerant (blend component), blowing agent, cleaning solvent, fluoropolymers, propellant and sterilization
134a	1,1,1,2-Tetrafluoroethane	CF_3CH_2F	Refrigerant (blend component), blowing agent and propellant
143a	1,1,1-Trifluoroethane	CF_3CH_3	Refrigerant (blend component)
152a	1,1-Difluoroethane	CHF_2CH_3	Blowing agent and propellant
227ea	1,1,1,2,3,3,3-Heptafluoropropane	CF_3CHFCF_3	Propellant for medical aerosol
236fa	1,1,1,3,3,3-Hexafluoropropane	$CF_3CH_2CF_3$	Refrigerant and fire extinguishant
245fa	1,1,1,3,3-Pentafluoropropane	$CF_3CH_2CHF_2$	Blowing agent
365mfc	1,1,1,3,3-Pentafluorobutane	$CF_3CH_2CF_2CH_3$	Blowing agent and cleaning solvent
43-10mee	1,1,1,2,3,4,4,5,5,5-Decafluoropentane	$CF_3CHFCHFCF_2CF_3$	Specialty solvent

Table 4.4 Highest consumption Hydrofluoroolefins (HFO).

ASHRAE code	HFO name	Chemical formula	Applications
1132a	1,1-Difluoroethylene	CF_2=CH_2	Monomer for fluoropolymers
1233zdE	*trans* 1-Chloro-3,3,3-trifluoroprop-1-ene	*trans*-CHCL=CH-CF_3	Refrigerant, cleaning solvent and blowing agent
1234yf	2,3,3,3-Tetrafluoropropene	CH_2=CF-CF_3	Refrigerant
1234ze	1,3,3,3-Tetrafluoropropene	CHF=CH-CF_3	Refrigerant, blowing agent and propellant
1243zf	3,3,3-Trifluoropropene	CF_3-CH=CH_2	Refrigerant (replacement for HFC-134a)
1327	1,1,3,3,4,4,4-Heptafluorobutene	CF_2=CH-CF_2-CF_3	Blowing agent
1336mzz-Z	*cis*-1,1,1,4,4,4-Hexafluoro-2-butene	CF_3-CH=CH-CF_3	Blowing agent, refrigerant, fire extinguishant and cleaning solvent.
1345zfc	3,3,4,4,4-Pentafluoro-1-butene	CH_2=CH-CF_2-CF_3	Blowing agent
1354	2,4,4,4-Tetrafluorobutene	CF_3CH=CF-CH_3	Blowing agent
HBFO-1233xfB*	2-bromo-3,3,3-trifluoropropene	CF_3CBr=CH_2	

*Hydrobromofluoroolefin.

Table 4.5 Consumption of fluorocarbons in the United States, Canada, and Mexico, 2014 [2].

Application	Percent consumed
Refrigeration and air conditioning	43
Intermediates for polymer production	36
Foam blowing agents	12
Aerosal propellants	8
Solvent cleaning and degreasing	1
All other uses	1
Total	100

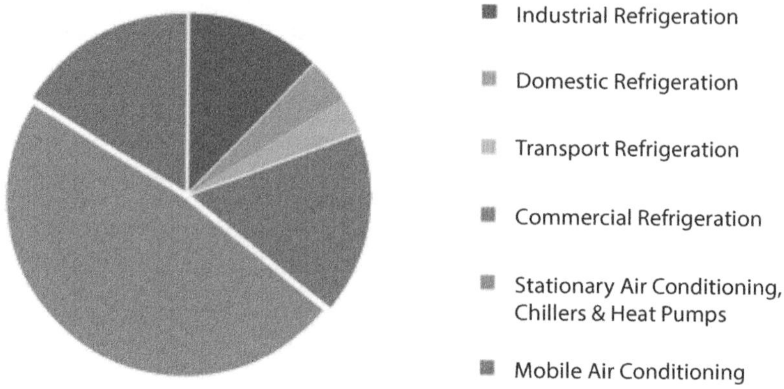

Figure 4.1 Global refrigerant market volume by application in 2016 [3].

Table 4.6 Classification of refrigerants according to the American Society of Heating, Refrigerating and Air Conditioning Engineers.

Refrigerants	Classification	Description
Fluorocarbons	1	Non-flammable
Methylene chloride or Ammonia	2	Flammable
Propane, Butane, Pentane	3	Highly Flammable

4.1.1 Refrigeration Applications

Refrigeration and air-conditioning end-uses typically use a refrigerant in a vapor compression cycle to cool and/or dehumidify a substance or space, like a refrigerator cabinet, room, office building, or warehouse. This section presents those applications according the descriptions provided by the US Environmental Protection Agency [4].

4.1.1.1 Chillers

Chillers typically cool water, which is then circulated to provide comfort cooling throughout a building or other location. Chillers can be classified by compressor type, including centrifugal and positive displacement. Chillers used to cool industrial processes are discussed under Industrial Process Refrigeration.

4.1.1.2 Cold Storage Warehouses

Cold storage warehouses store meat, produce, dairy products, and other perishable goods. The majority of cold storage warehouses in the United States use ammonia as the refrigerant in a vapor compression cycle, although some rely on other refrigerants.

Table 4.7 Breakdown of fluorocarbon consumption as refrigerants and potential substitutes prior to the development of HFOs [2].

Type of refrigeration system	Percent of refrigerant fluorocarbon consumption	Fluorocarbons sold for use current systems	Potential substitutes in these systems
Stationary air conditioning (cooler, chillers and heat pumps)	43	HFC-134a HFC-125 HFC-32 HFC blends	HFC-134a HFC-32/125 HFC-125/143a/134a HFC-32/125/134a HFO-1234yf Ammonia (vapor compression and absorption) Hydrocarbons
Commercial and industrial refrigeration	31	HFC-125 HFC-143a HFC-134a HFC-152a HFC blends	HFC-134a HFC-125/143a/134a HFC-125/143a HFC-32/125/134a HFO-1234ze Ammonia Hydrocarbons CO_2
Mobile air conditioning	24	HFC-134a HFO-1234yf	CO_2 HFC-134a HFO-1234yf
Household refrigerator/freezer	3	HFC-134a	Hydrocarbons Ammonia

4.1.1.3 Commercial Ice Machines

Commercial ice machines are used in commercial establishments (e.g., hotels, restaurants, convenience stores) to produce ice for consumer use. Ice machines produce ice in various sizes and shapes, and with different retrieval mechanisms (e.g., dispensers or self-retrieval from bins).

4.1.1.4 Household Refrigerators and Freezers

Household refrigerators and freezers are intended primarily for residential use, although they may be used outside the home. Household freezers only offer storage space at freezing temperatures. Products with both a refrigerator and freezer in a single unit are most common. Small refrigerated household appliances may also include chilled kitchen drawers, wine coolers, mini fridges, household beverage centers, ice makers that are part of a household refrigerator-freezer, and stand-alone ice makers for household use.

4.1.1.5 *Ice Skating Rinks*

Ice skating rinks are used by the general public for recreational purposes and also include professional rinks. These systems frequently use secondary loop refrigeration systems.

4.1.1.6 *Industrial Process Air Conditioning*

Industrial process air-conditioning units, which are distinct from commercial and residential air conditioning, provide comfort cooling for operators and protect process equipment. This end-use is often used when ambient temperatures approach 200°F (93°C) and corrosive conditions exist.

4.1.1.7 *Industrial Process Refrigeration*

Industrial process refrigeration systems cool process streams in industrial applications. The choice of substitute for specific applications depends on ambient and required operating temperatures and pressures.

4.1.1.8 *Motor Vehicle Air Conditioning*

Motor vehicle air-conditioning systems (MVACs) provide comfort cooling for passengers in light-duty cars and trucks, buses, trains, and other forms of transportation.

4.1.1.9 *Non-Mechanical Heat Transfer Systems*

Non-mechanical heat transfer systems include cooling systems that do not rely on a vapor compression cycle, such as those using convection to remove heat from an area. Two types of such systems are recirculating coolers, i.e., systems with fluid pumps, and thermosiphons, i.e., those that rely on natural convection currents. This end-use also includes Organic Rankine Cycle (ORC) devices that typically pump refrigerant to recover and utilize energy from lower-temperature heat rejected from other processes.

4.1.1.10 *Residential and Light Commercial Air Conditioning and Heat Pumps*

This end-use includes equipment that cools enclosed spaces in households and commercial industries, but excludes chillers – which include room air conditioning such as window units, packaged terminal air conditioners (PTAC) and heat pumps (PTHP), and portable air conditioners; central air conditioners (i.e., ducted); non-ducted systems (both mini and multi splits); packaged rooftop units; water-source and ground-source heat pumps; and other products. Residential and light commercial air conditioning and heat pumps are often distinguished from chillers by the fact that they condition the air directly, rather than cool (or heat) water that is then used to condition air.

4.1.1.11 *Residential Dehumidifiers*

Residential dehumidifiers are primarily used to remove water vapor from ambient air or directly from indoor air for comfort or material preservation purposes. While air-conditioning systems often combine cooling and dehumidification, this application serves only the latter purpose.

4.1.1.12 Refrigerated Transport

Refrigerated transport moves products (e.g., perishable goods) from one place to another by various modes of transportation while maintaining necessary temperatures, including refrigerated ship holds, truck trailers, railway freight cars, ships, and other shipping containers.

4.1.1.13 Retail Food Refrigeration

Retail food refrigeration, or commercial refrigeration, includes equipment designed to store, display, process, or dispense chilled or frozen goods for commercial sale. This end-use includes these categories of equipment: stand-alone equipment, remote condensing units, supermarket systems, and refrigerated food processing and dispensing equipment.

4.1.1.14 Vending Machines

Vending machines are self-contained units that dispense goods that must be kept cold or frozen.

4.1.1.15 Very Low Temperature Refrigeration

Very low temperature refrigeration systems require maintaining temperatures at approximately -80 degrees Fahrenheit (-62 degrees Celsius) or lower. Examples include medical freezers and freeze-dryers, which generally require extremely reliable refrigeration cycles to maintain low temperatures and must meet stringent technical standards that do not normally apply to refrigeration systems.

4.1.1.16 Water Coolers

Water coolers are self-contained units providing chilled water, and possibly heated water, for drinking. They may or may not feature detachable containers of water.

4.2 Oil in Refrigerants

The refrigeration compressor requires oil for lubrication of the moving parts. Ideally 100% of the oil in the compressor would be retained but in practice some oil leaks into the refrigerants. Oil types and strategies are used to circulate the oil through the system and return it back to the compressor on a regular and continuous basis.

A component called oil separators strips most of the oil from the discharge gas and return the oil to the compressor. These devices are used in larger systems but are still less than 100% effective by themselves.

In very large systems such as chillers, we are beginning to see oil free compressors equipped with magnetic bearings. An example can be seen in Figure 4.2 (TurboCor from Danfoss) – these pumps are still less common in the field.

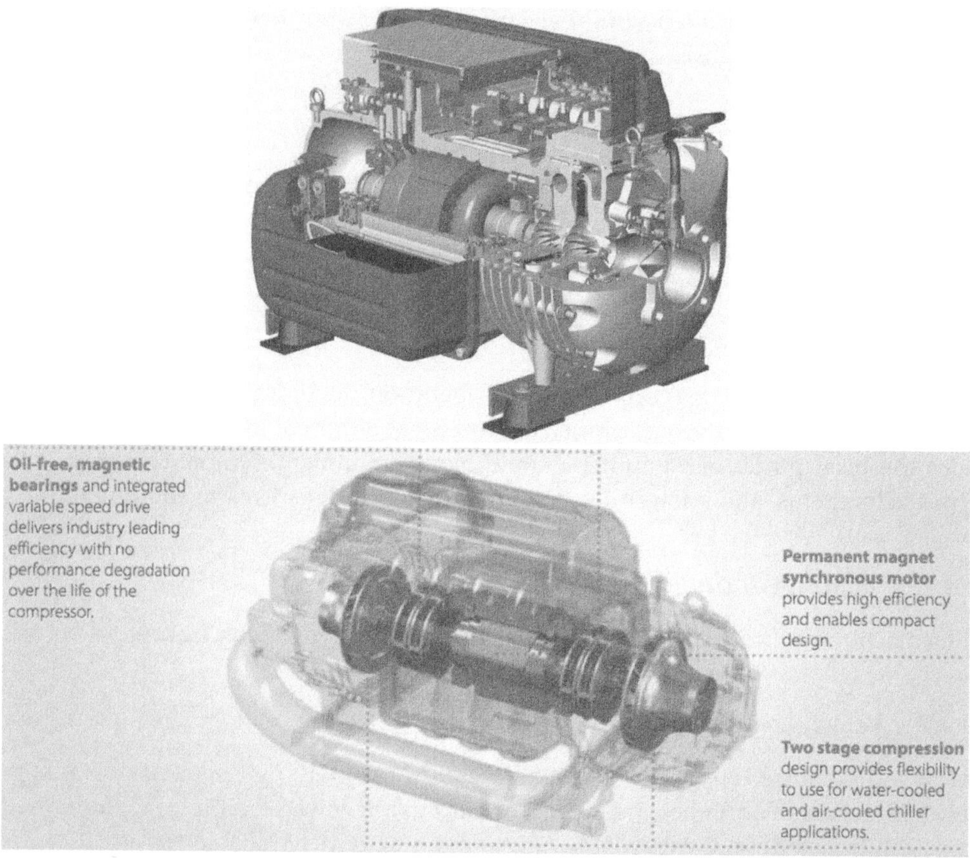

Figure 4.2 Example of large oil-free compressor [TurboCor® Compressor, Courtesy of Danfoss Corp, www.danfoss.com, 2019].

Consequently, oil circulates through the system and is returned to the compressor as a part of the air conditioning and refrigeration system during normal operation.

Liquid refrigerant should be prevented from entering the compressor because it can cause more rapid and potentially catastrophic oil loss. This phenomenon is called "flooding". It can occur while the system is running when the refrigerant superheat is allowed to stay at -17.8°C (0°F) as it enters the compressor which indicates the presence of liquid refrigerant mixed with the suction vapor.

A flooded start is flooding that occurs during startup when liquid refrigerant has been allowed to collect in the compressor, in the suction line or even in the evaporator. Both of these conditions can cause oil loss from the compressor as well as oil dilution which can result in rapid compressor wear [5].

Preventing flooding is a significant part of oil management and involves setting superheat properly and using other strategies such as crankcase heaters, non-bleed expansion valves and pump down on the off cycle to help keep liquid refrigerant out of the compressor.

Another factor in lubrication is oil breakdown that can occur at high temperatures. We should consistently monitor discharge temperatures exiting the compressor to ensure it doesn't exceed 107°C which equates to around 149°C at the compressor discharge valves (on a reciprocating compressor). This helps to ensure that the oil doesn't break down and "carbonize". This does vary based on the compressor type, system and oil type, but is a generally accepted rule in the absence of a more detailed guideline.

On a properly functioning compressor, the mass flow rate (amount of refrigerant moving through the compressor) and the suction gas temperature are the primary factors that impact compressor discharge temperature. Often high discharge temperatures occur when the suction pressure is low, superheat is low or compression ratio is high (high head pressure, low suction pressure) or some combination of these issues [5].

4.2.1 Oil Return

Once the oil has left the compressor it must circulate through the system and return to the compressor crankcase and there are a few key factors that impact oil return:

1. Oil/refrigerant miscibility (how well the oil mixes and moves with the refrigerant)
2. Oil Viscosity (oil thickness)
3. Refrigerant velocity throughout the circuit

The oil should be suited to the refrigerant type and of the proper viscosity for the compressor and the temperature application. Refrigerant velocity should be maintained according to manufacturer recommendations and low velocity will primarily be an issue in evaporator coils and suction lines when the suction pressure is lower than design due to improper tubing sizing, low evaporator load, metering device underfeeding or undercharge.

Naphthene based mineral oils are suitable for refrigeration systems using CFC or HCFC refrigerants and has been the standard oil used. Mineral oil works well with refrigerants that contained Chlorine but is not miscible with modern HFC and HFO refrigerants. Examples besides mineral oil include alkylbenzene, polyolester, polyvinyl ether, and polyalkylene glycol. Key oil variables are: miscibility, viscosity, hygroscopicity and propensity to hydrolysis [5].

4.3 Monomers and Intermediates

Consumption of fluorocarbons as intermediates for the production of fluorine-containing polymers (Table 4.8) and elastomers represents the second most important use of these compounds. These applications consumptive have been exempted from the regulations directed at ozone-depleting materials. These polymers find widespread use in industrial, commercial, and consumer applications in which their durability, chemical resistance, lubricity and dielectric properties are superior to less costly alternatives.

The primary raw materials used to produce fluoropolymers and fluoroelastomers are the hydrochlorofluorocarbons (HCFC), chlorodifluoromethane, chlorodifluoroethane (HCFC-22 and HCFC-142b) and difluoroethane (HFC-152a). While there are many polymer grades with differing compositions and properties, the largest volume fluoropolymer resins are polytetrafluoroethylene (PTFE), polyvinylidene fluoride (PVDF) and fluorinated ethylene propylene polymer (FEP). Fluoroelastomers are a relatively small part of the portfolio of fluorinated polymers but their growth rate is faster than the other fluoropolymers. Figure 4.3 shows a summary of the routes to selected fluoropolymers.

Table 4.8 Consumption of fluorocarbons as monomers for manufacturing fluoropolymers and fluoroelastomers [2].

Material	US consumption (thousands MT)
Polytetrafluoroethylene (PTFE)	26.3
Fluorinated ethylene-propylene (FEP)	12.5
Polyvinlyidine fluoride (PVDF)	11.4
Fluoroelastomers	9.0
Polyvinyl fluoride (PVF)	4.3
Perfluoroalkoxy polymers (PFA/MFA)	2.4
All other fluoropolymers (E-CTFE, ETFE, CTFE-VDF, THV PCTFE, amorphous)	6.7

4.4 Foam Blowing

A number of plastics, including polyurethane and polystyrene, have high thermal insulation properties. Those properties are utilized when the plastics are converted to foam. The process to convert them is called *foam blowing*. Foam blowing is most effectively achieved using fluorocarbon gases. The foam products exhibits high performance insulation properties at low-density which makes them desirable for many applications. HFCs replaced CFCs and partly replaced HCFCs in foam blowing. HFOs are most desirable for foam blowing because of no impact on ozone and low to no global warming potential.

The three main foam application areas are:

- Home appliances
- Building insulation
- Automotive insulation

Foam blowing agents work by imparting porosity to solid plastic bodies thus reduce the density and increase the thermal insulation properties of plastics such as polyurethanes, polystyrene, and polyolefin resins. For example, these foams are commonly used to insulate refrigerators and freezers, residential and commercial buildings, marine, automotive, and many other applications.

4.4.1 Foam Blowing Agents

Control of blowing conditions can result in the formation of closed cells that help retain the blowing agent within the foamed matrix. If the blowing agent has low thermal conductivity, the foamed products will have good insulating properties. Hydrocarbons (particularly cyclopentane), carbon dioxide/water systems, HFCs, and HFOs have replaced the fluorocarbons that have ozone depletion potential (CFCs and HCFCs).

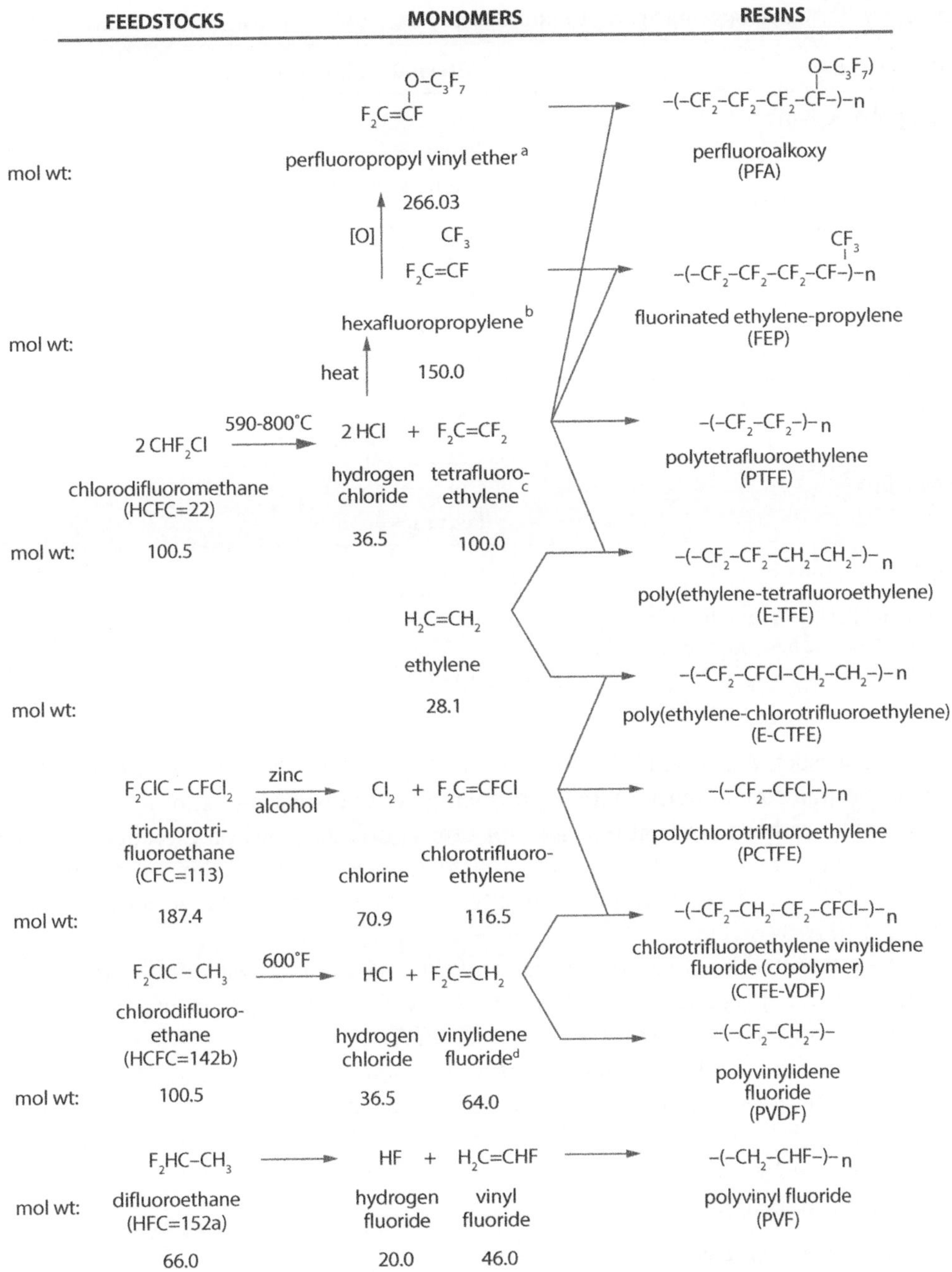

Figure 4.3 Routes to manufacturing of select fluoropolymers [6].

The type and amount of blowing agent that gives optimum insulating, flotation or cushioning performance to foams depend both on the polymer material used and the circumstances of its use. For example, where thicker foams can be used for insulation in architectural applications, hydrocarbons such as cyclopentane may be the blowing agent of choice for rigid polyurethane foams. Where thin, high-performance blown in place polyurethane foams are

Table 4.9 Alternative blowing agents to fluorocarbons for polymer foam manufacturing [2].

Polymer sytem	Alternative blowing agents[1]
Rigid polyurethane[2]	CO_2/H_2O Pentanes Butanes/pentanes with CO_2
Flexible polyurethane[2]	Liquid CO_2 CO_2/H_2O Acetone Polyols H_2O/Variables pressure foaming
Polystyrene	CO_2 Hydrocarbons
Polyolefins	CO_2 Pentanes
Polyisocyanurate[2]	Pentanes

[1]Other than HCFCs and products based on chlorine chemistry.
[2]These products are base on chlorine chemistry.

needed in items such as household refrigerators, HFCs or HFOs are preferred because they have lower thermal conductivity, thereby producing better insulating foams.

In other polyurethane applications, and for other polymers, carbon dioxide/water blowing agents might be preferred given the cost-performance tradeoffs. Depending on the type of polymer system, the major fluorocarbon alternative blowing agents for plastics are listed in Table 4.9 [2].

HFC's were targeted for phase out because of their high global warming potential. HFC fluorocarbon blowing agents—HFC-134a, HFC-245fa, and HFC-365mfc—will no longer be listed as acceptable substitutes under the EPA Significant New Alternatives Policy (SNAP) program because low global warming potential alternatives like HFOs and non-fluorinated alternatives will be used [7]. A number of alternatives to fluorocarbons have also been proposed because of their effectiveness even though they are less efficient than fluorocarbons (Table 4.9).

4.4.2 Foaming Process

Polymer foams have many advantages including lightweight, high strength, and effective sound attenuation, thermal insulation, and ability to absorb impact energy. Polymer foams are used nearly in every market and industrial segment such as construction, automotive, aerospace, agriculture, electronics, shipping and others. Polymer foams can be defined as porous material, generally speaking. Their properties and applications depend mainly on the physical and chemical properties of the polymeric matrix and the pore structure. Polymer foam is characterized by the pores' density, shape, size, anisotropy ratio and whether it is open-cell or closed-cell (Figure 4.4) [8, 9].

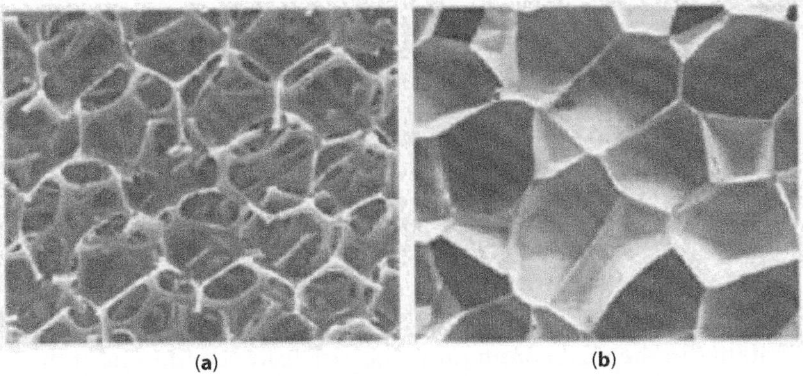

Figure 4.4 Examples of cellular solids: (a) open-cell polyurethane foam (b) closed-cell polyethylene foam [10].

In open cell foams, cells are connected with each other (Figure 4.2). They have softer and spongier appearance. Open cell foams are incredibly effective as a sound barrier in normal noise frequency ranges. It provides better absorptive capability than closed-cell foams. The disadvantage of the closed-cell foam is that it is denser, requires more material, and therefore, is more expensive. In closed cell foams, the foam cells are isolated from each other and cavities are surrounded by complete cell walls. Generally, closed cell foams have lower permeability, leading to better insulation properties. Absorb sound, especially bass tones. Closed cell foams are usually characterized by their rigidity and strength, in addition to the high R-value (Resistance to heat flow).

There are three ways a polymer may be foamed. They include mechanical, chemical and physical foaming. There are few applications for mechanical foaming. This method resembles making whipped cream during which air is blended with the bulk of melt. Mixing pulls (entrains) air into the polymer melt rendering the air into the foaming gas. An example of a use of this technique is making foam from vinyl plastisols (liquid dispersions of vinyl) for producing sheets for flooring applications.

In chemical foaming chemistry is the means of the foam formation process [11, 12]. In other words, chemistry controls the rate of formation of the polymer during which the viscous fluid is converted to a cross-linked (three-dimensional) structure. Chemistry also controls the rate of activation of the blowing agent, which is either by a drop in solubility in the monomer solution as the reaction proceeds, or by thermal degradation. The characteristics of the blowing agent determine the amount of gas generated, the rate of gas generation, the foaming pressure, and the net amount of gas retained in the cells. This process is not applicable to polymers that are not cross-linked.

Physical foaming process employs a *blowing agent* that when added to the plastic volatilizes during the melting process. Either gaseous or liquid agents can be used. Usually a nucleating compound is required to control the cell size. The foaming process conditions are determined by the chemical structure and composition of the polymer. Important variables include temperature, type of foaming agent, and the cooling rate of the expanded structure when dimensionally stabilizing the foam. The nature of the foaming agent and its concentration in the plastic determine the rate of gas evolution, gas pressure, gas retention in the cells, and heat absorption/release due to the degradation/activation of the blowing agent.

Figure 4.5 shows the evolution of the four classes of the fluorocarbon propellants. The trajectory of the developments has moved the fluorocarbons from agents harmful to ozone layer with high global warming potential to the fourth generation of hydrofluorocarbon olefins that are virtually benign on both fronts. Table 4.10 shows the polymer types and applications of foams blown using fluorocarbons. Table 4.11 shows the consumption of fluorocarbons and hydrocarbons in 2002 and 2015.

Every technique of foam manufacturing consists of the steps of cell initiation, growth, and stabilization. The usual method for classification of foaming methods is based on the cell growth and stabilization.

Initiation or nucleation refers to the formation of cells in the polymer. The nuclei are small discontinuities in the melt continuum. The expansion conditions of the foam lead to the growth of the nuclei. The drive force that determines the growth of each cell is the difference between the pressure inside and outside that cell. The pressure difference is calculated from the Eq. (4.1) [12].

$$\Delta P = \frac{2\gamma}{r} \tag{4.1}$$

Surface tension of the melt (γ) and the cell radius (r) are the variables determining the pressure difference (ΔP). Surface tension of the melt is a function of several variables including the type of polymer, temperature, pressure, and other additives present. The external pressure of the cell is the pressure that the melt is under, and the internal pressure of the cell is the pressure that is generated by the blowing/foaming agent. A gas or solid, where the blowing agent is accumulated, initiates foaming.

The process by which the cells grow is quite complex because the properties of the melt change during the cell growth phase. Few quantitative models have been proposed which completely describe the cell growth. Viscosity and pressure of the melt change during while

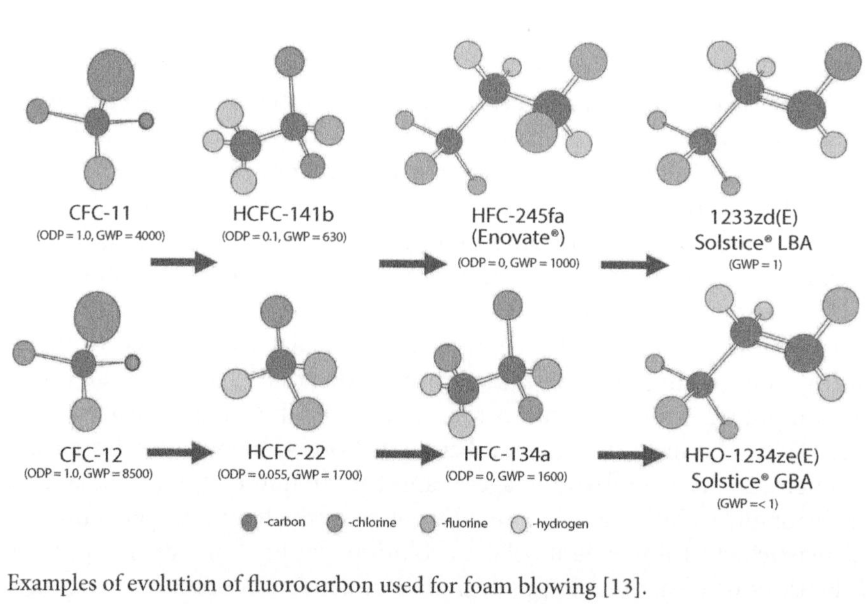

Figure 4.5 Examples of evolution of fluorocarbon used for foam blowing [13].

Table 4.10 Polymer types and applications of blown foams [14].

Foam type		Application area							
		Refrigeration & transport		Buildings & building services					
		Domestic appliances	Other appliances	Reefers & transport	Wall insulation	Roof insulation	Floor insulation	Pipe insulation	Cold stores
Polyurethane	Injected P+P	✓✓✓	✓✓	✓✓✓	✓			✓✓	
	Boardstock				✓✓✓	✓✓✓	✓		
	Cont. Panel			✓✓	✓✓✓	✓✓✓			✓✓✓
	Disc. Panel			✓✓	✓✓✓	✓✓✓			✓✓✓
	Cont. Block			✓✓		✓		✓✓✓	✓✓
	Disc. Block			✓✓				✓✓	✓✓
	Spray		✓✓		✓✓	✓✓✓		✓	
	One component				✓✓	✓			✓
Extruded Polystyrene	Board			✓✓	✓✓✓	✓✓✓	✓✓✓	✓	✓✓
Phenolic	Boardstock				✓✓✓	✓✓✓			
	One Panel				✓✓	✓			✓✓✓
	Disk Block							✓✓✓	✓✓
Polyethylene	Board						✓		
	Pipe							✓✓✓	
Mineral Fibre		✓	✓	✓	✓✓✓	✓✓✓		✓✓✓	✓

(Continued)

Table 4.10 Polymer types and applications of blown foams [14]. (*Continued*)

		Application area					
		Transport		Comfort		Packaging	Buoyancy
Foam type		Seating	Safety	Bedding	Furniture	Food & other	Marine & leisure
Polyurethane	Slabstock	√√		√√√	√√√	√√	
	Moulded	√√√		√√	√√		
	Integral Skin		√√√		√	√	
	Injected/P-I-P						√√√
	Cont. Block						√
	Spray						√
Extruded Polysterene	Sheet					√√√	
	Board						√√√
Polyethlene	Board					√√	√√

√√√ = Major use of insulation √√ = Frequent use of insulation √ = Minor use of insulation.

Table 4.11 Estimates of consumption of fluorocarbon blowing agents in 2002 and 2015 [15].

Blowing agent	CFCs	HGFCs	HFCs	Hydrocarbons
Consumption (metric tones) - 2002	11,300	128,000	11,200	79,250
Consumption (metric tones) - 2015	Nil	50,000	73,000	177,250

the cells are growing. Viscosity changes affect the rate of cell growth and the polymer flow. Pressure drop in the blowing agent, Eq. (4.1), is inversely proportional to the radius in contrast to volume. Pressure is higher in smaller cells than in larger cells. That leads to inter-cell gas diffusion or breaking of the cell walls.

4.4.3 Flexible Polyurethane Foams

Polyurethane (PU) foam is the most broadly used flexible polymer foam. A wide variety of thermal insulation, packaging materials, comfort cushions, bed mattresses, carpet backings and resilient floor coverings are manufactured with PU foams. Flexible polyurethane foams are open cell materials that allow free movement of air between the foam cavities. They are commonly available in density of 13-80 kg/m^3 [12].

In addition to polyalcohol and toluene diisocyanate (TDI), the monomers, other ingredients including solvent, additives and blowing agents are required for foam manufacturing. In the absence of a blowing agent hard polyurethane plastic is formed. Carbon dioxide that is the primary blowing agent is produced in the reaction mixture as water reacts with the isocyanate group. The PU chains crosslink with others via chemical reactions.

In practice, a number of additives are blended with the raw materials. A tin based additive is used to stabilize the foam while a few amine-based additives are added to control the extent of cross linkage and the rate of reaction. Dyes are also added to distinguish different grades and batches of foams.

There are three basic steps for producing PU foam described below and depicted in Figure 4.6 [16].

Step 1 - Mixing of the raw materials

During production (Figure 4.6), the raw materials (TDI), polyalcohol, blowing agents and additives) are pumped from their own storage tank to a common mixing chamber. Adequate dispersion can be achieved by the stirring of high-speed impeller installed in the mixer.

Step 2 - Foam forming and settling

The foam gradually solidifies when travelling up the settling chamber (Figure 4.6) by the action of paper conveyor. It is then cut into 2.2 m long blocks by an electric cutter after the foam is hardened.

Step 3 - Curing

The newly formed foam blocks are still very hot when transported to the storage area. They must be cured at room temperature for at least 18 hours before further processing.

Solid foam is formed when gas is blown through solidifying plastic. Depending on its ability to retain original shape after compression, solid foam can be classified as either flexible or rigid. Polyurethane foam is the most widely used flexible foamed plastic, being used

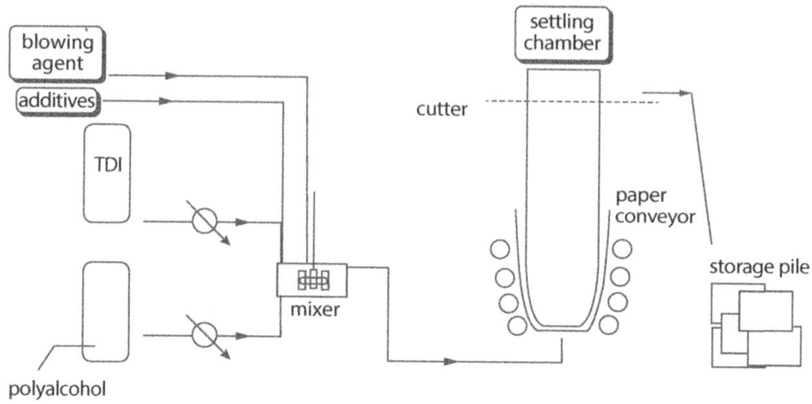

Figure 4.6 Schematic diagram of the Dunlop polyurethane flexible foam process [16].

for thermal insulation and packaging materials, cushions, bed mattresses, carpet backings and resilient floor coverings [16].

4.5 Aerosol Propellants

Aerosols can be divided into three main product categories:

1. Consumer aerosols including cleaning products, tire inflators, personal care products, spray paints, novelty aerosols, food products, safety horns
2. Technical aerosols including lubricant sprays, dusters, contact cleaners, pesticides, degreasers, mold release agents
3. Medical aerosols including metered dose inhalers (MDIs). There are also aerosols that deliver treatment for other medical purposes e.g., nasal and topical aerosol sprays.

The functions of fluorocarbon and other aerosol propellants are to expel and disperse the contents of a container in a controlled manner (Figure 4.7). This is done through selection of the proper container pressure, dispersing system design, and compatibility between the aerosol material and the compound being dispersed. When the plunger (1), in Figure 4.7, is pressed a hole in the valve (2) allows a pressurized mixture of product and propellant (3) to flow through the plunger's exit orifice.

Prior to the Montreal Protocol, CFCs, particularly CFC-11 and CFC-12, were used extensively in non-medical aerosols as propellants or solvents. In developed countries as early as in 2015, more than 98% of non-medical aerosols now use non-ozone-depleting (Table 4.12), low-GWP propellants (hydrocarbons, dimethyl ether, CO_2 or nitrogen). Some of the non-medical applications that previously used CFCs have also converted to not-in-kind technologies (spray pumps, rollers, etc.). These substitutions led to a total reduction of greenhouse-gas emissions of aerosol origin by over 99% between 1977 and 2001 [18].

HFCs in aerosol applications are approved substitutes under the EPA SNAP program because they have zero ozone depleting potential. However, HFC-134a, HFC-125, and HFC-227ea are being phased out over time as they have higher global warming potential than other alternatives. Despite the scheduled transition, consumers value their performance

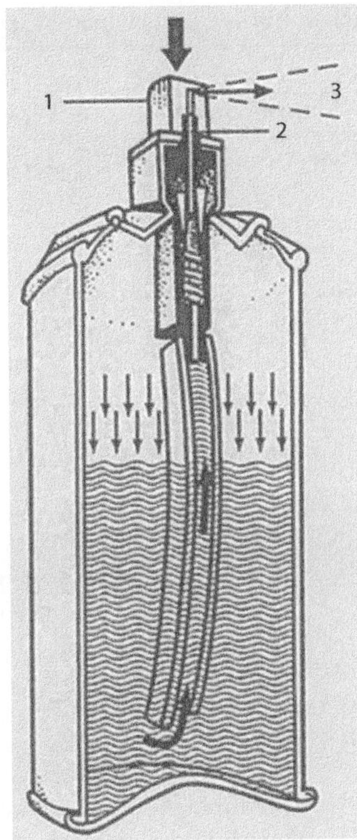

Figure 4.7 Schematic of the cut-away of the construction of aerosol can [17].

properties of inertness, vapor pressure, and nonflammability (with the exception of HFC-152a which is flammable)—advantages that have justified their higher price in certain applications [11].

HFOs, particularly HFO-1234ze (trans-1,3,3,3-Tetrafluoroprop-1-ene) and 1233zd(E) (trans-1-Chloro-3,3,3-Trifluoropropene) are replacing HFCs as aerosol propellants in many applications. Both compounds have ODP of zero and GWP<1. The major non-fluorocarbon alternatives are lower cost hydrocarbons—propane and butane—as well as a variety of other aerosol propellants, such as dimethyl ether, carbon dioxide, nitrogen, and the not-in-kind alternative, aerosol bladder (Bag-on-Valve) systems [20].

In terms of total volume, consumption of the alternative aerosol propellant technologies is much larger than that of fluorocarbons. However, use of the lower cost hydrocarbon alternatives may be constrained by both safety considerations, due to their flammability, and by limitations on the allowable release of volatile organic compounds (VOCs). Not-in-kind substitution may be possible in some applications through the use of alternative delivery systems such as aerosol bladders, hand pumps for liquids, and systems analogous to the dry powder inhalers (DPIs) used to deliver medications.

4.6 Fire Extinguishing Agents

Fire extinguishing systems are sometimes built into the building or other protected structures. Water sprinklers are by far the most common type of fixed system because they are

Table 4.12 Common commercial aerosol propellants beyond HFC's [19].

Propellants and alternatives	GWP	Flammability	Types of aerosols
Hydrocarbon blends (Propane, n-butane, iso-butane)	3	High	Aerosols that can use a highly flammable propellant
Dimethyl either (DME)	1	High	Aerosols that can use a highly flammable propellant; convenient to use with water-based formulations
HFO-1234ze	7	Low	Aerosols requiring a very low flammability propellant
Compressed gases -CO_2 -N_2 -Air -N_2O	 1 0 0 298	Non-flammable	} Some technical and consumer } aerosols } Some food products (e.g cream)
Not-in-kind - Pump sprays - Liquids - Roll-on liquids/sticks - Powders	0	Non-flammable where liquid dispensed is non-flammable	e.g cleaners, nasal sprays e.g lubricating oils e.g deodorants e.g dry powder inhalers

inexpensive, highly reliable and safe for people. But water damages building structures and their contents; it is sometimes ineffective; and it is impractical where weight and space are limited e.g., in an airplane. In these scenarios, fire extinguishers use different materials--ones that flood a protected compartment with a fire-fighting gas.

Carbon dioxide (CO_2) works well, but it is fatal at the concentrations necessary to extinguish a fire, thus cannot be used where people are present. Bromo trifluoromethane (CF_3Br, or Halon® 1301) has a relatively low boiling point and toxicity level--properties that made Halon® 1301 the firefighting chemical of choice for applications where sprinklers cannot be used. It is the most effective and common fluorocarbon for Handheld extinguishers. The choice of agent for this application had been bromo chloro difluoromethane (CF_2ClBr), referred to as Halon® 1211 until they were banned [21].

Halon® agents were widely used in fire suppression until alternatives, which have zero ozone depleting potential became available. The firefighting industry has developed disciplined practices that have made fire extinguishants nearly non-emissive, except in the case of fire. Fluorocarbons are generally used in occupied spaces where very low toxicity solutions are needed and where water is not appropriate due to the potential nature of the fire or limitations in system design [2].

Phase-out of Halon® has led to a scramble by government and industry researchers to find environmentally suitable replacements. None have been identified with all the positive

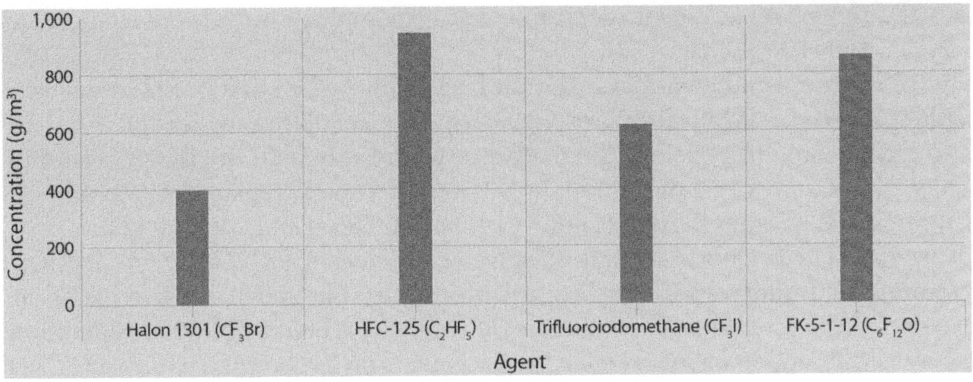

Figure 4.8 Comparison of fire-extinguishing and suppressing agents for engines [22].

qualities of Halon® 1211 and Halon® 1301. Bromine and chlorine atoms in the Halon® molecule--the very ones that are so damaging to the stratospheric ozone–are also incredibly aggressive scavengers of hydrogen atoms, which are key to maintaining a combustion chain reaction. Indeed, bromine and chlorine atoms are released as Halon® decomposes in the heat of the fire, establishing a catalytic cycle involving HBr and HCl; the cycle converts active hydrogen atoms to stable H_2 molecules, breaking the chain reaction.

Manufacturers have introduced new families of chemicals containing no chlorine or bromine, that is hydrofluorocarbons (HFCs), which have physical properties similar to the Halon® and no ozone depletion potential. But because of lacking Br or Cl atoms, unfortunately the HFCs cannot disrupt the combustion reaction to the same degree. HFCs extinguish fires in a manner similar to CO_2 or N_2--by absorbing heat and reducing the concentration of oxygen. Even so, several different companies are marketing HFCs like CHF_3, C_2HF_5, and C_3HF_7 for a variety of applications.

The replacements for Halon® are about 70% effective (Figure 4.8). This figure compares agents with concentration values published by the US Federal Aviation Administration (FAA) for minimum performance standard for engines. Significant increases in concentration (i.e., more agent) are required to demonstrate performance equivalency to that of Halon®. A challenge is presented to airframe manufacturers because any Halon® alternative system would be much larger and heavier, making integration into the airplane problematic [22].

Researchers are actively pursuing diverse materials--including iron- and phosphorous-containing compounds and hydrofluorocarbons-with the ability to inhibit flames. They are also developing better means of discharging more conventional chemicals, such as H_2O, N_2 and CO_2. For example, one idea is to use a solid propellant to generate an inert gas mixture--an approach identical to the systems in car air bags. Such a system, when activated, would extinguish a fire just as one blows out a candle [21].

4.6.1 Aerospace Fire Extinguishing

The most critical application of Halon® has been in aerospace in the aircraft engines, cargo compartment, cockpit, the cabin and the Auxiliary Power Unit. For almost 50 years, halogenated Halon® has been practically the sole fire extinguishing agents used in civil transport aircraft. The Halon® replacement agents found to be compliant include the HCFC Blend B, HFC-227ea, and HFC-236fa.

In the old airframes automatic systems released Halon® when heat or fire was detected in the cargo space. Hand-held Halon® extinguishers were available in the cockpit and the main cabin to put out any fires. Halon® alternatives are currently used in aircraft lavatories fire extinguishing systems while the search for a "drop-in" Halon® replacement continues. The ideal agent would be equivalent to Halon® with both ODP=0 and GWP=0 [23].

Boeing has stated: "efforts to find effective replacements for Halon® in airplane fire-extinguishing and suppression systems are promising, but much work remains for all stakeholders. Boeing continues to collaborate with industry groups and certification authorities to identify, certify, and implement Halon® replacements on its commercial airplanes" [22].

As of July 2016, the International Civil Aviation Organization (ICAO) has released the following statement that captures the status of Halon® replacement in aircraft applications:

1. "Halon replacement will continue to require the full cooperation and coordination of all stakeholders to advance the development of viable alternative agents. To this end, ICAO will continue to collaborate with all stakeholders to develop solutions.
2. "It is recommended that ICAO consider mandating the replacement of Halon® in cargo compartments of civil aircraft for which a new application for Type Certification [by FAA] has been submitted on or after 31 December 2024. This timeframe (Table 4.13) would allow enough opportunity for the manufacturers to transition to a non-Halon® alternative for new aircraft type designs and for States to certify such aircraft.
3. The procurement of Halon® is exclusively from recovery, reclaiming and recycling of Halon® and as Halon® reserves decline, it is expected that the rising cost of Halon® will provide a market incentive which, combined with the regulatory mandate, will drive industry to accelerate the transition to non-Halon® alternatives as well as to minimize waste of existing reserves. The continued industry-led efforts to achieve common solutions and realistic timeframes are worthy of recognition and support.

Most recently Boeing has found Halotron BrX™ to be a satisfactory replacement to Halon® 1211 in hand-held fire extinguishers. Halotron™ is a hydrobromofluorinated compound with the chemical formula of CH_2=CBr-CF_3 (2-bromo-3,3,3-trifluoro-1-propene). It has a GWP (CO_2=1, 100 years) of 0.26 and ODP (CFC11=1) of 0.0028

Table 4.13 Halon replacement cut-off dates [24].

Extinguishing systems in civil aircraft	ICAO SARP Cut-off date
Lavatory fire extinguishing systems	31 Dec 2011
Portable fire extinguishers	31 Dec 2018
Engine and APU fire extinguishing systems	31 Dec 2014
Cargo compartment fire extinguishing systems	28 Nov 2024

Table 4.14 Properties of halotron BrX™ [25].

Chemical Formula	$CF_3CBr{=}CH_2$ (stabilized with Proprietary additives)
Chemical Name	2-bromo-3,3,3-trifluoro-1-propene
Molecular Weight	174.9
Boiling Point @ 1 Atm.	93° (34°C)
Specific Gravity, Liquid	1.65
Cup Burner Extinguishing Concentration, heptane, % vol.	4.8
Atmospheric Lifetime	7 days
Ozone Depletion Potential (CFC-11=1.0)	0.0028[1]
Global Warming Potential (CO2=1.0, 100yr ITH)	0.26[1]
Cardio-Tox LOAEL, % vol.	1.0
Cardio-Tox NOAEL, % vol.	0.5
ASTM Standards for Quality Specification and Handling, Transportation and Storage	D8060, D8061

[1]*Journal of Geophysical Res*, vol 117, 2002.

(see Table 4.14). In 2016, ASTM established two Standards for Halotron BrX™. ASTM D8060 is the quality Standard Specification for Halotron BrX™. ASTM D8061 is the accompanying Standard Practice for Handling, Transportation, and Storage of the material [25].

Halotron BrX™ continues to be tested to various fire test standards. These include Underwriters Laboratories (UL), European Norm (EN), and International Standards

Table 4.15 Properties of Novec™ 1230 fire protection fluid [27].

Agent	Novec 1230	Halon 1301	HFC-125	HFC-227ea
Use Concentration[1]	4.5-6%	5%	8.7%-12.1%	6.7-8.7%
NOAEL[2]	10%[3]	5%	7.5%	9%
Safety Margin[4]	67-122%	Nil	Nil	3-34%
Global Warming Potential[5]	<1	6290	3170	3350
Ozone Depletion Potential[6]	0.0	12.0	0.0	0.0
SNAP[7] Approved	Yes	No	Yes	Yes

[1]Adjusted per 2012 NFPA 2001 requirements for minimum values (excluding halon): paragraph 5.4.2.4 and Table A.5.4.2.2(b).
[2]NOAEL for cardiac sensitization.
[3]NOAEL for acute toxicity, including cardiac sensitization.
[4]Safety Margin = (NOAEL = Use Concentration)/use Concentration.
[5]Intergovermental Panel on Climate (IPCC) 2013 Method, 100-year ITH (CO_2=1).
[6]World Meteorological Organization (WMO) 1998, Model-Derived Method.
[7]Signicant New Alternative Policy Program.

Organization (ISO) Standards. Tests against the UL711 standard include those for UL Class A, B, and C. Class A fires include fires involving common combustibles including wood. Class B fire test protocols require the repeated extinguishment of heptane fuel-in-depth pan fires with a portable fire extinguisher under controlled conditions. Class C test protocols ensure that if the agent is sprayed onto an energized electrical source that no electricity will flow back to the fire extinguisher operator. Halotron BrX™ easily passed the Class C test, as the agent is non-conductive [26].

Some of the other recent fire extinguishing candidates include Novec® 1230 Fire Protection Fluid by 3M. is a clean agent fire extinguishant which was developed as a halon replacement and an alternative HFCs. It belongs to a family of chemicals called halocarbons, a group which includes HFCs and fluoroketones. Novec 1230 fluid is a fluoroketone, while chemical clean agents like FM-200™ and ECARO-25® are HFCs (HFC-227ea, HFC-125). Novec™ 1230 fluid has a global warming potential (GWP) of less than one (Table 4.15) while these HFCs have a GWP of more than 3000. NovecTM 1230 fluid has the highest margin of safety for human occupancy among clean agents, including inert gas [26].

4.7 Cleaning and Drying Solvents

Fluorocarbons are used as solvents due to their specific properties that include: non-flammability, thermal and chemical stability, excellent dielectric properties, high material compatibility, low surface tension and viscosity, high liquid density, and low toxicity [28]. Fluorocarbon fluids are used for a wide variety of cleaning applications [29]. Their properties lend them to being excellent drying solutions after aqueous cleaning, eliminating the problems that commonly occur with oven drying, hot air knife drying or tumble drying. Some of applications include:

- Optics
- Electronic components
- Medical components
- Optical disks
- Hard disk drives

The most extensive use of specialty fluorocarbon fluids in the electronics industry is in the cleaning of printed circuit boards (PCB's) [30]. Cleaning PCB's has recently become more challenging, mainly because:

- Lead-free solder is used and the temperatures for soldering have increased, leaving behind a more stubborn residue to be cleaned.
- More is being packed onto one PCB, drastically reducing the gap between the parts that are mounted on the PCB's thus making it more difficult to clean.

An example of a suitable cleaning fluorocarbon is Vertrel™ MCA offered by Chemours Co. It is an azeotrope of 2,3-dihydrodeca-fluoropentane with trans-1,2-dichloroethylene. It is effective for use in vapor degreasing equipment. It has enhanced solvency power when compared to 2,3-dihydrodeca-fluoropentane alone, which makes it particularly effective for precision and specialty cleaning with difficult soils. Tables 4.16 and 4.17 summarize test

Tables 4.16 Plastic Compatibility with Vertrel™ MCA after immersion for 15 minutes at room temperature [31].

Compatible	
Polyethylene Polybinylchloride Polyester, PET, PBT Polyimide, PI, PEI, PAI Polyetherketone, PEK Polyaryletherketone, PEEK Polyarylsulfone, PAS Polypropylene Polyphenylene Sulfide, PPS	Acetal Epoxy Liquid Crystal Polymer Phenolic PTFE, ETFE Chlorinated PVC Ionomer ABS Polysulfone, PSO
Incompatible[a]	
Polystyrene Polyphenylene Oxide, PPO	Arcylic Cellulosic

[a]Material composition varies depending upon compounding agents, plasticizers, processing, etc. Specific materials should be tested for compatibility with solvent.

Tables 4.17 Elastomer compatibility with Vertex™ MCA after immersion for 1 week at 39°C [32].

Compatible	
Polysulfide (Thiokol FA) Chlorosulfonated PE Neoprene*	EPDM (Nordel*) Butyl Rubber*
Require Additional Testing	
Burna-N Urenethane Buna-S* Fluoroelastomers	Polychloroprene Silicone Natural Rubber

*Swelling, but with low extractables.

results on short-term exposures of unstressed plastics and elastomers simulating a typical cleaning cycle using Vertrel® MCA.

Vertrel™ MCA has "zero" ozone-depletion potential, and low global warming potential. It can replace CFC-113, methyl chloroform (1,1,1-TCA), hydrochlorofluorocarbons (HCFCs), and perfluorocarbons (PFCs) in many applications. Vertrel™ MCA is accepted by the U.S. Environmental Protection Agency (EPA) under the Significant New Alternatives Policy (SNAP) program, as a substitute for ozone-depleting substances.

Another example is Vertrel™ SDG by Chemours Co. intended to replace trichloroethylene (TCE) and n-propyl bromide (nPB) and perform in applications where maximum cleaning power is needed [33]. It can also be used as a substitute for other cleaners, such as HCFC-225 and its blends, HCFC-141b, HFEs, PFCs, CFCs, and aqueous cleaners, where safety, non-flammability and environmental concerns and/or floor space and cleanliness are paramount. It has low surface tension and strong solvency power for a wide range of soils, including oils, greases, waxes, and hydraulic fluids.

Vertrel™ SFR by Chemours Co. is an engineered mixture of nonflammable hydrofluorocarbons (HFCs), trans-1,2-dichloroethylene (t-DCE) and methanol. Vertrel™ SFR is designed to remove difficult to remove high temperature fluxes used in lead free and no clean solders. It is a strong solvent for a wide range of soils, including ionic soils. Its low surface tension and non-flammability render Vertrel™ SFR an effective ultrasonic vapor degreasing solvent. Vertrel™ SFR is a non-ozone depleter and can replace many solvents, such as trichloroethylene (TCE), n-propyl bromide (nPB), HCFC-225 blends, HCFC-141b, HFEs, PFCs, and CFCs. Vertrel™ SFR can also replace aqueous cleaners where floor space and cleanliness are at a premium.

Vertrel™ XF is HFC 43-10mee or 2,3-dihydrodecafluoropentane; empirical formula $C_5H_2F_{10}$ [34, 35]. It is a clear, colorless liquid with zero ODP and low GWP (Table 14.18) that is effective in several applications:

- Cleaning and rinsing agent
- Drying fluid
- Particulate remover
- Fluorocarbon lubricant carrier
- Solvent and dispersion media
- Heat transfer media
- Dielectric fluid
- Replacement for many HCFC, PFC, and CFC-113 applications

Another product by Chemours Co. is Opteon™ SF33 (HFO-1336mzz-Z) with a chemical formula of *cis*-1,1,1,4,4,4-Hexafluoro-2-butene). SF33 is a specialty fluid with zero ozone depletion potential and very low global warming potential of 2, as seen in Table 14.19. It's a clear, colorless, nonflammable, thermally stable, low toxicity, and environmentally benign fluid developed for cleaning, drying, and carrier fluid industrial solvent applications. The fluid has a boiling point of 33.4 °C and is appropriate for replacing PFCs, HCFCs, PFPEs, HFCs, and HFEs in industrial solvent applications [37].

Table 14.18 Environmental properties of Vertrel* XF [36].

Property	**Vertel™ XF**
Formula	$C_5H_2F_{10}$
Class	Hydrofluorocarbon (HFC)
Atmospheric Lifetime, yr	17.1
Ozone Depletion Potential (ODP)	0
Global Warming* Potential (GWP/100 yr ITH)	1650
Volatile Organic Compounds (VOC, g/L)	Exempt

*IPCC AR5 Fifth Assessment Report.

Table 14.19 Physical, environmental, and safety properties of Opteon™ SF33 [38].

Property	**Units**	**Opteon™ SF33**
Chemical Structure	–	Cis-$CF_3CH=CHCF_3$
Molecular Weight	g/mol	164
Boiling Point	°C (°F)	33.4 (92.1)
Freezing point	°C (°F)	-107 (-160.6)
Density at 25 °C (77 °F)	g/cm³	1.36
Viscosity at 25 °C (77 °F)	cP	0.38
Kb Value	–	11.3
Dipole Moment	D	2.9688
Hansen Solubility Parameters	$MPa^{1/2}$	∂_d13.9 ∂_p3.5 ∂_h2.1
Vapor Pressure at 25 °C (77 °F)	MPa	0.07
Flash Point, CC, ASTM D56	°C (°F)	None
Flash Point, OC, ASTM D1310	°C (°F)	None
Vapor Flammability, ASTM E681	%vol	None
Water Solubility	ppm	560
Critical Temperature	°C (F°)	171.3 (340.3)
Critical Pressure	MPa	2.9
Critical Density	g/cm³	0.471
Heat of Vaporization at BP	kJ/kg	166
Liquid Thermal Conductivity at 25 °C (77 °F)	W/m-k	0.077
Liquid Specific Heat at 25 °C (77 °F)	kJ/kg-k	1.2
Surface Tension	N/m	0.013
Dielectric Constant	–	32
Resistivity	ohm-cm	10^8
Break Down Voltage	kV	10
Global Warming Potential (GWP)	100-yr ITH	2
Ozone Depletion Potential (ODP)	–	0
Occupational Exposure Limit (OEL)	ppm	500

4.8 Carrier Fluids/Lubricant Deposition

Carrier and deposition solvents are among the most important materials used in lubricants, coatings and other formulations. Solvency, non-flammability, low surface tension and other physical properties impact your customer's manufacturing efficiency and safety, as well as the quality of their products [39]. They are widely used as replacements for PFC's in carrier fluid/lubricant deposition applications, especially in the hard disk drive and medical device industries [40]:

- Medical parts coating
- Hard disk drive coatings and lubricants
- Dry lubrication
- Lubrication with Krytox™ oil
- Automotive anti-squeak
- Conformal coating
    - Oil barrier
    - Flux barrier
    - Moisture barrier
    - Acid barrier

Specific applications of fluorocarbons in Carrier Fluid/Lubricant Deposition include [29]:

- Fluorolubricants deposition in the hard disk drive industry
- Silicone/PTFE-based lubricants deposition in the medical industry (Table 4.20).
- Spray adhesives carrier solvent in the adhesive industry

The benefits are several folds:

- Ideal replacement for PFCs, HCFCs, and nPB
- High solubility with various types of lubricants and adhesives
- Fast drying
- Low surface tension for coating of tight spaces
- Able to deposit a uniform layer of coating on a substrate

Table 4.20 Solubility of typical silicone fluids in Vertrel™ X-Si at room temperature (% oil loading in solvent) [41].

Oil	Vertrel™ X-Si	CFC-113	Hexane
DC-200*	14	19	25
DC-360*	21	24	46
DC-550*	33	39	58
DC-1107*	45	51	65
NuSil Med 4159	28	29	31

*As manufactured by Dow Corning Corp.

Vertrel™ X-Si by Chemours Co is a blend of Vertrel™ XF hydrofluorocarbon (2,3-dihydrodecafluoropentane) and hexamethyldisiloxane [42]. It is suited for use in medical applications as a solvent for cleaning or depositing silicone oil-based lubricants. It is also used as a swelling media for silicone rubber tubing. It has zero ozone depletion potential and low global warming potential and is exempted as a volatile organic compound. It is an alternative to environmentally regulated fluids, such as CFC-113, 1,1,1-trichloroethane (1,1,1-TCA or Methyl chloroform), HCFC-141b, and hexane.

4.9 Heat Transfer

Opteon™ SF10 is a clear, colorless, non-flammable, thermally stable fluorinated fluid developed in response to worldwide market demand for a low environmental impact heat transfer fluid, brine coolant, dielectric fluid, test fluid and other applications [43]. Characteristics of Opteon® SF10 are:

- Boiling point 110°C
- Zero ODP
- Very low GWP (2.5 GWP), 100-yr ITH

Vertrel™ Specialty Fluids are offered by Chemours Co. for heat transfer applications. They offer solutions to meet the complex challenges of heat management applications [44]. Fluorocarbon fluids offer an effective heat transfer fluid. In evaporative cooling, the fluid is enclosed in a heat cycle system. When it comes in contact with parts at high temperatures, it evaporates, cooling the parts. As the fluid moves to the cooler side of the heat exchanger system, it condenses back into liquid state. This fluid may be used to continuously cool down the parts that could be damaged due to excessive heat. Key characteristics of Vertrel® Specialty Fluids include:

- Zero ozone depletion potential (ODP)
- Nonflammable
- Solubility with a range of organic and fluorinated solvents
- An ability to be enhanced with additional components to form azeotropes designed for cleaning
- Low viscosity, low surface tension, and high density to penetrate tight or hard to reach spaces
- Safety for use
- Chemical and thermal stability
- Compatibility with most metals, plastics, and elastomers

Another example of a fluid is Opteon™ Sinera specialty fluid by Chemours Co. It is a clear, colorless, nonflammable, thermally stable, fluorinated fluid developed in response to worldwide market demand for a low environmental impact solvent for cleaning, as a carrier fluid, and many other applications. The fluid has a boiling point of 110 °C (230 °F), zero ozone depletion potential and a very low global warming potential. This fluid is appropriate to replace PFCs, PFPEs, HFEs, HFCs, and HCFCs in carrier fluid applications [45].

4.10 Etchants

The heart of chip manufacturing is lithography. It's like silk-screening, except instead of squeegeeing ink through a silk template onto a cotton T-shirt, you're shining ultraviolet light through a glass photomask onto a silicon substrate coated with an organic compound called photoresist. Where the UV light shines through, it chemically weakens the photoresist, leaving a pattern on the surface of the silicon. Then the wafer is sent through a chemical bath that etches trenches into the exposed substrate, while leaving the areas covered by the photoresist untouched. After removing the photoresist, other machines can fill those trenches with various materials, such as copper or aluminum that comprise the components of the processor [46].

4.10.1 What is Etching?

Dry or plasma etching is used for circuit-defining steps, while "wet" etching using chemical baths is aimed mainly to clean wafers. Dry etching is one of the most frequently used processes in semiconductor manufacturing – often with fluorocarbon gases. Before the start of etching a wafer is coated with photoresist or a hard mask (usually oxide or nitride) and exposed to a circuit pattern during photolithography. Etching removes material only from the pattern traces. This patterning and etching sequence is repeated multiple times during the chip building process [47].

Fluorocarbon-based plasma etching of SiO_2 and low di-electric constant films is an important processing step in integrated circuit manufacturing. One of the most widely employed plasma sources for this task is an inductively coupled plasma source. Process optimization in inductively coupled fluorocarbon plasmas has been found to be challenging due to the complexity of the fluorocarbon plasma etch mechanism [48].

Etch processes are referred to as conductor etch, dielectric etch, or polysilicon etch to indicate the types of films that are removed from the wafer. For example, dielectric etch is involved when an oxide layer is etched to leave "oxide isolators" separating devices from each other; polysilicon etch is used to create the gates in a transistor; dielectric etch is employed to etch *via holes* and trenches for metal conductive paths; and metal etch removes aluminum, tungsten, or copper layers to reveal the pattern of circuitry at progressively higher levels of the device structure [47].

Plasma etching begins by applying radio frequency electromagnetic energy to a gas containing a chemically reactive element, such as fluorine or chlorine. The plasma releases positively charged ions that bombard the wafer to remove (i.e., etch) materials and chemically reactive free radicals that react with the etched material to form volatile or nonvolatile byproducts. The electric charge of the ions directs them vertically toward the wafer. This produces the almost vertical etch profiles essential for the miniscule features in densely packed chip designs. Process chemistries differ depending on the types of films to be etched. Those used in dielectric etch applications are typically fluorine-based. Silicon and metal etch use chlorine-based chemistries [47].

4.10.2 Fluorocarbon Etchants

Fluorocarbon gases like C_2F_6, CHF_3, and c-C_4F_8 are important for plasma etching of SiO_2 in the semiconductor industry (Tables 4.21-4.23). During fluorocarbon plasma etching of

SiO_2, a polymer passivation layer is generally deposited on the surface of the wafer. The polymer layer regulates the etching by limiting the availability of activation energy and reactants and providing the fuel for removal of oxygen. They also contribute to global warming thus require consideration and evaluation of candidate gases with lower global warming impact [49]. Yet the candidates must perform satisfactorily in SiO_2 etching. Among them hexafluorobutadiene (C_4F_6) has been found to be effective. It has a GWP=0 and performs quite well as an etchant.

Hexafluoroethane is a versatile fluorocarbon etchant in semiconductor manufacturing (Table 4.21). It can be used for selective etching of metal silicides and oxides versus their metal substrates and also for etching of silicon dioxide over silicon. The primary aluminum and the semiconductor manufacturing industries are the major emitters of hexafluoroethane.

Hexafluorobutadiene (C_4F_6) is an unsaturated fluorocarbon with an alternating double bond (Figure 4.9). It exhibits very high performance for plasma, ion beam, or sputter etching in semiconductor device manufacturing with Xenon (Xe) and Argon (Ar) as carrier or dilution gas (Table 4.24). In using Argon as carrier gas, nitride selectivity can result in a

Table 4.21 Properties of hexafluoroethane (R116).

Property	**Unit**	**Value**
Formula		C_2F_6
GWP		9200
ODP		0
Physical State		Gas
Color		Colorless
Water Solubility		Slight
Toxicity		Toxic
Flammability		Non-flammable
Odor		Slight Ethereal
MW	g/mole	138.01
Boiling point	°C	-40.8
Freezing point	°C	-100.7
Critical Temperature	°C	19.67
Critical Pressure	mPa	2.98
Gas Density at @15°C	kg/m^3	5.84
Specific Volume @21.1 °C	m^3/kg	0.175

Table 4.22 Properties of fluoroform (R23).

Property	Unit	Value
Formula		CHF_3
GWP		14800
ODP		0
Physical State		Gas
Color		Colorless
Water Solubility	g/l	1
Toxicity		Toxic
Flammability		Non-flammable
Odor		Odorless
MW	g/mole	70.01
Boiling point	°C	-82.1
Melting point	°C	-155.2
Vapor Pressure @20°C	mPa	4.38
Gas Density @15°C	kg/m³	2.946
Specific Volume @15°C	m³/kg	0.339

decreasing function. The addition of Ar is thought to increase the importance of ion bombardment relative to neutral and direct reactive ion etching. Xenon as the other carrier gas has an increased selectivity. Therefore, mixtures of Ar/Xe are a good compromise [50, 51].

Like other fluorocarbon etchants it leaves a residual polymer on the dielectric surface. In the C_4F_6 molecule the C/F ratio is high enough to control the amount of polymer on the chamber surface and wafer surface with respect to F etching radicals. More importantly its intrinsic characteristics allow high selectivity to substrates or photo resists and wider process windows compared to e.g. C_4F_8. This behavior is particularly beneficial to address the need for geometries smaller than 0.25 μm [50].

4.11 Medical Applications

Even though small quantities of fluorocarbons are consumed in medical applications their role is quite important. A significant application of fluorocarbons in medicine is as anesthetics and analgesics. The compounds currently used for general anesthesia include isoflurane, desflurane, sevoflurane and methoxyflurane (Table 4.25).

In 1951, the British chemist Dr. Charles Suckling began an attempt to prepare the ideal inhalation anesthetic gas. His research led to the first synthesis of halothane (CF_3-CHClBr). Halothane was an excellent inhalation anesthesia except for a rare but severe detrimental

Table 4.23 Properties of octafluorocyclobutane (RC318).

Property	Unit	Value
Formula		c-C_4F_8
GWP		10300
ODP		0
Physical State		Gas
Color		Colorless
Water Solubility	g/liter	Slight
Toxicity		Toxic
Flammability		Non-flammable
Odor		Ethereal
MW	g/mole	200
Boiling point	°C	-5.8
Melting point	°C	-40.2
Critical Temperature	°C	114.8
Critical Pressure	kPa	270
Relative Gas Density	air =1	6.9
Specific Volume @21.1 °C	m^3/kg	0.175

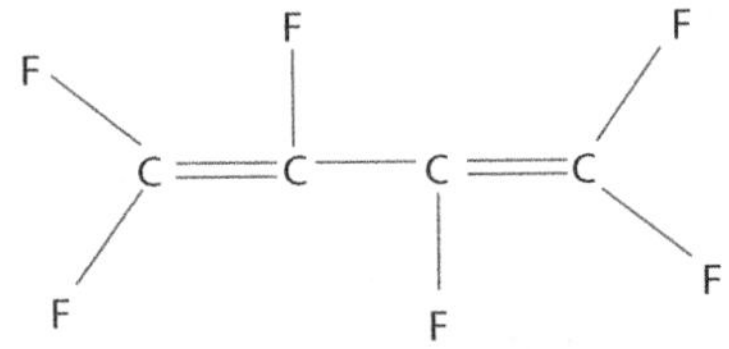

Hexafluoro-1,3-butadiene

Figure 4.9 Chemical structure of hexafluorobutadiene.

hepatic side effect. Its use as an anesthesia was discontinued in year 2000 because of the development of safer sevoflurane [53].

Dr. Suckling's pioneering work spurred the development of other fluorocarbons besides halothane including enflurane, isoflurane, desflurane, sevoflurane, and methoxyflurane. Every year, more than 46 million inpatient surgical procedures are performed in the United

Table 4.24 Properties of hexafluorobutadiene.

Property	Unit	Value
Formula		C_4F_8
GWP		0
Physical State		Gas
Color		Colorless
Odor		Odorless
Toxicity		Toxic
Flammability		Flammable
Explosive Limits	%	7-73
MW	g/mole	162.03
Boiling point	°C	6.5
Melting point	°C	-132
Critical Temperature	°C	139.6
Vapor Pressure	kPa	172.4
Vapor Density at 15 °C	air =1	6.79
Specific Volume	m^3/kg	0.040
Gas Density	kg/m^3	25

Table 4.25 Fluoro-ether anesthetics [52].

Anaesthatic name	Chemical structure	Introduction
Enflurane	CHF_2-O-CF_2-CHClF	1973
Isoflurane	CHF_2-O-CHCl-CF_3	1981
Desflurane	CHF_2-O-CHF-CF_3	1992
Sevoflurane	Ch_2F-O-$CH(CF_3)_2$	1994

States, a large number of which require general inhalation anesthetics [54]. Fluorocarbon agents play a major role as anesthetics, which have led to drastic reduction in the risks of using anesthesia.

4.11.1 Enfluorane

Enflurane is 2 chloro-1,1,2,-trifluorethyl-difluoro-methyl ether. Enflurane is a fluorinated ethyl methyl ether with a blood/gas coefficient of 1.8. Introduction of anesthesia is accomplished by using 3–4% enflurane in air or in oxygen or 1.5–3% in 65% nitrous oxide.

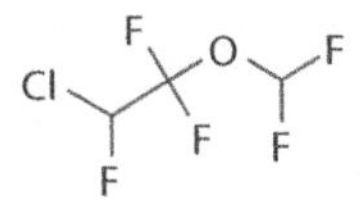

Enflurane is a potent and pleasant-smelling colorless liquid, stable in the presence of metals, alkali and indirect natural light, preservatives are unnecessary. Mixtures of enflurane and oxygen are neither flammable nor explosive in all proportions.

Enflurane is a maintenance rather than an induction agent. It enhances the effect of paralyzing agents more than other inhalation anesthetics. The impact of enflurane on arterial blood pressure reduction is dose-dependent because of its negative inotropic effects. Enflurane does not cause sensitivity of the heart. It also does not elicit *coronary artery steal* (a reduction in the blood flow directed to the coronary circulation). Enflurane can increase intracranial pressure (pressure inside the skull) and, especially in combination with hyperventilation, raises the risk of seizure. As in the case of other inhalation anesthetics, it can also trigger malignant hyperthermia (fatal rise in body temperature). Lucidity (clearance) from anesthesia with enflurane is slower than isoflurane [55, 56].

4.11.2 Isoflurane

Isoflurane is marketed under the trade name Forane®. It has a chemical composition of 1-chloro-2, 2,2-trifluoroethyl difluoromethyl ether, and its structural formula is:

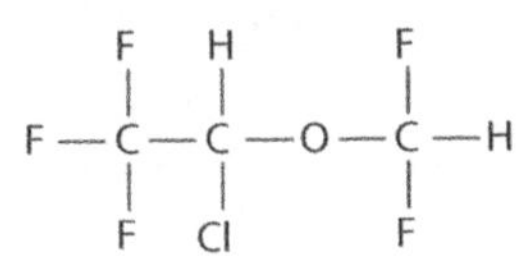

Isoflurane is a clear, colorless, stable liquid containing no additives or chemical stabilizers and has a boiling point of 48.5°C. Isoflurane has a mildly pungent, musty, ethereal odor. Samples stored in indirect sunlight in clear, colorless glass for five years, as well as samples directly exposed for 30 hours to a 2 amp, 115 volt, 60 cycle long wave ultraviolet light were unchanged in composition as determined by gas chromatography. Isoflurane placed in one normal sodium methoxide-methanol solution, a strong base, for over six months consumed essentially no alkali, indicative of strong base stability. Isoflurane does not decompose in the presence of soda lime (at normal operating temperatures), and does not attack aluminum, tin, brass, iron or copper.

Isoflurane is administered by vaporizing when applied as a general inhalation anesthetic drug. Induction of and recovery from isoflurane anesthesia are rapid. Isoflurane has a mild pungency, which limits the rate of induction, although excessive salivation or tracheobronchial secretions do not appear to be stimulated. The level of anesthesia may be changed rapidly with isoflurane, which is a profound respiratory depressant [57, 58].

4.11.3 Desflurane

Desflurane is marketed under the trade Suprane®. It has a chemical composition of 1,2,2,2-tetrafluoroethyl, and its structural formula is:

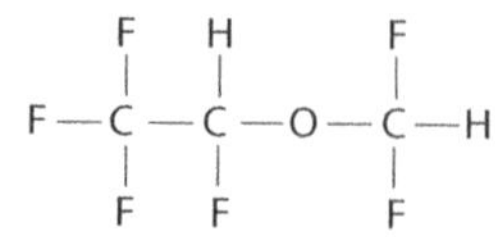

Desflurane is a colorless, volatile liquid below 22.8°C; it boils at 23.5°C. Data indicate it is stable when stored under normal room lighting conditions according to handling instructions. No discernible degradation occurs in the presence of strong acids. Desflurane does not corrode stainless steel, brass, aluminum, anodized aluminum, nickel plated brass, copper, or beryllium. Desflurane is chemically stable and its only known degradation reaction is through prolonged direct contact with soda lime producing low levels of fluoroform (CHF_3). The amount of CHF_3 obtained is similar to that produced by doses of Isoflurane in a monitored anesthesia care (MAC). American Society of Anesthesiologists define MAC as a planned procedure during which the patient undergoes local anesthesia together with sedation and analgesia [57].

Desflurane is administered by inhalation. The administration of general anesthesia must be individualized based on the patient's response. Hypotension and respiratory depression increase as anesthesia with Desflurane is deepened. In adults, some premedicated with opioid, a frequent starting concentration was 3% Suprane®, increased in 0.5-1.0% increments every 2 to 3 breaths. End-tidal concentrations of 4-11%, Suprane with and without N_2O, produced anesthesia within 2 to 4 minutes. When Suprane® was tested as the primary anesthetic induction agent, the incidence of upper airway irritation (apnea, breath holding, laryngospasm, coughing and secretions) was high [57].

4.11.4 Sevoflurane

Sevoflurane is marketed under the trade names Ultane® (by AbbVie Inc) and Sojorn® (Piramal Critical Care, Inc). It has a chemical composition of 2,2,2,-trifluoro-1-(trifluoromethyl) ethyl ether, and its structural formula is:

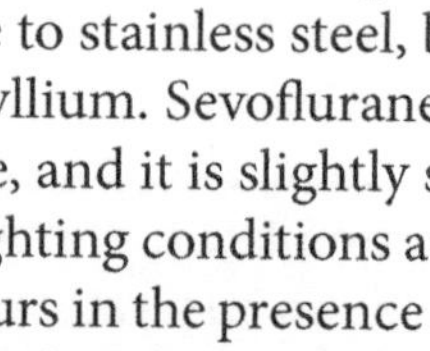

Sevoflurane is a clear, colorless, liquid containing no additives and has a boiling point of 58.1°C. Sevoflurane is not corrosive to stainless steel, brass, aluminum, nickel-plated brass, chrome-plated brass or copper beryllium. Sevoflurane is non-pungent, miscible with ethanol, ether, chloroform, and benzene, and it is slightly soluble in water. Sevoflurane is stable when stored under normal room lighting conditions according to instructions. No discernible degradation of Sevoflurane occurs in the presence of strong acids or heat. When in contact with alkaline CO_2 absorbents within the anesthesia machine, Sevoflurane can undergo degradation under certain conditions intensified by increase in temperature [57, 58].

Sevoflurane alkaline degradation occurs by two pathways. The first results from the loss of hydrogen fluoride with the formation of pentafluoroisopropenyl fluoromethyl ether, (PIFE,

$C_4H_2F_6O$), also known as Compound A, and trace amounts of pentafluoromethoxy isopropyl fluoromethyl ether, (PMFE, $C_5H_6F_6O$), also known as Compound B. The second pathway for degradation of Sevoflurane occurs primarily in the presence of desiccated CO_2 absorbents. Since the reaction of carbon dioxide with absorbents is exothermic, the temperature increase will be determined by quantities of CO_2 absorbed, which in turn will depend on fresh gas flow in the anesthesia circle system, metabolic status of the patient, and ventilation [57, 58].

Sevoflurane is an inhalational anesthetic agent for use in induction and maintenance of general anesthesia. Minimum concentration of Sevoflurane in oxygen for a 40-year-old adult is 2.1%. The low solubility of Sevoflurane facilitates rapid elimination via the lungs.

4.11.5 Methoxyflurane

Methoxyflurane was formerly marketed as Penthrane®. It is a halogenated ether in the form of a clear, colorless liquid that boils at 104.8°C. Its vapor has a strong fruity aroma. Methoxyflurane is miscible with ethanol, acetone, chloroform, diethyl ether, and fixed oils. It is soluble in rubber and has a chemical formula of $C_3H_4Cl_2F_2O$ and structure of:

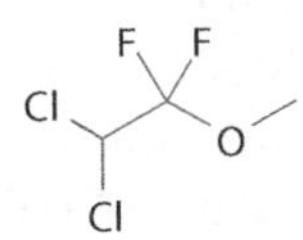

Methoxyflurane was in clinical use as a volatile inhalational anesthetic from its introduction in 1960 until the late 1970's. Methoxyflurane has a very high lipid solubility, which gives it very slow pharmacokinetics (induction and emergence characteristics); this is undesirable for routine application in the clinical setting.

The history of Methoxyflurane is fascinating, being widely used as an anesthetic long before it became clear that it caused dose-dependent permanent renal impairment. Anesthetic use ceased in the 1970's but analgesic use continued in Australian ambulances. In New Zealand, Methoxyflurane analgesia was withdrawn during the 1980's due to safety concerns. It is now supplied via the handheld Penthrox® inhaler and has recently become available in nearly all Australasian ambulances, hospital wards, medical centers, dental surgeries, ski fields and the military [59].

4.12 Usage of HCFCs and HFCs

There is insufficient data at the time of publication of this volume for the usage of replacement fluorocarbons such as HFOs. To provide the reader with an understanding of breakdown of consumption of fluorocarbons this section describes the usage of hydrofluorocarbons in various applications. Its content is based on a document published in 2016 on the occasion of a conference held by United Nations Environment Program (UNEP) in Kigali, Rwanda.

The Kigali Conference was held October 10-15, 2016 during which the Montreal Protocol was adjusted and a deal reached on a timetable that would mandate countries to phase down the production and usage of hydrofluorocarbons (HFCs) by the mid 21st Century. If implemented, global warming could be reduced by up to 0.5°C [60].

4.12.1 Introduction

The European FluoroCarbons Technical Committee (EFCTC) has published a detailed list of applications of fluorocarbons seen in Figure 4.10 without providing consumption quantities. The main applications for various fluorocarbons and blends and their quantitative contribution to global warming potential in 2012 were discussed in the 2016 Kigali document [60]. The five main application classes are seen below:

1. Refrigerants in RACHP (refrigeration, air-conditioning and heat pumps)
2. Propellants in aerosols
3. Blowing agents to manufacture insulating foam
4. Fire protection fluids
5. Solvents

The relative use of different fluorocarbons has changed significantly over the last three decades. Prior to the discovery of the ozone problem, the largest use of CFCs was for aerosols. The solvent market was also significant at that time. During the phase-out of CFCs, the applications structures have changed and large parts of the aerosol and solvent markets switched to not-in-kind (NIK) alternatives. The RACHP applications have grown significantly in which most CFC and HCFC consumption have switched to HFC products. Before the development of a strategy for reducing use of HFCs, it was important to understand the key sectors and sub-sectors that consume both HCFCs and HFCs [60].

The UNEP classifies the various fluorocarbons into four generations. There are minor variations among the generation classifications of UNEP and those by others. For instance, some authors focused on refrigerants have classified the substances used prior to the advent of CFC's, such as ammonia and sulfur dioxide, as first generation. The UNEP classification provides a convenient tool to discuss the differences among the fluorocarbons in all applications. The primary factors that characterize those classifications are ozone depletion (ODP) and global warming (GWP) potentials.

4.13 Breakdown of Fluorocarbons in Applications

The data in this section is based on the approximate breakdown of the global HCFCs and HFCs consumption in 2012. Data for 2012 has been used because represents a year in which there was little influence from the use of the 4th Generation Substances. Figure 4.10 indicates the majority of fluorocarbons are consumed by refrigeration, air-conditioning and heat pumps applications, expressed in metric tons. Figure 4.11 is "GWP-weighted", with the consumption expressed in tons of CO_2 equivalent. The RACHP uses are even more dominant in Figure 4.11 because they consume high GWP HFCs such as R-404A and R-410A in contrast to the aerosol and foam that use lower GWP HFCs.

Every application class consists of sub-applications, each of which has specific requirements that affect the selection of fluorocarbon. For example, in the RACHP applications, the type of equipment used is mainly based on a very similar technical process – the vapor compression cycle. However, the temperature of operation can vary considerably in different subsectors of this market. In the refrigerant uses evaporation temperature are defined for the most part according to the following:

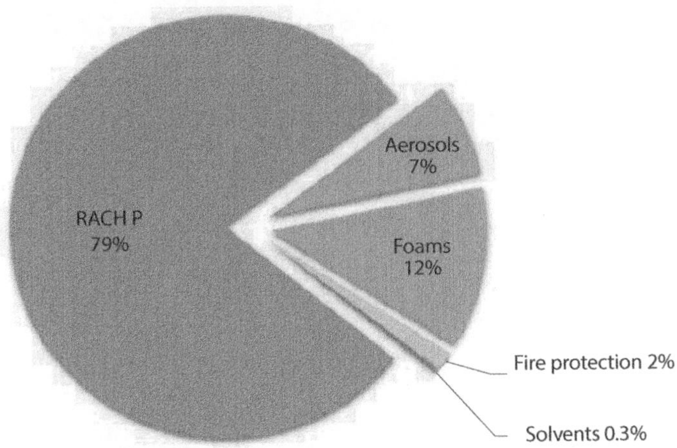

Figure 4.10 Breakdown of HCFCs and HFC consumption in applications by mass [60].

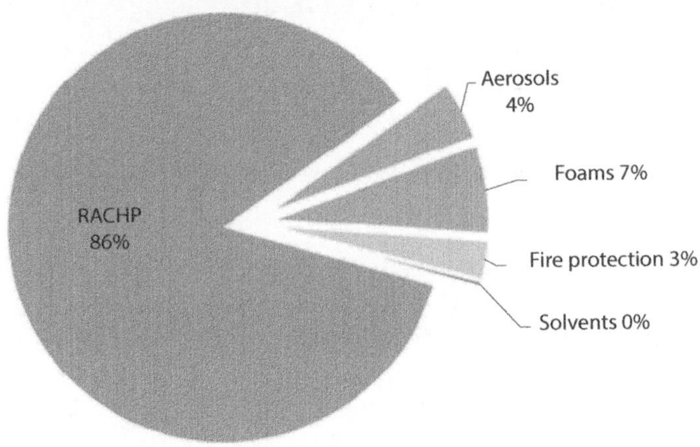

Figure 4.11 "GWP-weighted" breakdown of HCFCs and HFC consumption expressed in tons of CO_2 Equivalent [60].

-40°C for freezing ice cream
0°C for storing chilled foods
+10°C for air-conditioning
+30°C for a heat pump

The optimum thermodynamic properties for each of these applications vary quite a bit resulting in the selection of different refrigerants for each temperature value. The size and location of equipment can have an impact on the refrigerant selection. A large industrial refrigeration system may choose a refrigerant such as ammonia (which is toxic and slightly flammable). Small air-conditioning units in residential building require a fluorocarbon that is non-toxic and non-flammable to minimize risk to human and property.

The largest application of fluorocarbons in terms of GWP contribution comes from RACHP uses (86%, Figure 4.11), which will be the focus of this discussion. The combined use of HCFCs and HFCs is subdivided into different sub-segments of the largest main

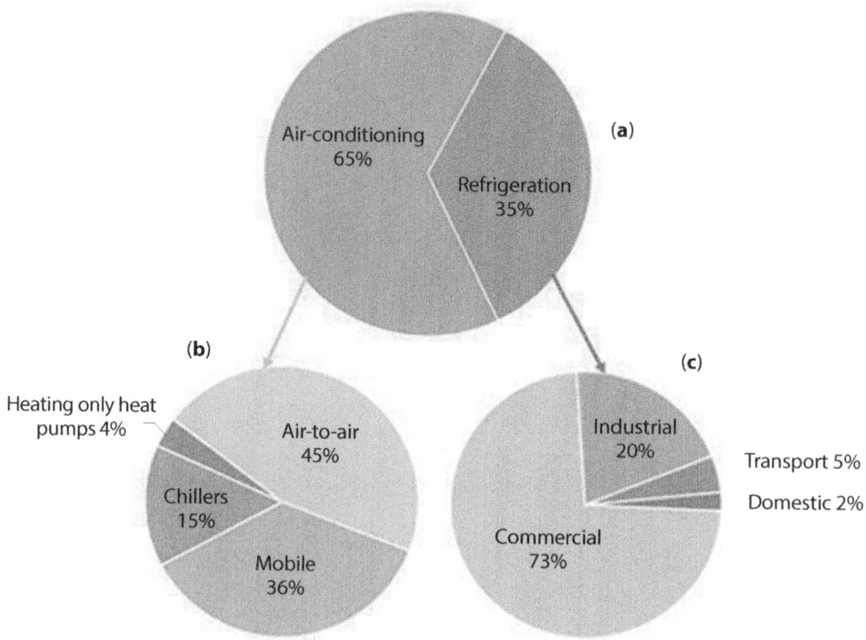

Figure 4.12 Breakdown of RACHP applications based on the depiction in Figure 4.11 [60].

applications. The percentage quantities shown in Figure 4.12 are global averages as opposed to the actual splits. The actual data vary from country to country. For example, the size of the air-conditioning market will be much larger in hot countries than the average displayed in Figure 4.12. Understanding the usage split in individual countries is a very important step in the development of a targeted HFC phase-down strategy (see Appendix II Kigali Fact Sheet No. 6 for more details).

The split between air-conditioning and refrigeration [Figure 4.12(a)] is 65% and 35%, respectively. A simple calculation shows globally over 56% of the total GWP is contributed by the use of fluorocarbons in air-conditioning applications. Refrigeration share amounts to 30% of the total GWP contribution.

Air-to-air air-conditioning [Figure 4.12(b)] use has a share of 45% of GWP contribution followed by mobile (automotive) at 36%, chillers at 15% and 4% by heating heat-pumps. To put those numbers in context 25% of the total global GWP stems from air-to-air air-conditioning followed by 20% total GWP contribution by mobile units.

4.14 Summary

Nearly three quarters (73%) of the refrigeration GWP is contributed by the commercial end-use, translating into 22% of the total global GWP. Industrial applications has a share of 20% of refrigeration GWP. Transport at 5% and domestic at an amazingly low 2% of the GWP of refrigeration round off the global warming potential contributions. Table 4.26 summarizes the contribution of various applications to global warming potential because of fluorocarbons use.

Table 4.26 GWP-weighted" breakdown of HCFCs and HFC consumption expressed in tons of CO_2 equivalent [60].

Application	Contribution to global warming potential, %
Air Conditioning	
Air-to-Air	86x0.65x0.45=25.2
Mobile	20.1
Chillers	8.4
Heating only heat pumps	2.2
Refrigeration	
Commercial	22
Industrial	6
Transport	1.5
Domestic	0.6
Aerosols	4
Foams	7
Fire Protection	3
Total	100.0

The data in Table 4.26 informs of the main causes the majority of global warming potential. Air-to-air and mobile air conditioning and commercial refrigeration contribute two thirds (67%) of global warming potential. Chillers, industrial refrigeration, aerosol and foaming applications together contribute a total of 25.4% to global warming potential. While pursuit of heat pumps, domestic refrigeration and fire protection applications are respectable, the net impact on the global warming potential is quite limited.

The challenge after development and commercialization of fluorocarbons with zero ODP was to reduce GWP of those gases. The good news is there has been a continuous decrease in the GWP of fluorocarbon gases as HFOs have been adopted an example of which can be seen in Figure 4.13. The average GWP of HFCs/HFOs placed on the EU market has decreased by about 23% since the phase-out of HCFCs at the end 2009 and has fallen every year since then, with a steeper decline following the 2014 F-Gas Regulation. The increasing use of extremely low GWP HFOs (and HCFOs) has contributed to this reduction and in 2017 these accounted for about 13% of the total HFCs/HFOs by weight, a significant increase from about 1% in 2014. The HFOs and HCFOs are being used alone and in blends with HFCs in a range of applications including mobile air-conditioning, refrigeration, stationary air-conditioning and heat pumps, foams and technical aerosols [28].

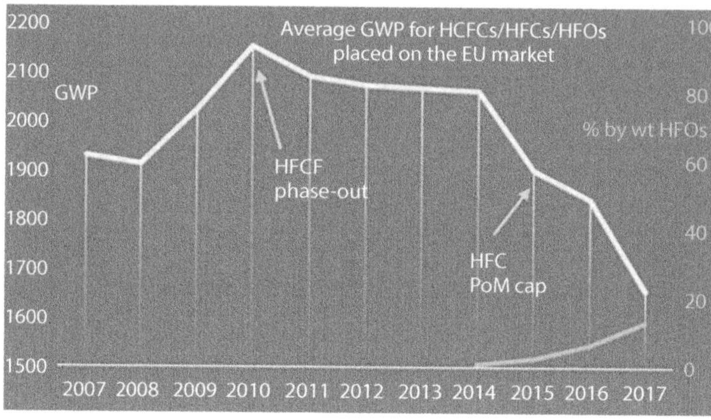

Figure 4.13 Decrease in the average GWP in the EU market – POM= placing on the market [28].

References

1. Booten, C., Nicholson, S., Mann, M., National Renewable Energy Lab, Abdelaziz, O., Oak Ridge National Lab, *Refrigerants: Market Trends and Supply Chain Assessment*, Clean Energy Manufacturing Analysis Center, Feb 2020.
2. *The Economic Benefits of Chlorine Chemistry in Fluorocarbons in the United States and Canada, IHS Chemical Consulting*, American Chemistry Council, May 2016.
3. *Refrigerant Market Volume by Application*, Grand View Research, www.grandviewresearch.com, Jan 2018.
4. *Substitutes in Refrigeration and Air Conditioning*, US Environmental Agency, www.epa.gov/snap/substitutes-refrigeration-and-air-conditioning, April 2020.
5. Orr, B., *Refrigerant Oil Basics*, HVAC School, www.hvacrschool.com/refrigerant-oil-basics, April 4, 2019.
6. Will, R.K., Kälin, T., Mori, H., Zhang, Y.E., Fluoropolymers, in: *Chemical Economics Handbook*, SRI Consulting, 2008.
7. Federal Register. 80, 138, available online at www.gpo.gov/fdsys/pkg/FR-2015-07-20/pdf/2015-17066.pdf, July 20, 2015.
8. Ma, Z.L. and Zhang, G.C., Microstructure and properties of foam plastics. *Eng. Plast. Appl*, 39, 11, 96–100, 2011.
9. Eaves, D., *Handbook of polymer foams*, RAPRA Technology Limited, Shrewsbury (UK), 2004.
10. Lee, L.J., Zeng, C.C., Cao, X., Han, X.M., Shen, J., Xu, G.J., Polymer Nanocomposites Foam. *Compos. Sci. Technol.*, 65, 2344–2363, 2005.
11. Suh, K.W., Foamed plastics, in: *Kirk-Othmer encyclopedia chemical technology*, 4th ed, vol. 11, pp. 730–83, John Wiley & Sons, New York, 1994.
12. Suh, K.W., *Foamed plastics, in on-line edition of Kirk Othmer encyclopedia chemical technology*, John Wiley & Sons, New York, 2000.
13. *Blowing Agents Technology*, Honeywell Corp, www.honeywell-blowingagents.com, December 2016.
14. *Guidance on the Process for Selecting Alternatives to HCFCs in Foams Sourcebook on technology options for Safeguarding the Ozone Layer and the Global Climate System*, United Nations Environment Program, 2010.
15. Ashford, P., Ambrose, A., Jeffs, M., *et al.*, Chapter 7: Foams, IPCC/TEAP Special Rep: Safeguarding the Ozone Layer and the Global Climate System, December 2016.

16. Yuen, D., *Based on the original article: Foams, Kirk-Othmer Encyclopedia of Chemical Technology*, 3rd ed, vol. 11, 88-89, March 2007.
17. *Fire Protection Handbook*, 19th ed, National Fire Protection Association (NFPA), January 2003.
18. Rand, S., Yamabe, M., Campbell, N., Hu, J. *et al.*, Chapter 10: Non-Medical Aerosols, Solvents, and HFC 23, in: *Safeguarding the Ozone Layer and the Global Climate System*, www.ipcc.ch/pdf/special-reports/sroc/sroc07.pdf, December 2016.
19. Tope, H., Energy International Australia. *Workshop on HFC Management: Technical Issues, Side Event*, Bangkok, 21st April 2015.
20. Douglas Fratz, D., *Aerosol Product Technology and Current Regulatory Definitions*, Consumer Specialty Products Association, Washington, DC, available at www.nist.gov/pml/wmd/upload/Doug-Fratz-CSPA-NIST-Meeting-1-9-14.pdf.
21. Scientific American, Grosshandler, W.L., www.scientificamerican.com/article/what-chemicals-are-used-i, leader of the Fire Sensing and Extinguishment Group in the Building and Fire Research Laboratory at the National Institute of Standards and Technology (NIST), accessed Dec 2016.
22. Bennett, R., *Replacing Halon® in Fire Protection Systems: a Progress report*, WWW.boeing.com/commercial/aeromagazine, 4th Quarter, 2011.
23. International Civil Aviation Organization, *Update on the development of Halon® Alternatives for Aircraft Fire Extinguishing Systems, A38-WP/36, TE/2*, June 6, 2013.
24. International Civil Aviation Organization (ICAO), SARPs in Annex 6 and 8, www.UnitingAviation.com, February 2020.
25. Halotron BrX™, American Pacific, www.Halotron.com, June 2018.
26. *Fire suppression-Novec 1230 Fire Protection Fluid*, 3M Corp, www.3m.com, February 2020.
27. 3M™, *Novec™ 1230 Fire Protection Fluid*, 3M Electronics Materials Solutions Division, Pub# 98-0212-3705-6, 2017.
28. European Fluorocarbons Technical Committee (EFCTC), www.fluorocarbons.org, March 2020.
29. *Vertrel™ Specialty Fluids*, TMC Industries, www.tmcindustries.com/vertrel.html, December 2016.
30. *Vertrel™ Specialty Fluids*, Chemours Co, www.chemours.com/businesses-and-products/fluoroproducts, December 2016.
31. *Vertrel™ MCA Specialty Fluid*, Precision Cleaning, No. K-04169-1, Chemours Co., www.Chemours.com, March 2020.
32. *Vertrel™ MCA Specialty Fluid*, Precision Cleaning, No. K-04169-1, Chemours Co., www.Chemours.com, May 2020.
33. *Vertrel™ SDG Specialty Fluid*, Precision Cleaning and Degreasing, No. C-11030, Chemours Co, www.Chemours.com, November 2016.
34. *Vertrel™ XF Specialty Fluid*, No. K-04159-2, Chemours Co., www.Chemours.com, September 2008.
35. *Vertrel™* XF Specialty Fluid, No. C-10968, Chemours Co., www.Chemours.com, November 2017.
36. *Vertrel™ X-Si Specialty Fluid*, No. C-10968, Chemours Co., www.Chemours.com, November 2020.
37. *Opteon™ SF33 Specialty Fluid*, No. C-11069, Chemours Co, www.Chemours.com, June 2018.
38. *Opteon™ SF33 Specialty Fluid*, No. C-11069, Chemours Co, www.Chemours.com, March 2020.
39. *Carrier solvents and lubricant deposition*, 3M™ Novec™ Engineered Fluids, www.3m.com, Pub No. 60-5002-0812-37, 2016.
40. *Vertrel™ Specialty Fluids*, Chemours Co, www.Chemours.com, March 2020.
41. *Vertrel™ X-Si Specialty Fluid*, Silicone Deposition, Removal, and Swelling, C-11058, Chemours Co, www.Chemours.com, November 2020.
42. *Vertrel® X-Si Specialty Fluid*, Silicone Deposition, Removal, and Swelling, C-11058, Chemours Co, www.Chemours.com, March 2020.

43. *Opteon™ SF10 Specialty Fluid*, Chemours Co, www.opteon.com/en/products/specialty-fluids/sf10, March 2020.
44. *Vertrel*™ Specialty Fluids, Chemours Co., www.chemours.com/en/brands-and-products/vertrel/products/xf, March 2020.
45. *Opteon™ Sinera Specialty Fluid, C-10840*, Chemours Co, www.Chemours.com, October 2016.
46. Snyder, J., *Wired Mag.*, www.wired.com/2010/10/inside-a-state-of-the-art-cleanroom, Oct 19, 2010.
47. Applied Materials Corp, www.appliedmaterials.com/semiconductor/products/etch/info, June 2018.
48. Li, X., Ling, L., Hua, X., Fukasawa, M., Oehrlein, G.S., *J. Vac. Sci. Technol.*, A21, 1, Jan-Feb 2003.
49. Takahashi, K., Itoh, A., Nakamura, T., Tachibana, K., *Thin Solid Films*, 374, 303, 2000.
50. *Sifren® 46*, Solvay Corp, www.solvay.com/en/markets-and-products/featured-products/sifren-46.html, December 2016.
51. Li, X., Hua, X., Ling, L., Oehrlein, G.S., Barela, M., Anderson, H.M., Fluorocarbon-based plasma etching of SiO_2: Comparison of C_4F_6/Ar and C_4F_8/Ar discharges. *J. Vac. Sci. Technol. A*, 20, 6, Nov/Dec 2002.
52. Rae, I.D., Ideal Gases: Anesthetics in the Heart of the Twentieth Century. *International Workshop on the History of Chemistry*, Tokyo, 2015.
53. Ebnesajjad, S., *Introduction to Fluoropolymers*, Elsevier, 2013.
54. *Tampa Bay Times*, www.tampabay.com, Sep 26, 2010.
55. Bowdle, T.A., Knutsen, L.J.S., Williams, M., Local and Adjunct Anesthesia, in: *Comprehensive Medicinal Chemistry II*, vol. 6, pp. 351–367, 2007.
56. Black, G.W., Enflurane. *Br. J. Anaesth.*, 51, 62, 1979.
57. www.drugs.com, December 2016.
58. https://pubchem.ncbi.nlm.nih.gov, December 2016.
59. Penney, J., *Methoxyflurane: Pharmacology And Clinical Applications of An inhaled Analgesic*, Christchurch Public Hospital, Christchurch, New Zealand, Sir Charles Gairdner Hospital, Christchurch, Australia, www.conference.co.nz/files/docs/penney.pdf, accessed Dec 2016.
60. Ozon Action Kigali Fact Sheet 2: Current Use of HCFCs and HFCs, UN Environment Program. *28th Meeting of the Parties 10-15 October 2016 in Kigali, Rwanda*, www.unenvironment.org/ozonaction/resources/factsheet/ozonaction-kigali-fact-sheet-2-current-use-hcfcs-and-hfcs.

5

Refrigeration Cycle and Refrigerant Selection: How Refrigerant Gases Work?

Majority of the fluorocarbons consumption is in refrigeration, which is the rationale for the inclusion of an introductory chapter on refrigeration cycle. The material in this chapter is basic information covered in thermodynamics courses in undergraduate engineering curriculums. It is included in this book to offer a refresher to readers with prior background in refrigeration technology and as a learning tool for others without it.

Refrigeration refers to a process of transferring heat from a low temperature location to a high temperature location under controlled conditions. To accomplish refrigeration expenditure of energy (or work) is required. The work of heat transport may be supplied by mechanical work, heat, electricity or another means. Without the work of heat transport the direction of heat transfer will be from the high to the low temperature location.

5.1 Refrigeration Cycle

In 1824 Nicolas Léonard Sadi Carnot described the operating principle of refrigeration cycle as a heat engine mathematically. The transfer of heat from a low-temperature to high-temperature requires a refrigerator or a heat pump. Refrigerators and heat pumps are basically the same device and only differ in their objectives. A refrigerator removes heat from a cold medium transferring it to a hot medium; a heat pump transfers heat/from a cold to a warm medium.

The Carnot cycle consists of the four steps when it is operating as a heat engine [1]. Those steps are displayed in Figures 5.1 and 5.2 and enumerated below.

(a) Reversible isothermal expansion of the gas in step 1-2 (in Figures 5.1 and 5.2) takes place at the high temperature, T_1. The gas expands and performs work on the surroundings. Gas temperature is held constant during the process (isothermal expansion). During the expansion the gas absorbs heat Q_1 from the high temperature medium. Consequently, the entropy of the gas increases by an amount shown in Eq (5.1).

$$\Delta S_1 = \frac{Q_1}{T_1} \tag{5.1}$$

Sina Ebnesajjad, PhD. Concise Handbook of Fluorocarbon Gases: Applications in Refrigeration and Other Industries, (87–98) © 2021 Scrivener Publishing LLC

(b) In step 2-3 (Figures 5.1 and 5.2), the gas continues to expand by a reversible adiabatic process at constant entropy work output. As a result of losing energy because of expansion the gas cools down to a low temperature, T_2.

(c) In step 3-4 (Figures 5.1 and 5.2) the gas is compressed reversibly isothermally at the low temperature, T_2. In this process the surroundings transfer work to the gas causing an amount of thermal Q_2 to transfer from the system to the low temperature reservoir. The entropy of the system thus decreases exactly by the same amount as was increased in step 1-2 as shown in Eq (5.1).

$$\Delta S_2 = \frac{Q_2}{T_2} \tag{5.2}$$

(d) In step 4-1 (Figures 5.1 and 5.2), the gas continues to be compressed by a reversible adiabatic process at constant entropy work output. As a result of gaining energy because of compression the gas heats up to a higher temperature. In this step, the surroundings do work on the gas, increasing its internal energy by compressing it. That in turn raises the temperature to T_1 due solely to the work added to the system, but the entropy remains unchanged. At this point the gas return to same state as at the start of step 1.

The most efficient heat engine cycle is the *Carnot Cycle*, consisting of two isothermal processes and two adiabatic processes (Figure 5.3). The Carnot cycle can be thought of as the most efficient heat engine cycle possible by physical laws. The second law of thermodynamics states that not all the supplied heat in a heat engine can be used to do work, the Carnot efficiency sets the limiting value on the fraction of the heat that can be so used.

Reversing the Carnot cycle does reverse the directions of heat and work interactions as seen in Figure 5.3. A refrigerator or heat pump that operates on the reversed Carnot cycle is called a Carnot refrigerator or a Carnot heat pump. The objective of a refrigerator is to

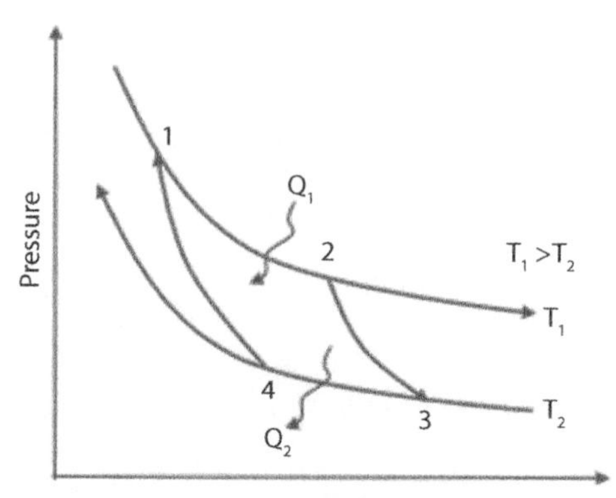

Figure 5.1 Illustration of a Carnot cycle on a PV diagram – depicting the work done.

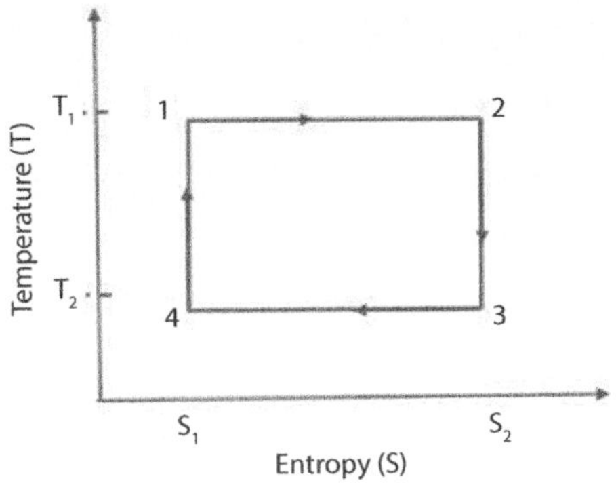

Figure 5.2 Illustration of a Carnot cycle operating as a heat engine on a temperature-entropy plot.

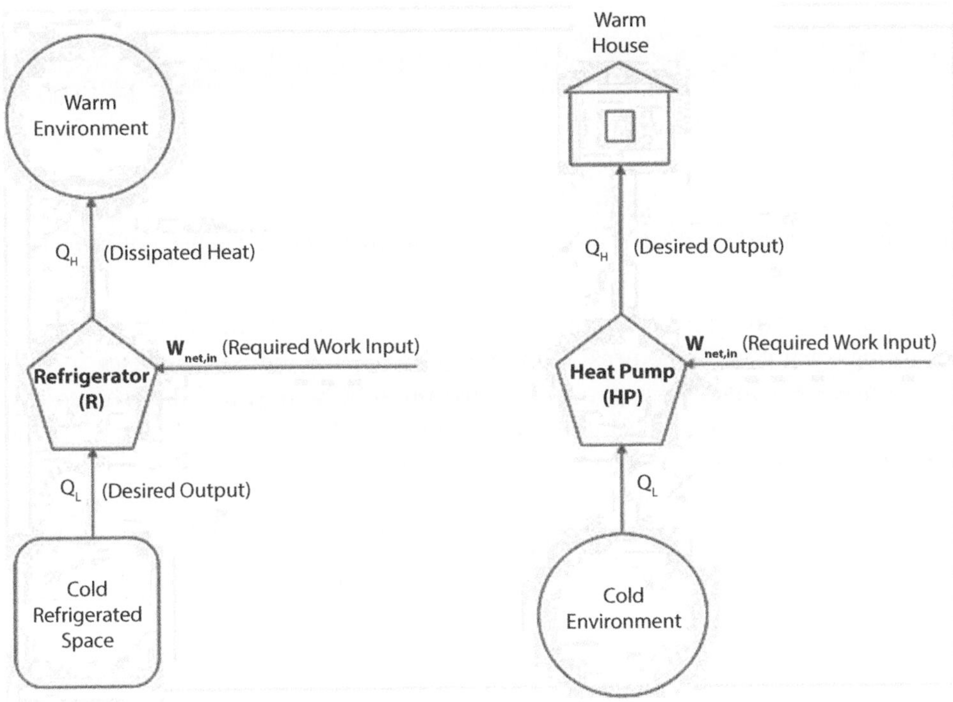

Figure 5.3 A comparison of operating steps of refrigerator (R) and a heat pump (HP) according to Carnot cycle.

remove heat (Q_L) from the cold medium; the objective of a heat pump is to supply heat (Q_H) to a warm medium.

5.1.1 Reversed Carnot Cycle

The reversed Carnot cycle is the most efficient refrigeration cycle operating between two specified temperature levels as shown in Figure 5.4. It sets the highest theoretical coefficient of performance (COP). The coefficient of performance for Carnot refrigerators and heat pumps are given by equation Eq (5.3) and Eq (5.4) [2].

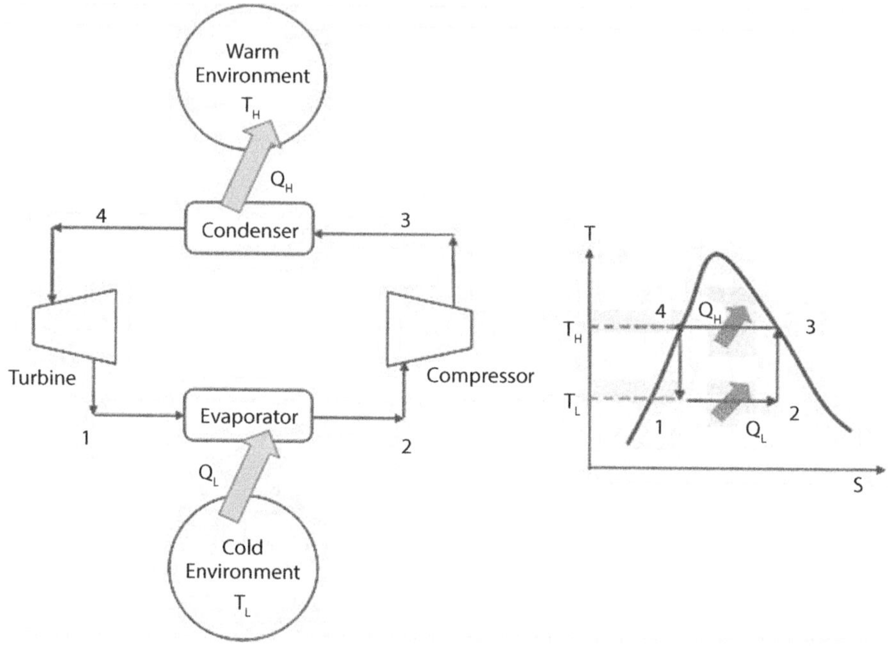

Figure 5.4 Major components of reversed Carnot cycle and temperature (T) vs entropy (S) diagram [2].

The Carnot cycle cannot be approximated in an actual cycle, because:

1- executing Carnot cycle requires a compressor that can handle two-phases

2- also process 4-1 involves expansion of two-phase flow in a turbine. The coefficient of performance of a Carnot refrigerator and heat pump (ideal) is given by Eq (5.4) and Eq (5.6) that are simplified forms of Eq (5.3) and Eq (5.5).

$$\text{refrigerator} \quad COP = \frac{Q_L)}{(Q_H - Q_L)} \tag{5.3}$$

$$\text{refrigerator} \quad COP_{R,Carnot} = \frac{1}{\left(\frac{T_H}{T_L} - 1\right)} \tag{5.4}$$

$$\text{heat pump} \quad COP = \frac{Q_H}{(Q_H - Q_L)} \tag{5.5}$$

$$\text{heat pump} \quad COP_{HP,Carnot} = \frac{1}{\left(1 - \frac{T_L}{T_H}\right)} \tag{5.6}$$

5.1.2 Ideal Vapor-Compression Refrigeration Cycle

The vapor-compression refrigeration seen in Figure 5.5 is the most widely used cycle for refrigerators, air conditioners, and heat pumps [1]. Assumptions for *ideal* vapor-compression cycle are:

1. Irreversibilities within the evaporator, condenser and compressor are ignored
2. No frictional pressure drops
3. Refrigerant flows at constant pressure through the two heat exchangers (evaporator and condenser)
4. Heat losses to the surroundings are ignored
5. Compression process is isentropic

5.1.3 Actual Vapor-Compression Refrigeration Cycle

There are significant differences between ideal and actual vapor-compression refrigeration. Most of them differences arise from the irreversibilities in various components. In a typical refrigerator the refrigerant enters the compressor as a slightly superheated vapor, as opposed to saturated vapor in the ideal cycle. The suction line connecting the evaporator to the compressor is quite long.

Leading to pressure drop and heat transfer to the surroundings. The compressor is not internally reversible thus increases the entropy. To reduce entropy growth, a multi-stage compressor can be used that is equipped with refrigerant cooling during the compression.

Another difference between ideal and actual cycle is the refrigerant leaves condenser as a sub-cooled liquid. The sub-cooling process not only increases the cooling capacity and but it prevents entry of any vapor (bubbles) to the expansion valve. Rejection and addition of heat in the condenser and evaporator do not take place at a constant pressure (and temperature) because of pressure drop of the refrigerant.

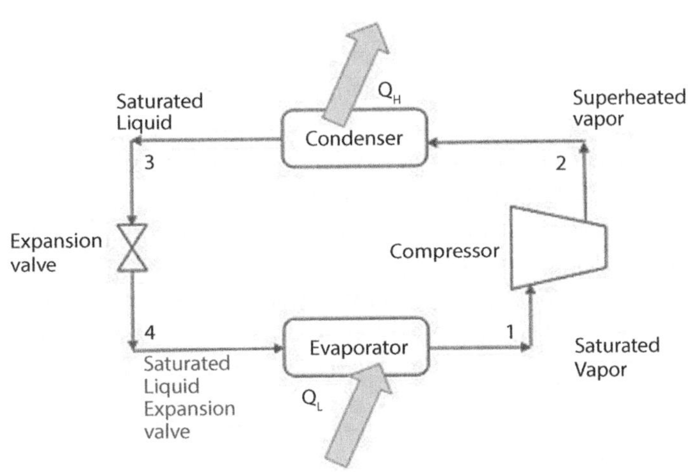

Figure 5.5 Schematic diagram of an Ideal Vapor-Compression Refrigeration Cycle.

5.2 Selection of Right Refrigerant

When designing a refrigeration system, there are several refrigerants from which to choose. An abbreviated list of refrigerant is given in Table 5.1. All, except R-404a, are single compounds, which is a blend of R-125/143a/134a. The right choice of refrigerant depends on the actual application and other factors. In addition to individual compounds blends of multiple refrigerants have been used since phase-out of chlorofluorocarbons (CFC) and hydrochlorofluorocarbons (HCFC). The latest generation of refrigeration agents is hydrofluorocarbons (HFC). They do not have ozone depleting potential (ODP) but contribute to global warming.

Each single-component refrigerant has its own pressure-temperature relationship and unique physical properties, such as density, heat of vaporization and heat-transfer coefficient. To match the properties of a single obsolete refrigerant with a blend (see Sec. 5.3), the individual components must be mixed in the right proportions [3].

The required or highly desirable characteristics of a widely used refrigerant are listed below:

1. Environmental acceptability
2. Chemical stability – The refrigerant must not attack the material it contacts.
3. Materials compatibility
4. Refrigeration-cycle performance
5. Adherence to nonflammable and nontoxic guidelines, per Underwriters Laboratories (UL)
6. Boiling point - The evaporator (typically a coil) must boil all of the liquid into a vapor before the end of the coil.

Several factors must be considered in selecting a refrigerant including ozone depletion potential, global warming potential, combustibility, leak detectability, low toxicity and cost and favorable thermodynamic properties. Desirable thermodynamic properties of refrigerants are listed below:

1. The heat of vaporization of the refrigerant should be high. The higher the enthalpy of vaporization (h_{fg}) the greater the refrigerating effect per unit weight of fluid circulated.
2. The specific heat of the refrigerant should be low. The lower the specific heat, the less heat it will pick up for a given change in temperature during the throttling or in flow through the piping, and consequently the greater the refrigerating effect per unit weight of refrigerant.
3. The specific volume of the refrigerant should be low to minimize the work required per unit weight of refrigerant circulated since evaporation and condenser temperatures are fixed by the temperatures of the surroundings.

The new refrigerants such as HFO-1234yf and HFO-1234ze have been considered as long-term replacements for HFC-134a to comply with the Kyoto protocol. In small size refrigeration systems, these refrigerants have poor performance than HFC-134a besides their minor flammability. Previous studies reveal that the addition of HFC-134a in small quantity can

Table 5.1 Examples of candidate refrigerants with low ozone depletion potential (ODP<<1) [4].

	Atmospheric life		Environmental		Safety				COP (efficiency)	
	Years	Days	ODP	GWP	Toxicity	Flammable	Burn velocity (BV) (cm/s)	Minimum ignition energy (MIE) (mJ)	Max theory 2-stage	Max theory 1-stage
R-11	45	16,425	1	4750	A	1	0	–	9.467	9.101
R-12	100	36,500	0.82	10,900	A	1	0	–	9.102	8.584
R-22	11.9	4,344	0.034	1760	A	1	0	–	8.965	8.483
R-123	1.3	475	0.12	79	B	1	0	–	9.401	8.949
R-134a	13.4	4,891	0	1300	A	1	0	–	9.055	8.472
R-410A	28.2	10,293	0	1924	A	1	0	–		7.988
R-245fa	7.7	2,811	0	858	A	1	0	–	9.305	8.778
R-32	5.2	1,898	0	677	A	2L	6.7	15		8.215
R-1233zd(E)	0.0712	26	0.0002	1*	A	1	0	–	9.345	8.854
R-1234yf	0.031	11	0	<1*	A	2L	1.5	5000	8.868	8.168
R-1234ze(E)	0.038	14	0	<1*	A	2L	0**	–	9.067	8.452
R-513A	5.9	2,158	0	572	A	1	0	–	8.934	8.280
CO_2	50	18,250	0	1	A	1	0	–		3.533
Ammonia	0.01	4	0	0	B	2L	7.2	100		8.774

GWP = Global Warming Potential.

improve the performance of a system without exceeding the global warming potential limit (150) specified by the European Union. However, this concept is not studied in small-scale refrigerators. Therefore, the performance of low and medium temperature systems of capacity 190 and 285 liters, respectively, working with different hydrofluoroolefin/hydrofluorocarbon (HFO/HFC) mixtures, has been studied by mathematical simulation [5].

The software tool 'MATLab' and the refrigerant property database 'REFPROP' have been used in the simulation. The results showed that the coefficient of performance (COP) of Hydrofluoroolefin (HFO) refrigerants and its mixtures decreases from 2.4 to 15.7% than that of HFC-134a. Among the mixtures, HFO-1234ze/HFC-134a (90/10) shows better performance and its predicted coefficient of performance (COP) and exergetic efficiency are 4–8.3% and 5.1–10.5%, respectively, higher than that of other mixtures considered in this study [5].

The numerical computations for predicting first law and second law performances for two vapor compression refrigeration systems have been carried out using R1234ze and R1234yf as fourth generation low GWP refrigerants for replacing third generation refrigerant (R134a). The comparison was carried out using HFC-134a for two systems (i.e. System-1: consisting of

Table 5.2 Comparison of first law efficiency in terms of COP of both systems [6].

Refrigerant	COP_System-I	COP_System-II
R-134a	2.464	3.095
R1234ze	2.42	3.090
R1234yf	2.255	2.981
R-227ea	2.056	2.889
R236fa	2.361	3.052
R245fa	2.634	3.182
R404a	2.009	2.864
R407c	2.084	2.809
R410a	2.259	2.925
R152a	2.652	3.156
R290	2.405	3.067
R600a	2.484	3.11
R600	2.627	3.17
R-125	1.785	2.74
R123	2.744	3.23
R32	2.416	2.917
R717	2.728	2.999
R507a	2.030	2.877

condenser, evaporator compressor and throttling valve and System-2 consisting of condenser, evaporator compressor and expander). It was discovered that system-2 using expander gives the best thermodynamic performances around 25.65% higher than system-1 using throttle valve System-1: consisting of condenser, evaporator compressor and throttling valve and System-2 consisting of condenser, evaporator compressor and expander (Table 5.2) [6].

While non-fluorinated refrigerants (sometimes called *natural* refrigerants) have some desirable characteristics in vapor compression cooling, the majority of the refrigeration industry prefers using fluorocarbon refrigerants. The reason is they offer the best balance of efficiency, safety and environmental impact. In fact, many of the new hydrofluoroolefin (HFO, GWP<1) refrigerant options have GWP values lower than non-fluorinated refrigerant alternatives. Non-fluorinated refrigerants will see increased usage in the coming years in applications where their unique characteristics are best suited—such as in low temperature refrigeration [7].

Policy makers, the public, and manufacturers face a complex picture with respect to refrigerant choices. Direct environmental concerns (e.g. ODP, GWP, leak rates), indirect environmental concerns (energy efficiency), safety and performance are among the complicating factors.

5.3 Refrigerant Blends

During most of the history of air conditioning and refrigeration industry, nearly all systems have been based on a single chemical compound or *pure* refrigerant. The search for single-equivalent replacements for chlorofluorocarbon and hydrochlorofluorocarbon refrigerants were unsuccessful. It proved simply impossible to develop single-component replacements that met all the requirements and desirable attributes of the CFCs and HCFCs. Those included:

- Environmental acceptability
- Chemical stability
- Materials compatibility
- Refrigeration-cycle performance
- Adherence to nonflammable and nontoxic guidelines, per UL
- Boiling point

Many of the refrigerant replacements were not pure, but instead were azeotropes[1], zeotropes[2] or near-azeotropes[3] because of being comprised of two or more compounds [8, 9]. A mixture's components were chosen based on the final desired characteristics examples of which included vapor pressure, transport properties, lubricant and material compatibility, thermodynamic performance, cost, flammability, toxicity, stability and environmental

[1]Azeotrope: A blend that, when used in refrigeration cycles, does not change volumetric composition or saturation temperature appreciably as it evaporates (boils) or condenses at constant pressure.

[2]Zeotrope: A blend that, when used in refrigeration cycles, changes volumetric composition and saturation temperatures to varying extents as it evaporates (boils) or condenses at constant pressure.

[3]Near-azeotrope: A zeotropic blend with a small temperature and composition glide over the application range and no significant effect on system performance, operation and safety.

Temperature glide – Because the composition of a zeotrope alters during a phase change, there is a slight change in evaporating and condensing temperature at constant pressure. This phenomenon is known as "glide."

properties. The concentrations of the components were selected based on the exact characteristics that were sought in the final product.

When an *azeotrope*, *near-azeotrope* or *zeotrope* is in the pure liquid or pure vapor state, the composition is totally mixed, and all properties are uniform throughout; however, when both liquid and vapor are present (such as in the evaporator, condenser or perhaps receiver), a mixture's behavior depends upon whether it is an azeotrope or zeotrope. The composition of the liquid and vapor of an azeotrope remain virtually identical when both liquid and vapor are present. If a leak occurs, there will not be a substantial change in composition of the refrigerant left in the system.

Fractionation refers to a change in the composition of a refrigerant blend by preferential evaporation of the more volatile component(s) or condensation of the less volatile component(s). To ensure fractionation does not occur during charging, it is recommended that zeotropic mixtures be liquid charged rather than vapor charged. Liquid must be removed from the refrigeration cylinder. It then can be flashed through a metering device and charged into the system in its vapor state. The refrigerant manufacturer's recommendation should be closely followed.

The composition of the vapor and liquid of a zoetrope are different when both liquid and vapor are present. If a leak occurs in this region of a system and only vapor leaks out, there can be a change in the composition of the refrigerant left in the system [10].

The options in Table 5.2 were valid for the selection of refrigerants until October 2016 when the 28th Montreal Protocol meeting took place in Kigali, Rwanda. At this meeting

Table 5.3 Examples of refrigerants types with low or zero ODP [9].

Refrigerant	Type	GWP
R-12	CFC	10900
R-502	CFC	4700
R-22	HCFC	1810
R-134a	HFC	1430
R-404A	HFC	3922
R-407A	HFC	2107
R-407C	HFC	1774
R-410A	HFC	2088
R-422A	HFC	3143
R-422D	HFC	2729
R-32	HFC	675
R-1234yf	HFO	4
HC-290	Propane	20
R-717	Ammonia	0
R-744	Carbon Dioxide	1

198 countries agreed to amend the protocol to signal the international community's commitment to practical action to implement 2015 Paris Agreement to limit global warming to 2°C, and the more ambitious target of 1.5°C by the end of 21st century. It was agreed to phase down the production and consumption of HFC's, which are potent greenhouse gases according to planned schedule. That leaves a few options at bottom four rows of Table 5.3 and a few others discussed later in this book.

5.4 Comparison of Refrigerator and Air Conditioning Systems

An interesting question is what are the differences between refrigeration and air conditions? Another one is how does those differences affect the selection of the cooling medium (refrigerant) for the two systems? This section discusses those differences.

Refrigeration refers to processes that remove thermal energy away from a cold location and transfer that energy to a higher temperature location. Naturally or spontaneously entropy always increases indicating thermal energy moves from higher to lower temperature locations. Consequently, refrigeration runs counter to the second law of thermodynamics and requires expenditure of work to allow the low to high temperature heat transfer to take place [11].

In a refrigerator, when the refrigerant enters the compressor, the refrigerant is compressed adiabatically thus raising the refrigerant temperature. Then, the hot refrigerant is sent into the condenser under high pressure, which is typically a tube outside the refrigerator. Here, the temperature of the refrigerant is higher than the temperature of the surroundings. So, the refrigerant gives off heat to the surroundings and gets condensed into a liquid. If you touch the tubes at the back of a refrigerator, it feels hot because you are feeling the heat given out by the condenser.

Next, the refrigerant goes through an expansion valve, where the pressure of the refrigerant is lowered. The refrigerant expands here, evaporating and cooling down. The refrigerant now enters the evaporator, which is a set of tubes inside the refrigerator. At this stage, the refrigerant is at a temperature lower than the temperature of the air inside the refrigerator. Now, the refrigerant draws in heat from inside the refrigerator. This is the mechanism of heat removal from the cold interior space of a refrigerator. Once the refrigerant goes through the evaporator, it is sent again into the compressor…and the cycle continues repeatedly.

Air conditioning is a type of refrigeration where thermal energy is taken away from the air in a large space such as a room or a vehicle. Air conditioners are fitted into rooms so that they cool the air inside them. Air conditioners also reduce humidity in rooms, because the water vapor in the room can condense around the colder parts of the air conditioner. The condensed water can be then drained off.

Air conditioners work in the same way as refrigerators. The primary difference between an air conditioner and a refrigerator is that air conditioners not only maintain the temperature of the air they also help to regulate humidity and filter the air. An air conditioning includes a fan in order to circulate air allowing the hot air to contact the evaporator at a faster rate [11].

Compressors are not independent standing units in refrigerators and air condition units. They play important continuous roles in both operations. If the compressor were removed, the refrigeration process would cease to exist.

(1) Most compressors are independent entities used for compressing gases. This implies they compress gases to supply to other applications or storage. They, therefore, do not have continuous direct connection with another piece of equipment. The refrigeration and the air conditioning compressors are indispensable parts of the complete vapor compression cycle that includes additional components such as the condenser expansion valve and the evaporator.

(2) The sole function of the other compressors is to compress a gas. The refrigerant compressor compresses the refrigerant to the pressure corresponding to the saturation pressure that is higher than the temperature of the naturally available air or water. This enables the cooling of the refrigerant in the condenser by the available atmospheric air or water.

The refrigerant compressor also circulates the refrigerant through the entire cycle. After leaving the compressor the refrigerant is forced into the condenser, the expansion valve, and the evaporator. Last it is pulled into the compressor for compression and recirculation though the entire cycle anew.

(3) The capacity of the refrigerating compressor determines the capacity of the entire refrigerating or air conditioning plant. The capacity of the air or gas compressor doesn't determine the capacity of the other plants. For instance, though the pneumatic tools depend on compressed air for its operation, it has its own capacity.

(4) The refrigerating compressor is an integral part of the cycle that includes other components of the cycle like condenser, expansion valve, and the evaporator. The working and proper functioning of each of these components depends on the working of other components. If the refrigeration compressor breaks down, the whole system will collapse. There is no such dependency in case of the other compressors. If the gas compressor breaks down, a standby compressor can be started easily [12].

References

1. Çengel, Y.A. and Boles, M.A., *Thermodynamics: An Engineering Approach*, 7th ed, McGraw-Hill, 2011.
2. Bahrami, M., *ENSC 461 course material, Refrigeration Cycle*, Simon Fraser University, 2016.
3. Lavelle, J., Understanding Refrigerant Blend Performance. *RSES J.*, 29–33, www.rses.org, January 2006.
4. Trane Climate Technology, www.Trane.com, Nov 2016.
5. Raveendran, P.S. and Sekhar, S.J., Energy and exergy analysis on hydrofluoroolefin/hydrofluorocarbon (HFO/HFC) refrigerant mixtures in low and medium temperature small-scale refrigeration systems. *Proceedings of the Institution of Mechanical Engineers, Part E: Journal of Process Mechanical Engineering, Sage Journals*, Institute of Mechanical Engineers, October 9, 2019.
6. Mishra, R.S., Thermodynamic performance of vapor compression refrigeration systems using HFO refrigerants for replacing HFC refrigerants. *Int. J. Res. Eng. Innov.*, 2, 5, 492–497, 2018.
7. *Considerations for Next Generation HVAC Refrigerants*, pub no ENV-APN001A-EN 0215, Trane, an Ingersoll Rand, www.trane.com, 2015.
8. Beeton, W., Buynacek, B., Bell, D., Gephart, J., Rajendran, R., Schemmel, R., *Refrigerants for Commercial Refrigeration Applications*, Copeland, Emerson Climate Technologies, 2005.
9. *Refrigerants for commercial refrigeration applications*, Emerson Climate Technologies, 2015.
10. The Air Conditioning, Heating and Refrigeration NEWS, www.achrnews.com, Nov 2016.
11. www.bayt.com, Dec 2016.
12. www.brighthubengineering.com, December 2016.

6

Preparation of Fluorocarbons

This chapter describes the basic preparation methods for commercial aliphatic (except cyclic) fluorocarbons. There is no repository for that type of information. The available sources include books, journal articles, government reports and publications and patents. Most of the technological developments for commercial fluorocarbons can be gleaned from the patent art. The coverage in this chapter, though not inclusive of every existing source, relies on important published sources especially patents. It follows a chronological path starting at the beginning of manufacturing of chlorofluorocarbons to the present state of art: hydrofluoroolefins.

The main focus in this chapter will be on commercial fluorocarbons with substantial and important applications. The following main groups of fluorocarbons agents will be discussed because of their commercial significance. This grouping is for the most part chronological. Numbers 1 to 4 indicate decreasing harm to the ozone layer. HCFC and HFC's have global warming potential, which has resulted in phase out of their use (HFC's as of October 2016). Hydrofluoroolefins neither impact the ozone layer nor contribute to global warming.

1. Chlorofluorocarbons (CFC's)
2. Hydrochlorofluorocarbons (HCFC's)
3. Hydrofluorocarbons (HFC's)
4. Hydrofluoroolefins (HFO's)
5. Specialty fluorocarbons – namely perfluorocarbons

6.1 Introduction

Refrigerators in the late 1800s and early 1900s used the toxic gases, ammonia (NH_3), methyl chloride (CH_3Cl), and sulfur dioxide (SO_2), as refrigerants. After a series of fatal accidents in the 1920s when methyl chloride leaked out of refrigerators, a search for a less toxic replacement begun as a collaborative effort of three American corporations-Frigidaire, General Motors, and Du Pont. CFCs were first synthesized in 1928 by Thomas Midgley, Jr. of General Motors, as safer chemicals for refrigerators used in large commercial applications. Frigidaire was issued the first patent, number 1,886,339, for the formula for CFCs on December 31, 1928 [1].

In 1930, General Motors and Du Pont formed the Kinetic Chemical Company to produce Freon (a Du Pont tradename for CFCs) in large quantities. By 1935 Frigidaire and its competitors had sold 8 million new refrigerators in the United States using Freon-12

Sina Ebnesajjad, PhD. Concise Handbook of Fluorocarbon Gases: Applications in Refrigeration and Other Industries, (99–154) © 2021 Scrivener Publishing LLC

(CFC-12) made by the Kinetic Chemical Company and those companies that were licensed to manufacture this compound. In 1932 the Carrier Engineering Corporation used Freon®-11 (CFC-11) in the world's first self-contained home air-conditioning unit, called the "Atmospheric Cabinet" [1].

Because of the CFC safety record for nontoxicity, Freon became the preferred coolant in large air-conditioning systems. Public health codes in many American cities were revised to designate Freon as the only coolant that could be used in public buildings. After World War II, CFCs were used as propellants for bug sprays, paints, hair conditioners, and other health care products. During the late 1950s and early 1960s the CFCs made possible an inexpensive solution to the desire for air conditioning in many automobiles, homes, and office buildings. Later, the growth in CFC use took off worldwide with peak, annual sales of about a billion dollars (U.S.) and more than one million metric tons of CFCs produced [1].

The net accomplishment of manufacturing fluorocarbons is controlled fluorination of a starting organic compound material for conversion to the desired product. The reaction medium of fluorination reactions could be either liquid or gaseous. The extreme reactivity of fluorine required closely controlled reaction conditions virtually always in the presence of a catalyst. One of the features of fluorination reactions is, that in addition to the desired products, they generate major amounts of by-products. Selectivity and increase in yields allow the actual fluorocarbon products to yield greater than 90% of the reaction's theoretical yield.

One little emphasized fact is the intrinsic role of chlorine chemistry in the production of fluorocarbons. First, he fluorination process control is improved by the presence of chlorine. Second the distribution of the reaction by-products is affected by the use of starting materials containing excess chlorine over what would be required in the final products. In those cases, the extra chlorine is converted into hydrochloric acid that is easily removed from the fluorocarbon products. This technique was commonly utilized for the production of CFCs and HCFCs [2].

Typically, the processes begin with starting materials that included methylene chloride, chloroform, carbon tetrachloride, trichloroethylene, perchloroethylene, and trichloroethane. A benefit of the presence of chlorine in a molecule was its favorable impact on the reaction rates thus improving commercial manufacturing. This approach has also been used in the production of HFCs with no chlorine in the molecule. The reaction ingredients included methylene chloride, chloroform, vinyl chloride, trichloroethylene, perchloroethylene, and trichloroethane. Products included trifluoromethane (fluoroform, HFC-23), difluoromethane (HFC-32), pentafluoroethane (HFC-125), 1,1,1,2-tetrafluoroethane (HFC-134a), 1,1,1-trifluoroethane (HFC-143a) and 1,1-difluoroethane (HFC152a) [2].

Multiple products, including HCFCs and HFCs, may be made from a single starting material by altering the extent of fluorine in the reaction or its location on the molecule. It is also possible to conceive of chemistries that can produce HFCs from chlorine-free starting materials. For example, HFC-152a has been produced by the selective fluorination of acetylene, although that process is no longer operated commercially in North America. Fluorocarbons are consumed as intermediates in the production of fluoropolymers, fluoroelastomers, and HFOs. When CFCs or HCFCs are used as the starting material, chlorine is removed during the production of the monomer and would not appear on the polymer, unless a chlorinated fluoropolymer is to be produced.

6.2 Classification of Fluorocarbons

Commercial fluorocarbons have been classified by their molecular chemical structures mainly grouped as CFC, HCFC, HFC and HFO compounds. Another classification method is by the common original compound from which the fluorocarbons have been driven. Further classifications are made by the application type, ozone depletion potential, global warming potential and others.

Most fluorocarbons appear to be derivatives of few aliphatic hydrocarbons. The chemical structures of these fluorocarbons indicate they are derivatives of methane, ethane or propane. This section describes the classification of fluorocarbons by the compound of origin. A convenient method of depicting fluorocarbons containing C, Cl, H and F is by use of a ternary composition diagram resembling phase diagram. Figures 6.1–6.3 show

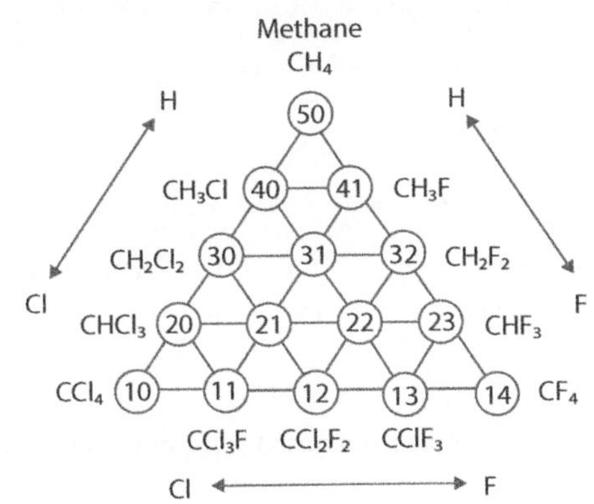

Figure 6.1 Ternary chemical composition diagram of methane-based chlorofluorocarbons.

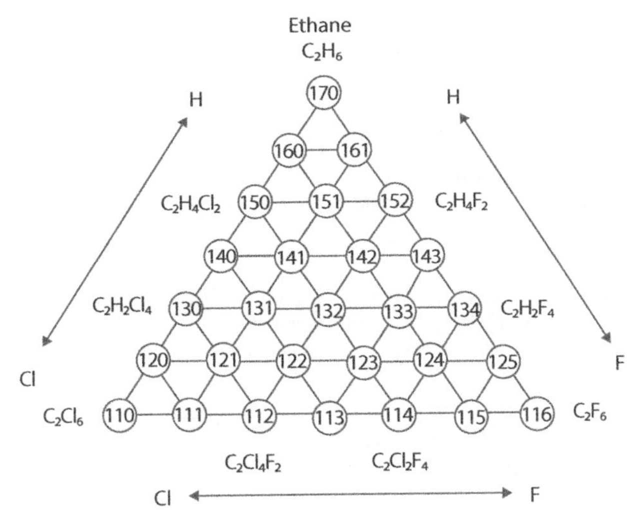

Figure 6.2 Ternary composition diagram of ethane-based chlorofluorocarbons.

the diagrams for methane, ethane and propane based fluorocarbons. Figures 6.4–6.6 display the chemical formulas of actual methane and ethane based fluorocarbons along with ASHRAE Codes for compounds currently used in fluorocarbons applications and the boiling points.

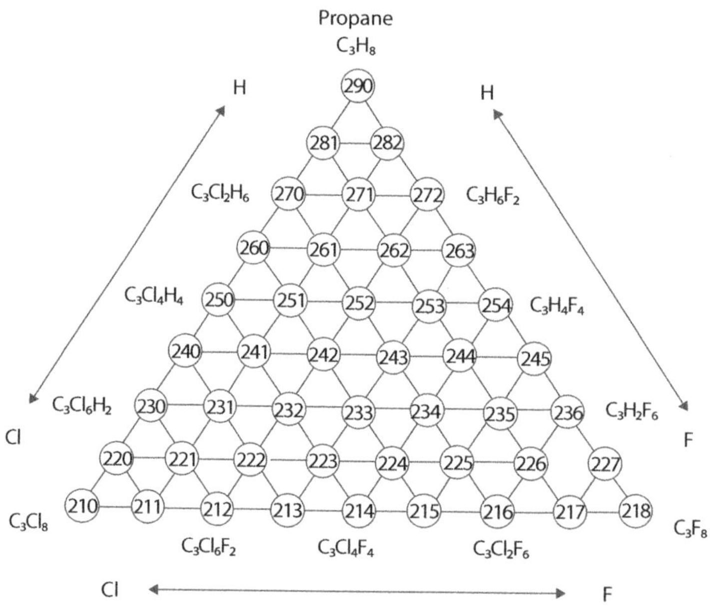

Figure 6.3 Ternary composition diagram of propane-based chlorofluorocarbons.

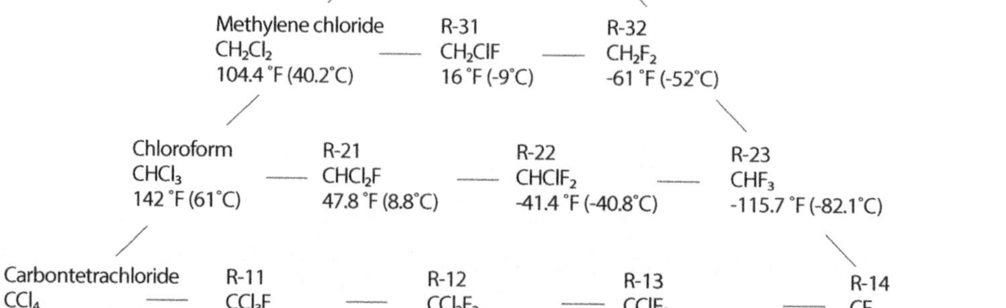

Figure 6.4 Chemical formulas of actual methane based fluorocarbons along with their ASHRAE codes and boiling points [3].

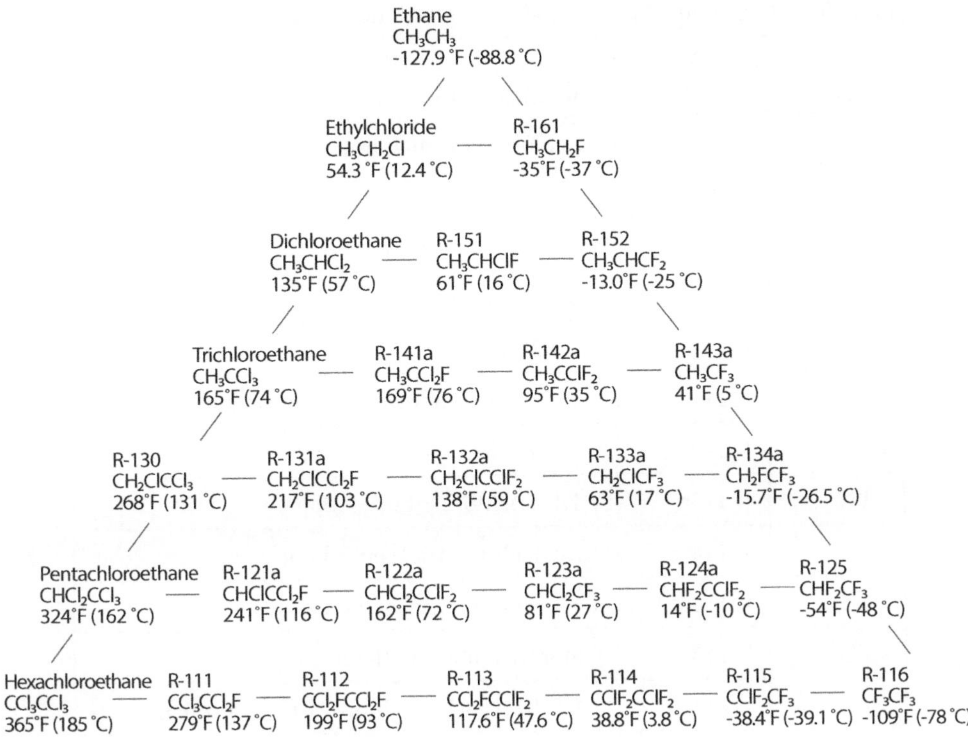

Figure 6.5 Chemical formulas of actual ethane based fluorocarbons along with their ASHRAE codes and boiling points [2].

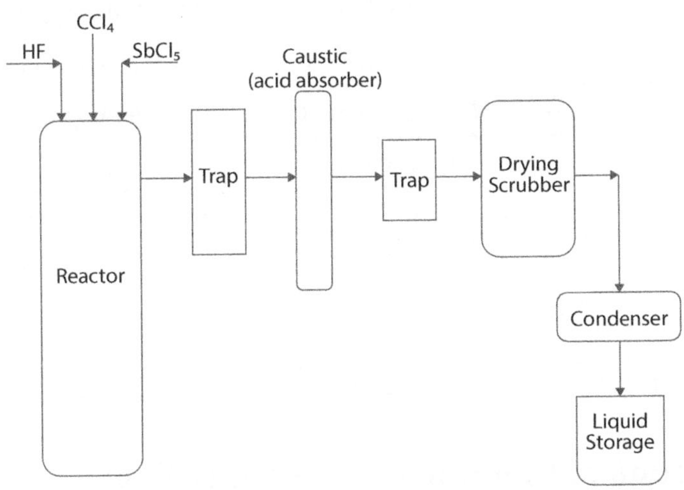

Figure 6.6 Schematic of the direct halogen exchange process for CFC preparation.

Tables 6.1 through 6.6 list the fluorocarbons according to the classification of chemical structure: CFC, HCFC, HFC and HFO. Phase out of HCFC and CFC's are well under way and in some cases complete. HFO compounds are being developed continuously to meet the needs of the industries for ozone and climate safe alternatives. Because of the proliferation of HFO fluorocarbons, it is likely some may be missing from the Table 6.6.

Table 6.1 Name and ASHRAE codes of chlorofluorocarbons [4].

Class	ASHRAE code	Chemical name
CFC	R-11	Trichlorofluoromethane
CFC	R-12	Dichlorodifluoromethane
CFC	R-13	Chlorotrifluoromethane
CFC	R-111	Pentachlorofluoroethane
CFC	R-112	1,1,2,2-Tetrachloro-1,2-difluoroethane
CFC	R-112a	1,1,1,2-Tetrachloro-2,2-difluoroethane
CFC	R-113	1,1,2-Trichlorotrifluoroethane
CFC	R-113a	1,1,1-Trichlorotrifluoroethane
CFC	R-114	1,2-Dichlorotetrafluoroethane
CFC	R-114a	1,1-Dichlorotetrafluoroethane
CFC	R-115	Chloropentafluoroethane
CFC	R-211	1,1,1,2,2,3,3-Heptachloro-3-fluoropropane
CFC	R-212	Hexachlorodifluoropropane
CFC	R-213	1,1,1,3,3-Pentachloro-2,2,3-trifluoropropane
CFC	R-214	1,2,2,3-Tetrachloro-1,1,3,3-tetrafluoropropane
CFC	R-215	1,1,1-Trichloro-2,2,3,3,3-pentafluoropropane
CFC	R-216	1,2-Dichloro-1,1,2,3,3,3-hexafluoropropane
CFC	R-216ca	1,3-Dichloro-1,1,2,2,3,3-hexafluoropropane
CFC	R-217	1-Chloro-1,1,2,2,3,3,3-heptafluoropropane
CFC	R-217ba	2-Chloro-1,1,1,2,3,3,3-heptafluoropropane
CFC	R-C316	Dichlorohexafluorocyclobutane
CFC	R-C317	Chloroheptafluorocyclobutane

6.3 Preparation of Chlorofluorocarbons (CFCs)

This section describes the basic processes that were developed for the manufacturing of chlorofluorocarbons. Those techniques were later modified for the production of HCFC's, HFC's and HFO's. This is why the important processed have been described in significant detail chronologically.

The earliest disclosure of the fluorocarbon's use as refrigerant dates back to a patent published in 1931 based on the late 1920's research at Frigidaire Corporation [5]. They disclosed

Table 6.2 Name and ASHRAE codes of hydrochlorofluorocarbons [4].

Class	ASHRAE code	Chemical name
HCFC	R-21	Dichlorofluoromethane
HCFC	R-22	Chlorodifluoromethane
HCFC	R-31	Chlorofluoromethane
HCFC	R-121	1,1,2,2-Tetrachloro-1-fluoroethane
HCFC	R-121a	1,1,1,2-Tetrachloro-2-fluoroethane
HCFC	R-122	1,1,2-Trichloro-2,2-difluoroethane
HCFC	R-122a	1,1,2-Trichloro-1,2-difluoroethane
HCFC	R-122b	1,1,1-Trichloro-2,2-difluoroethane
HCFC	R-123	2,2-Dichloro-1,1,1-trifluoroethane
HCFC	R-123a	1,2-Dichloro-1,1,2-trifluoroethane
HCFC	R-123b	1,1-Dichloro-1,2,2-trifluoroethane
HCFC	R-124	2-Chloro-1,1,1,2-tetrafluoroethane
HCFC	R-124a	1-Chloro-1,1,2,2-tetrafluoroethane
HCFC	R-131	1,1,2-Trichloro-2-fluoroethane
HCFC	R-131a	1,1,2-Trichloro-1-fluoroethane
HCFC	R-131b	1,1,1-Trichloro-2-fluoroethane
HCFC	R-132	Dichlorodifluoroethane
HCFC	R-132a	1,1-Dichloro-2,2-difluoroethane
HCFC	R-132b	1,2-Dichloro-1,1-difluoroethane
HCFC	R-132c	1,1-Dichloro-1,2-difluoroethane
HCFC	R-133	1-Chloro-1,2,2-Trifluoroethane
HCFC	R-133a	1-Chloro-2,2,2-Trifluoroethane
HCFC	R-133b	1-Chloro-1,1,2-Trifluoroethane
HCFC	R-141	1,2-Dichloro-1-fluoroethane
HCFC	R-141a	1,1-Dichloro-2-fluoroethane
HCFC	R-141b	1,1,-Dichloro-1-fluoroethane
HCFC	R-142	Chlorodifluoroethane

(*Continued*)

Table 6.2 Name and ASHRAE codes of hydrochlorofluorocarbons [4]. (*Continued*)

Class	ASHRAE code	Chemical name
HCFC	R-142a	1-Chloro-1,2-difluoroethane
HCFC	R-142b	1-Chloro-1,1-difluoroethane
HCFC	R-151	Chlorofluoroethane
HCFC	R-151a	1-Chloro-1-fluoroethane
HCFC	R-221	1,1,1,2,2,3-Hexachloro-3-fluoropropane
HCFC	R-222	Pentachlorodifluoropropane
HCFC	R-222c	1,1,1,3,3-Pentachloro-2,2-difluoropropane
HCFC	R-223	Tetrachlorotrifluoropropane
HCFC	R-223ca	1,1,3,3-Tetrachloro-1,2,2-trifluoropropane
HCFC	R-223cb	1,1,1,3-Tetrachloro-2,2,3-trifluoropropane
HCFC	R-224	Trichlorotetrafluoropropane
HCFC	R-224ca	1,3,3-Trichloro-1,1,2,2-tetrafluoropropane
HCFC	R-224cb	1,1,3-Trichloro-1,2,2,3-tetrafluoropropane

halofluoro derivatives of aliphatic hydrocarbon compounds as replacements for sulfur dioxide and ammonia because of their desirable boiling points low toxicity, nonflammability and stability. Halo-fluoro derivative of an aliphatic hydrocarbon (methane and ethane based) was defined to a derivative containing more than one fluorine atom with or without other halogen atoms, or one fluorine atom with one or more other halogen atoms. The general formula $C_nH_mF_pX_r$ was offered for suitable refrigerants in which n≥1, m≥0, p≥1, X represents a halogen other than F and the value of r may be zero when p≥1.

Some of the commercial developments of fluorocarbons in 1933 related to extinguishing fires using halogen derivatives of hydrocarbon compounds containing fluorine. Those compounds were viewed advantageous to carbon dioxide, carbon tetrachloride and other fire-extinguishing agents then used. Chief among the beneficial properties of halocarbons were their effectiveness as extinguishant and safety at the concentrations required to inhibit fires. Both methane and ethane based halocarbons could be effective, examples of which include CCl_2F_2 (CFC, R-12) and CHF_2-CH_2Cl (CFC, R-142). The degree of effectiveness of those compounds depended primarily on the ratio of hydrogen atom content to total halogen content. Preferable were compounds in which the hydrogen atom content is not any greater than the total halogen atom content [6, 7].

CFC's are commercially manufactured by substitution of fluorine for chlorine in chlorinated hydrocarbons (chlorocarbons) [8]. For example, R-12 one of the more promising CFCs was produced by the reaction of carbon tetrachloride (CTC) with a fluorinating agent in the presence of a catalytic agent in the reactor. CTC is a liquid at the room temperature

Table 6.3 Name and ASHRAE codes of hydrochlorofluorocarbons [4].

Class	ASHRAE Code	Chemical name
HCFC	R-224cc	1,1,1-Trichloro-2,2,3,3-tetrafluoropropane
HCFC	R-225	Dichloropentafluoropropane
HCFC	R-225aa	2,2-Dichloro-1,1,1,3,3-pentafluoropropane
HCFC	R-225ba	2,3-Dichloro-1,1,1,2,3-pentafluoropropane
HCFC	R-225bb	1,2-Dichloro-1,1,2,3,3-pentafluoropropane
HCFC	R-225ca	3,3-Dichloro-1,1,1,2,2-pentafluoropropane
HCFC	R-225cb	1,3-Dichloro-1,1,2,2,3-pentafluoropropane
HCFC	R-225cc	1,1-Dichloro-1,2,2,3,3-pentafluoropropane
HCFC	R-225da	1,2-Dichloro-1,1,3,3,3-pentafluoropropane
HCFC	R-225ea	1,3-Dichloro-1,1,2,3,3-pentafluoropropane
HCFC	R-225eb	1,1-Dichloro-1,2,3,3,3-pentafluoropropane
HCFC	R-226	Chlorohexafluoropropane
HCFC	R-226ba	2-Chloro-1,1,1,2,3,3-hexafluoropropane
HCFC	R-226ca	3-Chloro-1,1,1,2,2,3-hexafluoropropane
HCFC	R-226cb	1-Chloro-1,1,2,2,3,3-hexafluoropropane
HCFC	R-226da	2-Chloro-1,1,1,3,3,3-hexafluoropropane
HCFC	R-226ea	1-Chloro-1,1,2,3,3,3-hexafluoropropane
HCFC	R-231	Pentachlorofluoropropane
HCFC	R-232	Tetrachlorodifluoropropane
HCFC	R-232ca	1,1,3,3-Tetrachloro-2,2-difluoropropane
HCFC	R-232cb	1,1,1,3-Tetrachloro-2,2-difluoropropane
HCFC	R-233	Trichlorotrifluoropropane
HCFC	R-233ca	1,1,3-Trichloro-2,2,3-trifluoroprohane
HCFC	R-233cb	1,1,3-Trichloro-1,2,2-trifluoroprohane
HCFC	R-233cc	1,1,1-Trichloro-2,2,3-trifluoroprohane
HCFC	R-234	Dichlorotetrafluoropropane

(Continued)

Table 6.3 Name and ASHRAE codes of hydrochlorofluorocarbons [4]. (*Continued*)

Class	ASHRAE code	Chemical name
HCFC	R-234aa	2,2-Dichloro-1,1,3,3-tetrafluoropropane
HCFC	R-234ab	2,2-Dichloro-1,1,1,3-tetrafluoropropane
HCFC	R-234ba	1,2-Dichloro-1,2,3,3-tetrafluoropropane
HCFC	R-234bb	2,3-Dichloro-1,1,1,2-tetrafluoropropane
HCFC	R-234bc	1,2-Dichloro-1,1,2,3-tetrafluoropropane
HCFC	R-234ca	1,3-Dichloro-1,2,2,3-tetrafluoropropane
HCFC	R-234cb	1,1-Dichloro-2,2,3,3-tetrafluoropropane
HCFC	R-234cc	1,3-Dichloro-1,1,2,2-tetrafluoropropane
HCFC	R-234cd	1,1-Dichloro-1,2,2,3-tetrafluoropropane
HCFC	R-234da	2,3-Dichloro-1,1,1,3-tetrafluoropropane
HCFC	R-234fa	1,3-Dichloro-1,1,3,3-tetrafluoropropane
HCFC	R-234fb	1,1-Dichloro-1,3,3,3-tetrafluoropropane
HCFC	R-235	Chloropentafluoropropane
HCFC	R-235ca	1-Chloro-1,2,2,3,3-pentafluoropropane
HCFC	R-235cb	3-Chloro-1,1,1,2,3-pentafluoropropane
HCFC	R-235cc	1-Chloro-1,1,2,2,3-pentafluoropropane
HCFC	R-235da	2-Chloro-1,1,1,3,3-pentafluoropropane
HCFC	R-235fa	1-Chloro-1,1,3,3,3-pentafluoropropane
HCFC	R-241	Tetrachlorofluoropropane
HCFC	R-242	Trichlorodifluoropropane

and boils at 76.7°C. From the safety standpoint CTC is considered a dangerous compound because of sever hepatotoxicity, being a depressant and attacking the central nervous system. It also has high ozone depletion and global warming potential. Carbon tetrachloride is a potent solvent that used to be popular in the industry. Because of its adverse health effects its consumption has been curtailed.

The product of fluorination of carbon tetrachloride dichlorodifluoromethane underwent condensation and separation before being removed. In one scheme carbon tetrachloride was introduced into the reactor. A fluorinating agent like SbF_3 along with a catalyst such as $SbCl_5$ was also charged into the reactor. The mixture was heated up using a steam jacket to commence and speed the reaction.

Table 6.4 Name and ASHRAE codes of hydrochlorofluorocarbons [4].

Class	ASHRAE code	Chemical name
HCFC	R-243	Dichlorotrifluoropropane
HCFC	R-243ca	1,3-Dichloro-1,2,2-trifluoropropane
HCFC	R-243cb	1,1-Dichloro-2,2,3-trifluoropropane
HCFC	R-243cc	1,1-Dichloro-1,2,2-trifluoropropane
HCFC	R-243da	2,3-Dichloro-1,1,1-trifluoropropane
HCFC	R-243ea	1,3-Dichloro-1,2,3-trifluoropropane
HCFC	R-243ec	1,3-Dichloro-1,1,2-trifluoropropane
HCFC	R-244	Chlorotetrafluoropropane
HCFC	R-244ba	2-Chloro-1,2,3,3-tetrafluoropropane
HCFC	R-244bb	2-Chloro-1,1,1,2-tetrafluoropropane
HCFC	R-244ca	3-Chloro-1,1,2,2-tetrafluoropropane
HCFC	R-244cb	1-Chloro-1,2,2,3-tetrafluoropropane
HCFC	R-244cc	1-Chloro-1,1,2,2-tetrafluoropropane
HCFC	R-244da	2-Chloro-1,1,3,3-tetrafluoropropane
HCFC	R-244db	2-Chloro-1,1,1,3-tetrafluoropropane
HCFC	R-244ea	3-Chloro-1,1,2,3-tetrafluoropropane
HCFC	R-244eb	3-Chloro-1,1,1,2-tetrafluoropropane
HCFC	R-244ec	1-Chloro-1,1,2,3-tetrafluoropropane
HCFC	R-244fa	3-Chloro-1,1,1,3-tetrafluoropropane
HCFC	R-244fb	1-Chloro-1,1,3,3-tetrafluoropropane
HCFC	R-251	Trichlorofluoropropane
HCFC	R-252	Dichlorodifluoropropane
HCFC	R-252ca	1,3-Dichloro-2,2-difluoropropane
HCFC	R-252cb	1,1-Dichloro-2,2-difluoropropane
HCFC	R-252dc	1,2-Dichloro-1,1-difluoropropane
HCFC	R-252ec	1,1-Dichloro-1,2-difluoropropane
HCFC	R-253	Chlorotrifluoropropane

(*Continued*)

Table 6.4 Name and ASHRAEcodes of hydrochlorofluorocarbons [4]. (*Continued*)

Class	ASHRAE code	Chemical name
HCFC	R-253ba	2-Chloro-1,2,3-trifluoropropane
HCFC	R-253bb	2-Chloro-1,1,2-trifluoropropane
HCFC	R-253ca	1-Chloro-2,2,3-trifluoropropane
HCFC	R-253cb	1-Chloro-1,2,2-trifluoropropane
HCFC	R-253ea	3-Chloro-1,1,2-trifluoropropane
HCFC	R-253eb	1-Chloro-1,2,3-trifluoropropane
HCFC	R-253ec	1-Chloro-1,1,2-trifluoropropane
HCFC	R-253fa	3-Chloro-1,3,3-trifluoropropane
HCFC	R-253fb	3-Chloro-1,1,1-trifluoropropane
HCFC	R-253fc	1-Chloro-1,1,3-trifluoropropane
HCFC	R-261	Dichlorofluoropropane
HCFC	R-261ba	1,2-Dichloro-2-fluoropropane
HCFC	R-262	Chlorodifluoropropane
HCFC	R-262ca	1-Chloro-2,2-difluoropropane
HCFC	R-262fa	3-Chloro-1,1-difluoropropane
HCFC	R-262fb	1-Chloro-1,3-difluoropropane
HCFC	R-271	Chlorofluoropropane
HCFC	R-271b	2-Chloro-2-fluoropropane
HCFC	R-271d	2-Chloro-1-fluoropropane
HCFC	R-271fb	1-Chloro-1-fluoropropane

At the conclusion of the reaction the R-12 was separated from the mixture. By this method we have been able to obtain over 90% yield of CCl_2F_2 based on the amount of antimony fluoride used. Other products produced along with R-12 included other derivatives of carbon tetrachloride including $CClF_3$ (R-13) and CCl_3F (R-11). Midgely *et al.* demonstrated the possibility of using CH_2ClF (R-31) as a refrigerant in 1934 [9].

The Midgely *et al.* process [8] required antimony trifluoride (SbF_3) as a fluorinating agent. The method of manufacturing difluoro dichloro methane consisted of progressively replacing chlorine atoms of carbon tetrachloride with fluorine by interacting CCl_4 with SbF_3, removing the difluoro dichloro methane by fractionation as formed, and returning from the fractionation for further interaction with SbF_3, methane derivatives with less fluorine content than difluoro dichloro methane.

Table 6.5 Name and ASHRAE codes of hydrofluorocarbons [4].

Class	ASHRAE code	Chemical name
HFC	R-23	Trifluoromethane (Fluoroform)
HFC	R-32	Difluoromethane
HFC	R-41	Fluoromethane
HFC	R-125	Pentafluoroethane
HFC	R-E125	Pentafluorodimethyl ether
HFC	R-134	1,1,2,2-Tetrafluoroethane
HFC	R-134a	1,1,1,2-Tetrafluoroethane
HFC	R-E134	Bis(disfluoromethyl)ether
HFC	R-143	1,1,2-Trifluoroethane
HFC	R-143a	1,1,1-Trifluoroethane
HFC	R-143m	Methyl trifluoromethyl ether
HFC	R-E143a	2,2,2-Trifluoroethyl methyl ether
HFC	R-152	1,2-Difluoroethane
HFC	R-152a	1,1-Difluoroethane
HFC	R-161	Fluoroethane
HFC	R-227ca	1,1,2,2,3,3,3-Heptafluoropropane
HFC	R-227ca2	Trifluoromethyl 1,1,2,2-tetrafluoroethyl ether
HFC	R-227ea	1,1,1,2,3,3,3-Heptafluoropropane
HFC	R-227me	Trifluoromethyl 1,2,2,2-tetrafluoroethyl ether
HFC	R-236cb	1,1,1,2,2,3-Hexafluoropropane
HFC	R-236ea	1,1,1,2,3,3-Hexafluoropropane
HFC	R-236fa	1,1,1,3,3,3-Hexafluoropropane
HFC	R-236me	1,2,2,2-Tetrafluoroethyl difluoromethyl ether
HFC	R-FE-36	Hexafluoropropane
HFC	R-245ca	1,1,2,2,3-Pentafluoropropane
HFC	R-245cb	Pentafluoropropane
HFC	R-245ea	1,1,2,3,3-Pentafluoropropane

(Continued)

Table 6.5 Name and ASHRAE codes of hydrofluorocarbons [4]. (*Continued*)

Class	ASHRAE code	Chemical name
HFC	R-245eb	1,1,1,2,3-Pentafluoropropane
HFC	R-245fa	1,1,1,3,3-Pentafluoropropane
HFC	R-245mc	Methyl pentafluoroethyl ether
HFC	R-245mf	Difluoromethyl 2,2,2-trifluoroethyl ether
HFC	R-245qc	Difluoromethyl 1,1,2-trifluoroethyl ether
HFC	R-254cb	1,1,2,2-Tetrafluoropropane
HFC	R-254pc	Methyl 1,1,2,2-tetrafluoroethyl ether
HFC	R-263	Trifluoropropane
HFC	R-272	Difluoropropane
HFC	R-281	Fluoropropane
HFC	R-329ccb	1,1,1,2,2,3,3,4,4-Nonafluorobutane
HFC	R-338eea	1,1,1,2,3,4,4,4-Octafluorobutane
HFC	R-347ccd	1,1,1,2,2,3,3-Heptafluorobutane
HFC	R-347mmc	Perfluoropropyl methyl ether
HFC	R-347mmy	Perfluoroisoprophyl methyl ether
HFC	R-365mfc	1,1,1,3,3-Pentafluorobutane

The inventors described a detailed process for preparation of SbF_3 by fluorination of antimony trichloride ($SbCl_3$) in 1935. The invention comprised of interacting hydrofluoric acid with antimony trichloride to replace the chlorine with fluorine in the antimony compound. Hydrofluoric acid was passed through a molten mass of antimony trichloride contained in a steam heated lead receptacle, or pot. The hydrofluoric acid converted the antimony trichloride into antimony trifluoride, the replaced chlorine atoms. The hydrogen atoms produced hydrochloric acid leaving the reactor as a vapor. The reaction was represented by the following equation:

$$3HF + SbCl_3 \rightarrow SbF_3 + 3HCl \tag{6.1}$$

As the conversion of the antimony trichloride into antimony trifluoride proceeds, it crystallizes out progressively. The mixture of antimony trifluoride and antimony trichloride was a satisfactory fluorinating agent. All the fluorine fixed on the antimony was used during the subsequent fluorination processes. The resulting antimony trichloride was recycled by retreating it with hydrofluoric acid. The conversion to antimony trifluoride could be carried

Table 6.6 Name and ASHRAE codes of hydrofluoro-olefins [4].

Class	ASHRAE code	Chemical name
HFO	R-1132a	1,1-Difluoroethylene
HFO	R-1141	Fluoroethylene (vinyl fluoride)
HFO	R-1225ye(E)	(1E)-1,2,3,3,3-pentafluoroprop-1-ene
HFO	R-1225ye(Z)	(1Z)-1,2,3,3,3-pentafluoroprop-1-ene
HFO	R1225yc	1,1,2,3,3-pentafluoroprop-1-ene
HFO	R-1225zc	1,1,3,3,3-pentafluoroprop-1-ene
HFO	R-1234yc	1,1,2,3-tetrafluoroprop-1-ene
HFO	R-1234zc	1,1,3,3-tetrafluoroprop-1-ene
HFO	R-1234ye(E)	(1E)-1,2,3,3-tetrafluoroprop-1-ene
HFO	R-1234ye(Z)	(1Z)-1,2,3,3-tetrafluoroprop-1-ene
HFO	R-1234ze(E)	(1E)-1,3,3,3-tetrafluoroprop-1-ene
HFO	R-1234ze(Z)	(1Z)-1,3,3,3-tetrafluoroprop-1-ene
HFO	R-1234yf	2,3,3,3-Tetrafluoropropene
HFO	R-1234ze	1,3,3,3-Tetrafluoropropene
HFO	R-1243yc	1,1,2-trifluoroprop-1-ene
HFO	R-1243ye(E)	(1E)-1,2,3-trifluoroprop-1-ene
HFO	R-1243yf	2,3,3-trifluoroprop-1-ene
HFO	R-1243zc	1,1,3-trifluoroprop-1-ene
HFO	R-1243ze(E)	(1E)-1,3,3-trifluoroprop-1-ene
HFO	R1243zf	3,3,3-trifluoroprop-1-ene

out to completion but in later disclosure Midgely *et al.* expressed preference for a use of antimony trichloride and fluoride mixture as a fluorinating agent [10].

Other fluorinating agents have also been reported. As a Specific example, when 100 parts of SiF_4, 154 parts of CCl_4 and 7 parts of $SbCl_5$ were brought together, the first two reacted, in the presence of the third as a catalyst, to form either or both CCl_2F_2 and CCl_3F. The Silicon tetrafluoride and carbon tetrachloride can be brought into contact by bubbling the SiF_4 through the $SbCl_5$ in the presence of a catalyst at room temperature. Heating up the materials above the room temperature increased the reaction rate. Degradation reaction of SiF_4 produced a number of silicon by-products that were removed reaction products by washing with Water [11].

Yet a different reaction scheme to produce CFCs involves the interaction of a metal fluoride, a halogen derivative of an aliphatic hydrocarbon, and an antimony salt. When a pentavalent antimony salt was used it also acted a catalyst. If the antimony salt was not a catalyst another substance acted as a catalyst. Eq(6.2) and Eq(6.3) show examples of such reactions

$$2KF + SbCl_5 + CCl_4 \rightarrow CCl_2F_2 + 2KCl + SbCl_5 \qquad (6.2)$$

$$3CaF_2 + Sb_2(SO_4)_3 + CCl_4 + Cl_2 \rightarrow 3CCl_2F_2 + 3CaSO_4 + 2SbCl_3 \qquad (6.3)$$

In the reaction of Eq(6.2) the antimony compound is pentavalent and apparently serves the dual function of an antimony salt, which enters into the reaction and as a catalyst for the reaction. In the reaction of Eq(6.3) the antimony salt entered into the reaction and chlorine acted as the catalyst. However, the two reactions are essentially the same [12].

Preparation of fluorinating agents like antimony trifluoride presents an added step on the way to preparation of the chlorofluorocarbons. It was preferable to eliminate as many steps as possible to enhance the economics of a process. The next development represents significant progress in the preparation of CFCs because of direct hydrogen exchange. In summary a non-cyclic aliphatic hydrocarbon containing one or more atoms of chlorine, bromine or iodine was reacted with hydrofluoric acid in the presence of a catalyst such as an antimony compound.

For example, 500 parts gaseous anhydrous hydrofluoric acid was passed rapidly and steadily into reactor over a period of twenty-five hours (Figure 6.6) [13]. The reactor contained 600 parts of antimony-penta-chloride ($SbCl_5$) that was maintained at a temperature of 60° C. During this time 1925 parts (12% mols) of carbon-tetra-chloride was charged into the reactor. The products were condensed in a vessel maintained at -60° C followed by fractional distillation. The combined yield (based on CCl_4) of fluoro-tri-chloro-methane ($CFCl_3$) and difluoro-di-chloro-methane (CF_2Cl_2) was about 90%. The product composition consisted of 60% of CF_2Cl_2 and 40% of the $CFCl_3$ methane.

In another example, a mixture of 75 parts of $SbCl_5$, 10 parts of $SbCl_3$ and 10 parts of chloroform ($CHCl_3$) was charged to the reactor maintained at about 75° C. Anhydrous hydrogen fluoride gas was added at the rate of about 10 parts (by weight) per hour, and the chloroform at the rate of about 70 parts per hour. The products were condensed at -60° C. followed by fractional distillation to separate difluoro-chloro-methane and fluoro-di-chloromethane. The former boils at -15° C under a pressure of 2.03 bars; the latter boils at about +8.7° C at 1.01 bar pressure. The fractional distillation allowed the separation of any chloroform carried through the reactor for recycling. The yield of fluoro-di-chloro-methane is about 80-90% and that of di-fluoro-chloro-methane is about 10–5% both based on the chloroform consumption.

The processes of the previous two examples could be operated continuously or in batch mode. Improvements in the product yield and refinements of the direct halogen exchange process were reported in a series of patents in 1935-1936 [6, 13–18].

The industrial production of chlorofluorocarbon was achieved by substitution of fluorine for chlorine in chlorocarbons via halogen exchange reaction using hydrofluoric acid. Most of the CFCs were made by a liquid-phase reaction in presence of an antimony pentachloride catalyst. Alternatively the antimony catalyst could be added as $SbCl_3$ and converted to pentachloride by adding chlorine to the reaction mixture. The reaction proceeds by the reaction of hydrofluoric acid with antimony pentachloride resulting in chlorofluorides, the actual catalysts of the reaction. The halogen exchange takes place between the chlorocarbon and the chlorofluorides. The reaction scheme is seen in Eq(6.4) through Eq(6.8) showing the reaction that produce the various CFCs.

$$SbCl_5 + nHF \rightarrow SbnCl_{5\text{-}n} + nHCl \tag{6.4}$$

$$nRCl + SbF_nCl_{5\text{-}n} \rightarrow nRF + SbCl_5 \tag{6.5}$$

$$CCl_4 + HF \rightarrow CFCl_3 + HCl \tag{6.6}$$

$$CFCl_3 + HF \rightarrow CF_2Cl_2 + HCl \tag{6.7}$$

$$CFCl_3 + 2HF \rightarrow CHF_2Cl + 2HCl \tag{6.8}$$

As the reaction schemes in Eq(6.4) to Eq(6.8) indicate fluorine exchange takes place in steps. In commercial operations temperature is used to control the number of fluorine substitutes. In commercial reactors reactions temperature is in the range of 100-150°C and pressure 10.13-30.4 bars.

To produce ethane-based CFCs the starting chlorocarbon is hexachloroethane (HCE). It is a solid at the room temperature with a melting point around 183°C. Surprisingly HCE is not very toxic even if ingested though it is considered a skin irritant [20]. Hexachloroethane is a byproduct of some of the chlorination processes. It can be produced directly via a sequential chlorine substitution reaction ethane can be converted to HCE, seen in Eq(6.9). The most convenient way to work with HCE is to produce it *in-situ* by the reaction of tetrachloroethylene (CCl_2=CCl_2) with chlorine according to Eq(6.10) [12].

$$C_2H_6 + 6\ Cl_2 \rightarrow C_2Cl_6 + 6\ HCl \tag{6.9}$$

$$C_2Cl_4 + Cl_2 \rightarrow C_2Cl_6 \tag{6.10}$$

The high melting point of HCE is an impediment. Attempts are, therefore, made to reduce the hexachloroethane melting point before fluorinating it in liquid phase by reaction with HF. A specific example clarifies the processing method. First 75 parts of antimony pentachloride is added to 213 parts of hexachloroethane. Alternatively 57 parts of antimony trichloride and 17 parts of chlorine could be added to 213 parts of HCE. The combination has a significant impact on lowering the melting point of HCE from 183°C. to around 130° C. After the compounds are mixed in the specified proportions and slightly heated, the mixture will become a liquid. The C_2Cl_6 is fluorinated by the reaction of HF with the liquid mixture [21–24].

In one example the desired compound was 1,2-difluorotetrachloride (R-112, $C_2Cl_2F_4$). To maximize the yield of R-112, process steps and conditions were set such that temperature was held at 5°C and pressure at 1 bar. Compounds such as $C_2Cl_2F_4$, $C_2Cl_4F_2$, C_2Cl_5F, and any unreacted upon C_2Cl_6 were returned to the reactor where they were further fluorinated to obtain R-112. The reactions that take place in the reactor are displayed in Eq(6.11) and Eq(6.12):

$$C_2Cl_6 + SbCl_5 + 4HF \rightarrow C_2Cl_2F_4 + 4HCl + SbCl_5 \quad (6.11)$$

$$C_2Cl_6 + SbCl_3 + Cl_2 + 4HF \rightarrow C_2Cl_2F_4 + 4HCl + SbCl_3 + Cl_2 \quad (6.12)$$

In practice the process for R-112 production could be operated continuously by feeding C_2Cl_6 and HF to the reaction chamber and by recovering the $C_2Cl_2F_4$ and HCl from the top of the separator. The reaction conditions can be adjusted to maximize the yield of any single co-product of R-112. For example favorable yields of $C_2Cl_3F_3$ was obtained by maintaining a temperature of 48°C and pressure of 1 bar [22].

In another development fluorination aliphatic hydrocarbons took place without a catalyst but by only using the fluorinating agent itself [25]. A specific compound is pentavalent antimony salt containing three fluorine and two chlorine atoms with a formula of SbF_3Cl_2. This compound was prepared by allowing chlorine gas and antimony trifluoride to react in a vessel kept at a temperature above 70°C. The antimony trifluorodichloride was a sufficiently active fluorinating agent to the point of fluorinating hydrocarbons without a catalyst according to Eq(6.13) and Eq(6.14).

$$3C_2Cl_6 + 3SbF_3Cl_2 \rightarrow C_2Cl_2F_4 + C_2Cl_3F_3 + C_2Cl_4F_2 + 3SbCl_5 \quad (6.13)$$

$$3CCl_4 + 3SbF_3Cl_2 \rightarrow C_2Cl_2F_4 + CCl_2F_2 + C_2Cl_4F_2 + 2SbCl_5 \quad (6.14)$$

The fluoro and/or halo-fluoro derivatives of aliphatic hydrocarbons including methane, ethane, ethylene, etc., may be produced by this method. For example compounds from the various series such as $CHClF_2$, CCl_2F_2, $C_2Cl_2F_4$, C_2F_6 and $C_3H_3F_3Br_3$ may be obtained from the proper initial materials.

A method for converting perfluorochloroethane starting materials into perhalogenated ethane products of a different type than the starting materials which consisted of the step of contacting a perfluorochloroethane with at least one fluorine atom and at least one chlorine atom at a temperature between 150°C and 600°C. with the aid of a catalyst [26]. An example illustrates the procedure for catalyst preparation.

An activated alumina was employed in the form of 3 mm x 3 mm cylindrical pellets containing >99% (H_2O free basis) of alumina and low in sodium, iron and silica (0.03% Na_2O; 0.08% Fe_2O_3; 0.22% SiO_2). Before drying it has a 26% weight loss on ignition at 1000°C, and a surface area of 231m^2/g. The above activated alumina was dried by heating to 500°C. while sweeping with nitrogen for about one hour resulting in the loss of 9.7% by weight of water, based on dry weight.

After drying, the activated alumina, in an electrically heated tube was treated with $CF_2ClCFCl_2$ vapors at a space velocity of 180 volumes of $CF_2ClCFCl_2$ vapor (at standard conditions of temperature and pressure) per hour per volume of alumina for a period of 0.8 hour. The bed was maintained at an average temperature of 300°C during the treatment. At the outset a hot zone approximately 50°C hotter, than the average bed temperature, formed at the inlet to the bed and moved progressively down the bed toward the exit as the activation treatment proceeded [26].

During the activation treatment the principal products were a mixture of carbon dioxide and carbon monoxide at a ratio of 2:1 along with smaller amounts of tetrachloroethylene. After 0.8 hour, the evolution of carbon oxides suddenly ceased simultaneously with the appearance of the hot zone at the exit end of the bed. During this treatment the activated alumina in creased in weight by about 16%, based on dry weight.

The catalysts prepared were highly active for the conversion of perfluorochloroethanes having from 1 to 5 fluorine atoms into more highly fluorinated ethanes through disproportionation reactions. Such disproportionation reactions are represented by equations Eq(6.15) through Eq(6.19).

$$2C_2Cl_5F \rightarrow C_2Cl_4F_2 + C_2Cl_6 \tag{6.15}$$

$$2C_2Cl_4F_2 \rightarrow C_2Cl_3F_3 + C_2Cl_5F \tag{6.16}$$

$$2C_2Cl_3F_3 \rightarrow C_2Cl_2F_4 + C_2Cl_4F_2 \tag{6.17}$$

$$2C_2Cl_2F_4 \rightarrow C_2ClF_5 + C_2Cl_3F_3 \tag{6.18}$$

$$2C_2ClF_5 \rightarrow C_2F_6 + C_2Cl_2F_4 \tag{6.19}$$

Conversion of perfluorochloroethanes with 3-4 F-C bonds to those with 4-5 bonds was of particular interest because conventional methods using hydrogen fluoride or a fluorine containing metal salt required relatively drastic reaction conditions. Moreover, the reactions described here had high conversions and yields [26].

The disproportionation reactions of the invention may occur between like molecules, for example between two molecules of $CF_2ClCFCl_2$ or two molecules of CF_3CFCl_2 as shown in Eq.(6.20) and Eq.(6.21).

$$2CF_2ClCFCl_2 \rightarrow C_2Cl_2F_4 + C_2Cl_4F_2 \tag{6.20}$$

$$2CF_3CFCl_2 \rightarrow C_2ClF_5 + C_2F_3Cl_3 \tag{6.21}$$

On the other hand, mixed disproportionation reactions could take place occur between two isomers, or between two molecules containing different numbers of fluorine. Examples of mixed disproportionation can be seen in Eq.(6.22) and Eq.(6.30).

$$CF_3CCl_3 + CF_2ClCFCl_2 \rightarrow C_2Cl_2F_4 + C_2Cl_4F_2 \quad (6.22)$$

$$CCl_3CClF_2 + CCl_2FCCl_3 \rightarrow C_2Cl_3F_3 + C_2Cl_6 \quad (6.23)$$

$$CCl_3CF_3 + CCl_2FCl_2F \rightarrow C_2Cl_2F_4 + C_2Cl_5F \quad (6.24)$$

$$CCl_3CF_3 + CCl_3CClF_2 \rightarrow C_2Cl_2F_4 + C_2Cl_5F \quad (6.25)$$

$$CCl_3CF_3 + CCl_2FCCl_3 \rightarrow C_2Cl_2F_4 + C_2Cl_6 \quad (6.26)$$

$$CCl_2FCF_3 + CClF_2CCl_2F \rightarrow C_2ClF_5 + C_2Cl_4F_2 \quad (6.27)$$

$$CCl_2FCF_3 + CCl_2FCCl_2F \rightarrow C_2ClF_5 + C_2Cl_5F \quad (6.28)$$

$$CCl_2FCF_3 + CCl_3CClF_2 \rightarrow C_2ClF_5 + C_2Cl_5F \quad (6.29)$$

$$CCl_2FCF_3 + CCl_2FCCl_3 \rightarrow C_2ClF_5 + C_2Cl_6 \quad (6.30)$$

In addition to the disproportionation reactions mentioned above, rearrangement reactions may be also carried out in accordance to this process. The molecules that are symmetrically arranged with respect to the distribution of chlorine and fluorine are converted to asymmetric forms. Typically, Eq.(6.31) and Eq.(6.32) represent the rearrangement reactions [26].

$$CF_2ClCFCl_2 \rightarrow CF_3Cl_3 \quad (6.31)$$

$$CF_2ClCF_2Cl \rightarrow CF_3CFCl_2 \quad (6.32)$$

The rearrangement and disproportionation reactions described could occur simultaneously. For example, when $C_2Cl_4F_2$ was passed over the catalyst it converted to $C_2Cl_3F_3$ (C_2Cl_5F). And $C_2Cl_3F_3$ could disproportionate to the next higher fluorinated ethane $C_2Cl_2F_4$($C_2Cl_4F_2$). When $CF_2ClCFCl_2$ was treated/rearrangement to $C_2F_3Cl_3$ and mixed disproportionation between $CF_2ClCFCl_2$ and the rearrangement product CF_3Cl_3 occured simultaneously together with disproportionation of $CF_2ClCFCl_2$ with itself and of CF_3CCl_3 with itself.

The various rearrangement and disproportionation reactions undergone by the perfluorochloroethanes in accordance with the invention occurred at different rates depending upon the reaction temperature and other conditions. By controlling the reaction conditions, e.g. temperature, a desired disproportionation and/or rearrangement reaction could be made to predominate. In general, better conversions to disproportionation products higher in fluorine content were obtained at higher temperatures.

The catalytic reaction was run on $CF_2ClCFCl_2$ according to eleven different sets of conditions. In each experiment the catalyst employed consisted of 172 grams 3 mm x 3 mm cylindrical pellets of activated alumina prepared as previously described. It was treated with $CF_2ClCFCl_2$ at a temperature of 300°C until CO and CO_2 were no longer produced. The reaction temperatures reported were average catalyst bed temperatures. The results of eleven runs in which the catalyst temperature varied from 250° C. to 400°C are summarized in Table 6.7 [26].

A number of processes are known for the production of fluorinated and chlorofluorinated organic compounds. They include reactions carried out in a liquid or vapor phase. When the liquid phase technique is used, the reaction is influence by boiling a mixture of hydrofluoric acid and a halogenated organic compound in the presence of antimony halides with a reflux under pressure. In the vapor phase process a vaporized mixture of hydrogen fluoride and a halogenated organic compound is passed over catalysts that are usually based on aluminum, chromium, zirconium, thorium and other fluorides. The raw materials are halogenated organic compounds such a carbon tetrachloride, chloroform, perchloroethylene, etc., produced by thermal chlorination of hydrocarbons.

The process for preparing fluorinated and chlorofluorinated organic compounds using reacting hydrogen fluoride, chlorine and methane requires the presence of a catalyst and a mixture of continuously recycled hydrocarbons. This process has high yields and a rapid reaction time but it has the disadvantages of requiring catalysts, which exhibit a relatively short life, requiring frequent reactivations and periodic renewals thus interrupting the continuity of the process.

A process was developed for the preparation of fluorinated and chlorofluorinated organic compounds, that did not require frequent catalyst regeneration or replacement of the catalyst mass. One major advantage of this technique was a process that required an extremely reduced catalytic contact time so that the output of the whole process was increased. The process involves successive passing a mixture of chlorine, hydrogen fluoride and methane together with one or more halogenated hydrocarbons through two catalytic reaction zones. One of the zones was maintained at a higher temperature than the other zones. Specifically, the process consisted of passing the reactants through two catalytic zones. The temperature in one zone was kept between 350° and 500°C while the other zone's temperature ranged from 200 to 350°C [27].

The same inventors reported further progress in the fluorination process [28]. A continuous process was developed to convert methane into $CFCl_3$ and CF_2Cl_2 by passing the mixture of CH_4, Cl_2 and HF along with a recycle mixture over a solid catalyst bed. The recycle stream consisted of CCl_4, $CFCl_3$, $CHCl_3$, CH_2Cl_2, $CHCl_2F$ in which CCl_4 and $CFCl_3$ formed $\geq$ 60% of the mixture. The molar ratio of HF to CH_4 was 2:1 or higher. The bed temperature was in the range of 180-700°C and the contact time 0.1-30 seconds.

Table 6.7 Product composition of conversion of chlorofluorocarbons vs reaction temperature and space velocity [22].

Temp. °C.	Space velocity per hour	Product composition, mole percent									
		CF_3CF_2Cl	CF_3CFCl_2	CF_2ClCF_2Cl	CF_3CCl_3	$CF_2ClCFCl_3$	CF_2ClCCl_3 [a]	$CFCl_2CCl_3$	C_2Cl_4	C_2Cl_6	CF_3Cl_3 plus CF_3CFCl_2 plus CF_2ClCF_2Cl
250	345	0.1	20.0	0.8	50.1	14.1	12.2	1.5	0.5	0.7	71
275	350	0.4	21.4	1.1	48.1	12.8	12.4	2.1	0.8	0.1	71
300	175	1.0	20.8	1.8	54.9	6.1	9.9	2.6	1.3	1.6	78
300	355	0.7	21.9	1.8	47.9	10.5	12.5	2.5	1.2	1.1	72
300	700	0.5	21.0	1.6	41.7	18.7	12.7	2.0	0.6	1.2	64
300	1,280	0.4	19.6	1.0	31.8	33.1	11.9	1.3	0.6	0.3	52
350	350	2.3	23.1	2.8	46.1	8.1	10.7	2.8	2.7	1.5	72
350	690	1.7	23.2	2.6	42.7	12.0	11.4	2.6	2.6	1.1	69
350	1,230	1.1	22.8	2.2	38.7	18.2	12.1	2.2	0.8	1.9	64
400	680	3.6	25.1	3.6	39.3	8.0	10.7	2.7	5.7	1.3	68
400	1,230	2.3	24.6	3.4	39.2	14.0	11.4	2.5	4.7	0.9	64

A fluid-bed reactor that contained 218 ml of catalyst (consisting of fluorinated alumina impregnated with 10% by weight of ThF_4. A mixture of halogenated hydrocarbons was recycled continuously and a mixture of hydrogen fluoride, chlorine and methane was fed. At the reactor outlet the products were separated continuously. The reaction temperature was 470°C with a contact time of 1.99 seconds. The composition of the mixture of halogenated hydrocarbons which was recycled continuously was approximately the following in mole percent:

CCl_4	80.4
$CHCl_3$	6.0
CH_2Cl_2	0.8
$CFCl_3$	12.7

The molar ratios of the reactants and of the mixture of halogenated hydrocarbons to be recycled were as follows:

$$\text{Recycled mixture: } Cl_2\text{: HF: } CH_4 = 5.9\text{: }4.4\text{: }1.7\text{: }1$$

The conversions of the reactants were according to the following (%):

CH_4	97.4
Cl_2	85.3
HF	98.9

The obtained products and the yields computed on the converted CH_4 were as given below (%):

CF_3Cl	9.5
CF_2Cl_2	73.5
$CFCl_3$	15.9

The choice of catalyst included oxides chromium, nickel, cobalt and others [28].

6.3.1 Longevity of Process Catalysts

Antimony catalysts are quite effective but become less active after use like other catalysts. The recovery process is fairly simple and requires oxidation with chlorine. Longer term, formation of carbonaceous and tarry matter limits the activity of the catalyst beyond the reactivation capability of chlorine oxidation [21]. The options left are purification or disposal as waste. Environmental concerns about disposal have for some time spurred development of research into detoxification and rejuvenation of antimony catalysts [29–31].

A review of some of the past developments is presented here. US Patent 3,784,671, proposed the following method for regeneration of antimony catalyst. The following steps were taken: 1. $SbCl_5$ in the catalyst solution was thermally dissociated into $SbCl_3$ and Cl_2; 2. Extraction of the organic contamination with halogenated hydrocarbons; 3. Oxidizing the $SbCl_3$ contained in the residue with chlorine to $SbCl_5$; and 4. Distilling off the recovered $SbCl_5$ in a vacuum from the inorganic impurities. This method also has the disadvantage of being uneconomical because of high temperatures are necessary to achieve the thermal decomposition of $SbCl_5$ to $SbCl_3$ and Cl_2 [29].

The procedure for carrying out this thermal reaction as well as the following chlorination of the $SbCl_3$ requires a pressure vessel made of nickel alloy. The utilization of steel was not recommended for safety reasons, thermal stress in the presence of chlorine. At the high temperatures of the thermal decomposition, there is a possibility that a portion of the $SbCl_3$ undergoing sublimation, as well as part of the high melting organic compounds, would be pulled out with the large amounts of chlorine escaping from the apparatus, resulting in the possibility of encountering dangerously high pressures because plugage of the chlorine gas escape passageway.

Further, separation of $SbCl_3$ from the organic impurities requires great amounts of carbon tetrachloride and other solvents as well as a large expenditure of time and energy. The corresponding separation in this process by way of recrystallization of the $SbCl_3$ has the disadvantage that it must be repeated several times to enable recovery of sufficient quantities of antimony trichloride. Because great quantities of dissolved $SbCl_3$ are separated also with the solvent, the $SbCl_3$ must be further recovered by distillation of the solvent from the organic extracts. A distillation recovery of the dissolved $SbCl_3$ from the combined solvents can only be successful with a costly distillation apparatus because, for example, hexachloroethane boils only 35° C lower than antimony pentachloride, and tends to sublime at this temperature. Considering the required quantities of solvents by the process and the quantity of antimony trichloride recovered from this extraction material, this method is uneconomical.

In U.S. Patent 3,806,589 antimony halogen salt catalysts were separated through extraction with water from the halogenated hydrocarbons. For purification, the antimony chlorides were reduced, for example, with sulfur dioxide in the presence of alkali iodide to $SbCl_3$. For the removal of the heavy metal ions and the other foreign ions present, the $SbCl_3$ was converted with a concentrated ammoniacal solution into SbO_3. After reconversion of the filtered-off antimony oxides into $SbCl_3$, very pure $SbCl_3$ was obtained through fractional distillation. According to this procedure, the reduction of $SbCl_5$ requires only a relatively modest expenditure with a low cost reduction material. The method, however, has the same disadvantage as in the previously discussed processes that generally a reduction of the $SbCl_5$ $SbCl_3$ must be performed in the process.

In the exhausted catalyst solutions the antimony exists for the most part in a form as active $SbCl_5$. The inactivity of the catalyst solutions rests, to a considerable measure, upon the dilution of the antimony pentachloride by high boiling organic compounds. The removal of the high boiling organic compounds from the $SbCl_5$ is extremely difficult to perform on an efficient economical basis. According to earlier known methods great quantities of Sb^{5+} are converted into Sb^{3+}, which thereafter must be reoxidized [29].

Fernschild *et al.* [25] reported development of A process for the recovery of antimony pentachloride from catalyst solutions used in the fluorination of chlorinated hydrocarbons. The catalyst solutions contained exhausted antimony ions and high boiling halogenated hydrocarbons which were chlorinated to convert them to Sb^{3+} to Sb^{5+}. Treatment of the solution with hydrogen fluoride followed to generate $SbCl_xF_y$ compounds in which x is 4-2 and y is 1-3, the sum of x+y=5. The high boiling halogenated hydrocarbons were converted to lower boiling halogenated hydrocarbons. They could be removed because they had a lower boiling temperature than the boiling point of $SbCl_xF_y$. The remaining $SbCl_xF_y$ compound was converted to $SbCl_5$ using carbon tetrachloride. In the last step antimony pentachloride was separated from the remaining organic compounds and inorganic residue.

Lee proposed a method for detoxification of spent catalyst from a fluorination process included the step of hydrolyzing such spent catalyst in the presence of aqueous calcium chloride to produce insoluble compounds that could be separated from the aqueous medium [30]. The spent catalyst included antimony halide, such as $SbCl_5$, SbF_5, $SbCl_3$ and/or SbF_5, usually with some halogenated hydrocarbons. Some of the halocarbons included chlorinated methane, partially chlorinated ethane and/or chlorofluorocarbons (C1 to C5). Usually present were of arsenic compounds, such as $AsCl_3$, hydrofluoric acid, and antimony chlorofluorides, such as $SbCl_4F$.

The spent catalyst was directly treated with the aqueous calcium chloride solution preferably containing small amounts of a transition metal ion, such as Fe^{++} or Fe^{+++}, or aluminum ion (Al^{+++}). Next the emitted gases were treated. An organic liquid phase was separated from an aqueous liquid phase containing dispersed precipitated insoluble compounds of antimony and arsenic (when arsenic was present), such as calcium salts of acids, oxides, hydroxides and/or oxyhalides resulting from the hydrolysis reaction. The insoluble materials were separated from the aqueous liquid medium by filtration and the filtrate was neutralized with lime.

The neutralized filtrate, containing calcium chloride resulting from the neutralization reaction, could be returned to the hydrolysis step and/or be further treated by a sulfide or hydrosulfide and a water-soluble iron salt. Examples included aqueous solutions of sodium sulfide or sodium hydrosulfide, and ferric chloride, to co-precipitate compounds of antimony and arsenic. The precipitates were removed from the aqueous medium, together with precipitated compounds of heavy metals that were present. Next the filtrate was discharged to a sewer system as long as toxic materials content was within the permissible range. The antimony compounds that were removed during the process were recovered and converted to useful form like chloride for fluorination catalysts [30].

A process for treating spent catalyst from chlorofluorocarbon production in which the catalyst includes antimony halides [31]. It underwent hydrolysis in an aqueous medium to produce insoluble antimony compounds and fluoride ions. The process involved hydrolyzing the catalyst in an aqueous solution of ferric chloride with a sufficient concentration of ferric ions to complex all of the fluoride ions produced by hydrolysis of the catalyst. After hydrolysis of the catalyst, and complexing of the fluoride ions, the reaction mass was neutralized by reacting it with an aqueous suspension of calcium or magnesium hydroxide. The insoluble antimony compounds were then separated from the neutralized reaction.

Vapor phase halogen exchange techniques have been developed that circumvent the catalyst disposal problem. Typically, in these processes vapor of chlorocarbons and excess hydrofluoric acid are passed over the catalyst at temperatures in the range of 250 to 400°C. Increasing the catalysts bed temperature increases fluorination though it still takes place in a stepwise manner. The common catalyst is CrF_3 or [32] Cr_2O_3 usually supported by alumina (Al_2O_3) [33, 34]. Improvements were made in the chromium catalyst during the research for CFC substitutes [35]. Vapor phase fluorinations are industrially quite popular. Process pressure is a modest 10 bar. One drawback to the process is formation of CFC-HF azeotropic mixture that complicates the downstream purifications.

A reported development described preparation of high surface area trivalent chromium oxides and hydroxides that are useful as catalysts for a variety of processes including fluorination [35]. The method for preparing a chromium oxide or hydroxide aerogel consists of the following steps:

(a) A solution comprising chromium (VI) oxide is made in a solvent that is reducing to chromium (VI) oxide;
(b) The solution is heated to a temperature at which chromium (VI) oxide was reduced by the solvent to a solid oxide or hydroxide form, producing a slurry or gel;
(c) The slurry/gel was heated continuously to above the critical temperature of the solvent; and,
(d) The solvent was vented and the oxidation products consisted of a chromium (III) oxide or hydroxide aerogel with surface area at least about 200 m^2/g.

6.4 Fluorocarbon Replacements of CFCs

The Montreal Protocol mandated the reduction of CFCs production by 50% by the end of 1998 because of their deleterious effect on the ozone layer. The deadline gave producers and consumers of CFCs a short decade to find substitutes. The allotted time was considered short taking into account the massive penetration of CFCs in numerous, some critical, applications across the globe. For example, before the Protocol, CFC-12, best known by its DuPont brand name Freon® 12 was the refrigerant of every home refrigerator and automobile air conditioning. CFC-12 was as close as one could get to a perfect refrigerant in those applications. A suitable substitute had to be found to replace this ubiquitous refrigerant.

The enactment of the Protocol created a crisis for the fluorocarbon consumer and manufacturer industry requiring a rush to develop substitutes for the CFC. Not only the replacement materials had to meet the Montreal Protocol's requirements but had to be useable in existing equipment. Chapter 10 provides a description of the evolution of the fluorocarbon replacements over the years.

The original Montreal Protocol was signed in 1987 and called for 50% reduction in production of CFCs 1998. Within a year of signing National Aeronautics and Space Administration (NASA) released new research findings indicating the ozone hole problem was much more sever than previously thought. Consequently, the pace of CFC phase-out came under questioning. A subsequent update to the Montreal Protocol in 1990 changed the timetable to full phase-out to 2000. CCl_4, $CHCl_3$ and CH_3CCl_3 were also added to the

Table 6.8 HCFC and HFC replacements for CFCs [21].

Application	CFC grade	Replacement
Refrigerant	12 11	HFC-134a, HCFC-123, blends, azeotropes
Foaming Agent	11	HFC-134a, HFC-152a, HCFC-124, HCFC-141b, blends, azeotropes
Cleaning Agent	113	blends, azeotropes

list of CFCs and Halon®'s to be banned. The acceleration of complete phase-out left the producers and suppliers of CFCs in a mad rush searching for alternatives.

The substituents for CFCs included HCFCs and HFCs that were developed at highly accelerated paces (Tables 6.1–6.5). The presence of hydrogen in HCFCs reduced their stability in the atmosphere. Degradation of the HCFCs and HFC took place mostly below the stratosphere leading to reduced ozone depletion and global warming potentials. The manufacturers focused on C1-C3 compounds. Many hydrochlorofluorocarbon and hydrofluorocarbon compounds were developed and evaluated. Just in the C2 group one could identify eighty-eight compounds with combinations of C, H, F and Cl atoms, excluding their stereoisomers.

An approach was developed to predict atmospheric lifetime, flammability and toxicity of potential compounds. Culling out those compounds that had unacceptable properties reduced the list of compounds.

Table 6.8 provides a list of important HCFC and HFC replacements for CFCs and their applications.

6.5 Substitutes for CFCs: HCFC and HFC

One of the issues in preparing replacements for CFC was mixture of isomers of HCFCs and HFCs was produced. Some of the isomers are more desirable than others and had to be purified and separated from its isomer mixture. However, the boiling points of isomers are usually so close to each other that purification by the distillation process (which is industrially the most generally used separation method) is not feasible or if feasible is not economical. Therefore, methods would have to be found to enhance the yield of the desirable isomers.

Downing *et al.*, dating back to 1947, reported one of the earliest HFC developments. Little did anyone know about the importance of HFC's decades later. They started with halogenated hydrocarbons and produced fluoroalkyl chains by pyrolysis. One of the products was $C_3F_6H_2$ (R-245fa) prepared by pyrolysis of CHF_2Br in a platinum-lined tube at 615°C. At the time this compound was a new chemical compound that can be purified by fractional distillation. Downing *et al.* stated $C_3F_6H_2$ could be useful as refrigerant in absorptive or adsorptive systems because of boiling point (15.3°C), volatility (123 kPa at 20°C) and low toxicity [36].

6.5.1 Preparation of Hydrochlorofluorocarbons (HCFCs)

A HCFC (2-chloro-1,1,1,2 tetrafluoroethane) was prepared by a process that consisted of contacting 1,1-dichloro 1,2,2,2-tetrafluoroethane with an active-hydrogen source selected from the group consisting of formamide and alcohols containing from 1 to 3 carbon atoms. The reaction took place in the presence of an active metal catalyst consisted of zinc, magnesium, and calcium. The reaction temperature was from 50° to 180°C when the catalyst was calcium and magnesium. The temperature ranged from 90° to 180°C when the metal was zinc [37].

For example, isopropyl alcohol in the presence one of zinc, magnesium or calcium reacted at 110°C to 130°C under autogenous pressure yielding 2-chloro-1,1,1,2 tetrafluoroethane [Eq(6.33)]. There were three product fractions: 0 CF_2=CHF, CF_3CHClF and CF_3CFCl_2. The organic recovery was 81.4 percent, conversion 56.8 percent, and 71.5 percent yield of 2-chloro-1,1,1,2-tetrafluoroethane.

$$CCl_2F\text{-}CF_3 + CH_3CHOHCH_3 + Zn \rightarrow CF_2{=}CHF + CF_3CHClF + CF_3CFCl_2 + ZnCl_2 \quad (6.33)$$

Ballard *et al.*'s method described selective preparation of 1-chloro1,2,2,2-tetrafluoroethane ($CH_2Cl\text{-}CF_3$). A mixture of 1,1-dichloro-1,2,2,2-tetrafluoroethane and 1,2-dichloro-1,1,2,2-tetrafluoroethane was heated to a temperature in the range form —35° to 30°C with an alkali metal amalgam in an active hydrogen containing liquid medium. A solvent was present for the dichlorotetrafluoroethanes. In the last step 1-chloro-1,2,2,2-tetrafluoroethane was separated from unreacted 1,2-dichloro-1,1,2,2-tetrafluoroethane [38].

Methanol and distilled water (solvent) were placed in a flask equipped with reflux and a cold trap and placed in an ice bath. The flask was purged with nitrogen and allowed to cool to 0° C. A mixture of 1,2-dichlorotetrafluoroethane and 1,1-dichlorotetrafluoroethane was charged into the flask forming a homogeneous mixture with the aqueous alcohol. The contents were stirred vigorously while a sodium amalgam was added to the mixture over a period of 35 minutes. The stirring was continued for an additional 40 minutes. The flask was removed from the ice-bath and warmed gently to aid distillation of the volatiles. After 2.25 hours, the contents of the cold trap were transferred by evaporation into an evacuated flask that was then let down to atmospheric pressure. The contents of the flask were quantitatively analyzed for starting materials and products by gas chromatography. Several similar experiments were run following this procedure at different reactant and solvent compositions [38].

Manzer and Rao [39] reported a process for the preparation of 1,1,1-trifluorodichloroethane and 1,1,1,2-tetrafluorochloroethane by fluorination of a tetrahaloethylene, $C_2Cl_{4-x}F_x$, in which x=0 to 3. The process required reaction of tetrahaloethylene and HF in the gas phase at a temperature in the range of 300°C to about 450°C. The process catalyst was a metal in an oxidation state greater than zero from the group including of chromium, manganese, nickel, and cobalt, and their mixtures. The catalyst was supported by mixture of aluminum, oxygen, and fluorine in such proportions that the fluorine content of AlF_3 content was ≥90% by weight of the catalyst. To ensure fluorine content was ≥90% the AlF_3 was pretreated with HF.

The procedure for fluorination was followed using 19.8 g. (30 cc) of $NiCl_2/Al_2O_3$ (2% $NiCl_2$) as the initial catalyst charge. The results of the reaction of HF with CCl_2=CCl_2 over the prepared catalyst are given in Table 6.9.

Gumprecht *et al.* made reported a process for the preparation of 1,1,1-trifluorodichloroethane (and/or 1,1,1,2-tetrafluorochloroethane, CF_3-CHFCl) by fluorination of a tetrahaloethylene, $C_2Cl_{4-x}F_x$ [40]. The value of x was from 0 to 3. Another compound was a pentahaloethane, $C_2HCl_{5-x}F_x$, in which x=0 to 2. The preparation method entailed contacting the gaseous tetrahaloethylene and/or pentahaloethane with HF and Cr_2O_3. The Cr_2O_3 catalyst was prepared by pyrolysis of $(NH_4)_2Cr_2O_7$.

The product stream of the catalytic reaction contained 1,1,1-trifluorodichloroethane and/or 1,1,1,2-tetrafluorochloroethane and pentafluoroethane. The amount of 1,1,1-trifluorodichloroethane and/or 1,1,1,2-tetrafluorochloroethane was greater than the amount of pentafluoroethane. In the final step 1,1,1-trifluorodichloroethane and/or 1,1,1,2-tetrafluorochloroethane were recovered by purification from the product stream. Table 6.10 shows examples of the reaction under several conditions and their impact on conversion and the composition of the product.

US Patent 5,030,372 describes a catalytic process for interconverting (converting one compound to the other) the chlorofluorohydrocarbons CF_3CHClF (HCFC-124) and $CClF_2CHF_2$ (HCFC-124a), using either a liquid or gas phase reaction. The catalyst for the reaction was an aluminum halide or halogenated alumina. The equilibration constant for the interconversion varied with temperature over the range of 50°C to 475°C. Formation of CF_3CHClF was more favorable as the temperature decreased [41].

Table 6.9 Examples of reaction conditions for HCFC preparation and product composition [39].

	Example	
	1	2
Temp.°C.	325	350
HF/C_2Cl_4 (mol ratio)	5/1	6/1
Contact Time (sec)	45	30
Conversion %	58.7	61.2
	Area Percent	
CF_3CHCl_2	70.0	60.6
CF_3CHClF	12.4	20.3
CF_3CHF_2	0.5	1.3
CF_3CHCl_2 plus CF_3CHClF	82.4	80.9

Table 6.10 Examples of reaction conditions and their impact on conversion and product composition [40].

	Example						
	1	2	3	4	5	6	7
Temperature	300°	300°	250°	225°	225°	225°	250°
HF/C_2Cl_4 mol ratio	4/1	4/1	4/1	6/1	10/1	6/1	6/1
Contact time (sec)	15	30	60	60	60	90	90
Conversion (percent)	79.8	89.4	97.8	94.9	90.4	98.3	99.5
	Area percent						
CF_3CHCl_2	65.3	57.9	76.1	73.1	74.2	78.3	68.3
CF_3CHClF	11.8	13.2	14.5	10.0	10.0	12.4	20.2
CF_3CHF_2	10.7	15.1	5.4	0.1	0.1	0.1	8.0
CF_3CHCl_2+ CF_3CHClF	77.1	71.1	90.6	83.1	84.2	90.7	88.5

*Area Percent refers to the area under the graph in gas chromatography.

A gas phase process for increasing in a composition the mole ratio of one compound selected from the group consisting of CF_3CHClF and $CClF_2CHF$. If the selected compound was CF_3CHClF, the mole ratio of CF_3CHClF to $CClF_2CHF_2$ was less than the equilibrium ratio at 50°C. If the selected compound was $CClF_2CHF_2$, the mole ratio of $CClF_2CHF_2$ to CF_3CHClF in the gaseous composition was less than the equilibrium ratio at 475°C. The gas mixture was passed over a catalyst consisting of a halide aluminum oxide, an aluminum halide or their mixtures. The reaction time was sufficiently long to allow substantial interconversion between CF_3CHClF and $CClF_2CHF_2$ to take place. The catalyst contact temperature was in the range of 50°C to 475°C. The reaction could also be conducted via a liquid phase process.

The mole ratio of CF_3CHClF to CF_2HCF_2Cl at equilibrium was significantly greater than 1.0 throughout the temperature range of 50°C to 475°C. This process could be employed for increasing the mole ratio of CF_3CHClF to CF_2HCF_2Cl in a composition. Another mixture of gases was CF_3CH_2F and CF_2HCF_2H in which the mole ratio of CF_3CH_2F to CF_2HCF_2H was increased in the product mixture by this process.

HCFC-124 and HCFC-124a can be respectively converted to CF_3CH_2F (HFC-134a) and CF_2HCF_2H (HFC-134) by hydrogenolysis; and this process provided a means for achieving desirable mixtures of HFC-134a and HFC-134 [41].

A process to increase isomerization of HCFC-123a to HCFC-123 involved a reaction at 100° to 200°C in the presence of a catalysts [42]. Examples of the catalyst of isomerization included alumina catalysts, chlorinated and fluorinated by HF. However, this method has the shortcoming in that catalytic activity declines quite rapidly. The known method to prevent the decline of the catalytic activity entails isomerization by contact between the mixture of a perhalogenated carbon and HCFC-123a with the chlorinated and fluorinated alumina. The problem with that technique was excessive processing steps required to separate HCFC-123 and the perhalogenated carbon after the isomerization reaction. Additionally, the perhalogenated carbon that was to prevent the decline of the catalytic activity was itself a banned substance because of its detrimental effect on the ozone layer.

Investigation of catalysts with high activity for isomerization turned up partially fluorinated alumina catalyst compounds that had long life. An excellent catalyst for the isomerization of HCFCs was obtained by calcinating $Al_2O_3.nH_2O$ (n=0 to 3) containing fibril of polytetrafluoroethylene. The catalyst wss obtained by calcinating $nAl_2O_3.nH_2O$ (n=0 to 3) containing fibrils of polytetrafluoroethylene at a temperature of at least 400° C. Next a specific example is presented [42].

About 1,700 g of the pure water was heated to 100°C and 70 g of aluminum isopropoxide was added while agitation and hydrolysis continued for 3 hours. The solution was allowed to stand for one day before being rinsed and filtered to recover a filter cake (aluminium hydroxide). The filter cake was converted into mortar by adding 0.875g of polytetrafluoroethylene of easy fibrillability and mechanically mixing them for about 3 minutes followed by drying for 60 min at 100° C. The blend was kneaded, while wet and warm 3 to 4, times for 60 minutes.

The filter cake containing polytetrafluoroethylene fibrils was dried for a day at 120°C and pulverized to the powder of uniform 22 to 40 mesh. An amount of 1.2 g of the powder was placed into the reactor tube and calcinated for 3 hours at 600°C under N_2 sweep to get fluorinated alumina catalyst. The elemental analysis of aluminum, fluorine and oxygen in the catalyst were determined by X-ray photoelectron spectroscopy analysis (Table 6.11).

Then, by using HCFC-123 containing 10% of HCFC-123a as the raw material of the isomerization, HCFC-123a was isomerized to HCFC-123 at 120 ml/min (gas) of the feed rate of the raw material. The reaction temperature was 60°C for the described catalyst. The conversion rate of the isomerization of HCFC-123a to HCFC-123 after 1 hour and 3 hours is shown in Table 6.12 [42].

Another development provided a process for the preparation of hydrogen-containing chlorofluoropropanes represented by the formula $C_3HCl_2F_5$ [43]. In this process monofluorodichloromethane was contacted with tetrafluoroethylene in the presence of a modified aluminum chloride catalyst at a temperature of 0 to 150° C yielding and rhydrogen-containing chlorofluoropropanes. The modified aluminum chloride was prepared by contacting anhydrous aluminum chloride with chlorofluorocarbon with 1 to 2 carbon atoms followed by the removal of liquid products. The product was 2,2-dichloro 1,1,1,3,3-pentafluoropropane.

Sievert et al found that certain isomeric hydrogen-containing chlorofluoropropanes with the formula $C_3HCl_2F_5$ could be prepared readily in good yields by the reaction of monofluorodichloromethane with tetrafluoroethylene in the presence of a modified aluminum chloride catalyst. The reaction may be represented by Eq(6.34).

$$CHCl_2F + CF_2{=}CF_2 \rightarrow C_3HCl_2F_5 \quad (6.34)$$

"$C_3HCl_2F_5$" in Eq(6.34) represents an isomeric mixture of hydrogen-containing chlorofluoropropanes. The reaction of $CHCl_2F$ with $CF_2{=}CF_2$ to produce hydrogen-containing

Table 6.11 Elemental analysis of the catalyst by x-ray photoelectron spectroscopy analysis [42].

Element	Case 1	Case 2	Case 3*
Al, %	35.5	37.3	32.4
F, %	8.3	3.9	23.2
O, %	56.2	58.8	44.2
Cl, %	–	–	0.2

*In the Case 3 a mixture of CFC-13 (Chlorotrifluoromethane, $CClF_3$) and N_2 was used.

Table 6.12 Conversion rate of the isomerization of HCFC-123a to HCFC-123, [42].

Example	After 1 hour	After 3 hours
1 (Case 1)	99.9	85.0
2 (Case 1)	95.0	88.0
3 (Case 2)	99.5	80.0
4 (Case 2)	98.0	65.0
Case 3	85.0	0.0

chlorofluoropropanes has been reported by [44, 45] in which the catalyst used was ordinary aluminum chloride. The disadvantage of the above process, using ordinary aluminum chloride as the catalyst, was the extensive halogen exchange reaction that took place. At least one aspect of such exchange was the production of a large amount of chloroform ($CHCl_3$) from $CHCl_2F$, which meant considerable loss of valuable reactant. The other disadvantage arising from extensive halogen exchange was that in this particular reaction the chloroform produced forms azeotropes with most of the $C_3HCl_2F_5$ isomers which makes recovery of $C_3HCl_2F_5$ by conventional industrial means, e.g., distillation, difficult and expensive.

Sievert *et al.* found a solution to the above problem by using modified aluminum chloride catalyst instead of ordinary aluminum chloride. Not only the yields of hydrogen-containing chlorofluoropropanes were higher, formation of chloroform from monofluorodichloromethane either eliminated or greatly reduced, but an unknown isomeric hydrogen-containing chlorofluoropropane, i.e., $CHF_2CCl_2CF_3$, was also obtained in appreciable quantities. This new isomeric dichloropentafluoropropane should be particularly desirable from structural considerations since the hydrogen atom is located on one of the terminal carbon atoms of the propane molecule and both chlorine atoms are located on the middle carbon atom. Such structure should ensure greater ease of dehydrochlorination in the atmosphere so that none of the compounds would survive to reach the ozone layer. This isomer would be less damaging to the ozone layer than the other isomers [43].

Gas phase and liquid phase processes were disclosed that used interconversion of CF_3CHFCl and CF_2HCF_2Cl in contact with a catalyst [46]. It consisted of essentially of halogenated aluminum oxide and/or aluminum halide. The reaction at elevated temperatures increased the mole ratio of one of said compounds in a composition relative to the other. Interconversion occurred along with hydrogenolysis in processes for producing mixtures of CF_3CH_2F and CF_2HCF_2H from mixtures consisted of CF_2HCF_2Cl. The mole ratio of CF_3CH_2F to CF_2HCF_2H, in the product mixtures produced from the mixtures, could be increased or decreased.

In the procedure for Liquid Phase Equilibration, the catalyst was loaded into a 10 ml Hastelloy® B nickel alloy shaker tube. The tube was cooled to -78° C. and evacuated. The desired amount of chlorofluoroethane was transferred into the tube. The sealed tube was placed in a shaker in a barricade and heated at the indicated temperature and time. At the end of the reaction the tube was once again cooled to -78° C. and the contents removed. The recovered liquid was analyzed using gas chromatography. Table 6.13 presents the results of heating in a shaker tube 2 g of HCFC-124 containing 5.5% HCFC-124a with $AlCl_3$. This example demonstrates the liquid phase process with $AlCl_3$ catalyst. The reaction time and temperature and the catalyst content have strong influence on the conversion of HCFC-124 to HCFC-124a.

In an example of gas phase reaction the reactor was loaded with 10.7 g Al_2O_3, and activated at 350° C with CFC-12 (gas at 5 ml/minute). The feed was a mixture of HCFC-124 and HCFC-124a charged at 3 ml/minute. The result of the ratio of HCFC-124/HCFC-124a at various temperatures is shown in Table 6.14. This example exhibits a method for maximizing HCFC-124 content plus the effect of changing reaction temperature by a vapor phase process.

6.5.2 Preparation of Hydrofluorocarbons (HFCs)

Manzer and Rao devised a catalytic method to convert HFCs to their more thermodynamic stable isomer [47]. By more thermodynamically stable isomers is meant those isomers with the lowest free energy of formation, as calculated using Benson's Group Contribution Method [48]. The products of their invention exchanged atoms such that the highest number of fluorine atoms existed on terminal carbon atoms ($-CF_3$). For example, in the case

Table 6.13 Conversion rate of HCFC-124a to HCFC-124 using a catalytic liquid-phase reaction [46].

Temp. °C.	Time hour	$AlCl_3$ Wt. (g)	124/124a	123*/124	C_2Cl_4
150	4	0.05	3040	ND	0.056
150	1	0.05	2860	ND	0
250	1	0.5	38	0.69	0
100	2	0.5	3300	0.14	0
100	2	0.15	75	3×10^{-4}	0

Table 6.14 Ratio of HCFC-124/HCFC-124a in a catalytic gas phase process [46].

Temp.	124/124a*
Start	18
102	960
85	1600
70	2400
62	3250
52	4900
46	9800

*124/124a = CF_3CHClF/CHF_2CClF_2 area percent ratio.

of saturated hydrofluorocarbons with two carbon atoms, 1,2-difluoroethane (HFC-152) to 1,1-difluoroethane (HFC-152a); 1,1,2-trifluoroethane (HFC-143) to 1,1,1,-trifluoroethane (HFC-143a); and 1,1,2,2-etrafluoroethane (HFC-134) was isomerized to 1,1,1,2-tetrafluoroethane (HFC-134a):

HFC-134: CHF_2CHF_2 HFC-134a: CH_2FCF_3

The catalyst composition utilized contained an aluminum fluoride (AlF_3) or fluorinated alumina. Fluorinated alumina refers to a high fluorine-content composition of aluminum, oxygen, and fluorine. The total fluorine content of the catalyst ≥50% (by weight), most effective at 80 weight percent. A. To prepare the catalyst $Al(OH)_3$ (38.98 g) was dissolved in 48% aqueous HF (100 ml). The solution was evaporated at room temperature over 72 hour. The resulting white residue was dried at 110°C for 48 hours. The solid was then heated in air at 5°C/min to 500°C and held at this temperature for 3 h.

The reactor was charged with $Al(OH)_3$ at the temperature of 200°C. A mixture of HF and N_2 gas (3 molar ratio) was passed through the reactor. The nitrogen flow was decreased with time until neat HF was being passed through the reactor. At this point the temperature was gradually raised to 450°C and maintained (15 to 300 minutes). The fluorine content of the catalyst composition corresponded to an AlF_3 content, exclusive of any added metal, of at ≥50%. After temperature stabilized flow of the hydrofluorocarbon was started. The gases exiting the reactor were analyzed by gas chromatography. For example a one to one molar ratio mixture of CHF_2CHF_2 (HFC-134) with air was passed over the AlF_3 catalyst at 425°C with a contact time of 60 seconds. After about 30 minutes of operation a 99.0% conversion of HFC-134 to HFC-134a with selectivity of 94.6% was observed [47].

A 1990 patent by Moore and Massey [49] disclosed an invention about a process for the manufacture of 1,1,1,2 tetrafluoroethane (HFC-134a), which is a useful refrigerant, aerosol propellant, blowing agent and solvent. There are a numbers of methods for preparation of this HFC including a process for making tetrafluoroethanes. One route is by hydrogenation of dichlorotetrafluoroethanes in the temperature range 350°–500° C. Another approach manufactures 1,1,1,2-tetrafluoroethane by reacting trifluoroethylene in vapor phase with hydrogen fluoride. A chromium oxide is used in the process at a temperatures of 200° to 500° C. Operation of both of these processes is costly. To overcome the adverse economics Moore and Massey proposed preparation of 1,1,1,2-tetrafluoroethane at high yields and high selectivity by isomerisation of 1,1,2,2-tetrafluoroethane in presence of a fluorination catalyst.

Effective fluorination catalysts included aluminium fluoride, sodium fluoride, gamma alumina and, especially, chromia (Cr_2O_3). Even though unmodified chromia may be used but it is more effective to use a prefluorinated chromia that has been used in other fluorination reactions. After use, the catalyst could be regenerated by treatment with air at 400-420° C.

The isomerisation temperature should be sufficiently high to convert the 1,1,2,2,-tetrafluoroethane but not so high to result in excessive generation of by-products such as pentafluoroethane. Temperatures in the range of 300 to 900°C were effective. Lower temperatures, 400-450°C, can be employed when chromia is the fluorination catalyst. At those temperatures, the 1,1,2,2-tetrafluoroethane, optionally diluted with an inert gas such as nitrogen, was contacted with the catalyst for times from 1 to 50 seconds [49, 50].

In an example of the technology hydrogen (at 1000 ml/min) and tetrafluoroethylene (at 100 ml/min) were passed over 120 g of 5% palladium/alumina at ambient temperature in a glass reactor. The off-gases were collected in two chilled traps. There was an immediate exothermic event causing the temperature to rise to 250°C. After 210 min, the hydrogen and tetrafluoroethylene flows were turned off and the reactor and the lines were purged with nitrogen. The material condensed in the traps was weighed and analyzed. The product was 84.4 g (94% yield) of 1,1,2,2-tetrafluoroethane with a purity of 98.5%.

To isomerize 1,1,2,2-Tetrafluoroethane (and diluent nitrogen) was passed through a Hastelloy tubular reactor packed with 80 ml of chromia while the reactor was heated in a furnace. The gases leaving the reactor were passed through an empty tube for cooling. In the next step the gas was scrubbed using a 25% potassium hydroxide solution to remove hydrogen fluoride. Finally the gaseous products were condensed in a chilled trap containing trichloroethylene and CO_2 ice. Gas samples were analyzed by gas chromatography.

Details of a number of runs using pre-fluorinated chromia, which had been dried at 400° C. for 4 hours and then "activated" by passage of air at 200–300 ml/min at 400° C. overnight, are given below in Table 6.15. A flow rate of 50 ml/min A134 and 100 ml/min nitrogen corresponds to a catalyst contact time of approximately 12 sec. Table 6.16 gives details of experiments using the same catalyst after regeneration with 400 ml/min air at 400°C for 16 hours. The data indicate the highest conversions of R-134 to R-134a were obtained with the used (fluorinated) chromia.

An important development came in 1994 [51]. Manzer reported on an improved process for the manufacture of all-important 1,1,1,2-tetrafluoroethane (HFC-134a). HF reacted was reacted with either trichloroethylene or 2-chloro-1,1,1-trifluoroethane (HCFC-133a) in the presence of a catalyst at elevated temperature to form a mixture of 2-chloro-1,1,1-trifluoroethane (HCFC-133a) and 1,1,1,2-tetrafluoroethane (HFC-134a) and other organic by-products.

The improvement was in conducting the reaction in a single reaction zone in the presence of a catalyst composition comprised of one or more metals selected from the group of chromium, Group VIII, Group VIIB, Group IIIB, Group IB and metals with an atomic number from 58 to 71. The metals were supported by aluminum fluoride or carbon. HFC-134a was recovered from the mixture and HCFC-133a was recycled. This process produces

Table 6.15 Results of Conversion of HFC R-134 to HFC R-134a using a fluorinated chromia catalyst [49, 50].

Catalyst volume ml	Temperature °C.	A134 flowrate ml/min	Nitrogen flowrate ml/min	Concentration in off gas mole %			
				A134	A134a	3FE	A125
80	288	50	100	93.5	1.0	0.1	0.2
80	325	50	100	91.4	3.0	0.1	0.4
80	420	50	100	69.9	21.5	0.2	0.7
80	460	50	100	59.4	34.1	0.3	1.8
80	497	50	100	52.6	34.6	0.2	8.5
80	550	50	100	48.1	28.1	0.3	27.9
80	510	50	100	56.0	26.0	0.6	12.7
80	465	50	100	73.4	21.0	1.0	2.7
80	410	50	100	88.5	9.1	0.6	0.4
80	470	50	100	74.9	19.4	0.2	3.8

A134 = 1,1,2,2-tetrafluoroethane (R-134), A134a = 1,1,1,2-tetrafluoroethane (R-134a), 3FE = trifluoroethylene, A125 = pentafluoroethane (R-125).

Table 6.15 Results of conversion of HFC R-134 to HFC R-134a using a fluorinated chromia catalyst after regeneration with air [49, 50]. (*Continued*)

Catalyst volume ml	Temperature °C.	A134 flowrate ml/min	Nitrogen flowrate ml/min	Concentration in off gas mole %			
				A134	A134a	3FE	A125
80	420	50	100	29.1	4.2	–	60.2
80	343	50	100	66.9	25.7	0.4	3.7
80	328	50	100	75.5	21.1	–	1.1
80	375	50	100	43.6	48.8	0.1	5.4
80	388	50	200	52.3	43.0	0.2	2.8
80	400	50	300	46.3	48.1	0.1	3.9
80	412	50	400	49.8	44.3	0.1	4.4
80	410	50	200	40.3	50.2	0.9	7.1
80	430	50	300	47.1	43.6	1.6	6.2
80	440	50	400	53.1	36.8	2.2	6.5
80	445	50	100	28.4	52.9	0.5	16.6

(*Continued*)

Table 6.15 Results of conversion of HFC R-134 to HFC R-134a using a fluorinated chromia catalyst after regeneration with air [49, 50]. (*Continued*)

Catalyst volume ml	Temperature °C.	A134 flowrate ml/min	Nitrogen flowrate ml/min	Concentration in off gas mole %			
				A134	A134a	3FE	A125
80	450	50	150	41.4	45.1	0.7	11.1
80	455	50	100	40.6	42.5	0.6	14.7
80	455	50	100	47.9	42.2	0.8	7.9

A134 = 1,1,2,2-tetrafluoroethane (R-134), A134a = 1,1,1,2-tetrafluoroethane (R-134a), 3FE = trifluoroethylene, A125 = pentafluoroethane (R-125).

Table 6.16 Results of fluorination of trichloroethylene with HCFC-133a Recycle [51].

Ex.	Hrs	Temp	% CF_3CH_2F	% CF_3CH_2Cl	% CF_3CF_2H	%CFCl=CHCl
2	2	380° C.	29.3	65.6	0.8	0.1
3	5	360	28.5	68.9	0.3	0.0
4	17	350	26.2	72.3	0.0	0.0
5	50	350	15.3	83.5	0.0	0.0
6	94	370	18.5	79.5	0.1	0.1
7	120	400	18.4	74.7	1.0	0.2

relatively little by-products, of the order of 5%. These organic by-products in the mixture could be recycled with trichloroethylene in a molar amount at least equal to the molar amount HFC-134a recovered from the mixture plus the molar amount of other organic by-products. Additional HF from 3 to 30 times the molar amount of trichloroethylene was added to the reactor along with the recycle stream.

This process produced tetrafluoroethane as the major product at greater than 99% trichloroethylene conversion with a reduction in the number of process steps for producing HFC-134a from trichloroethylene via HCFC-133a. An actual example of the process is described next [51].

The reactor (a 0.5 inch ID, 12 inch long Inconel® alloy pipe) was charged with the amount of catalyst as described in the following examples, and placed in a sand bath. The bath was gradually heated to 400° C. while N_2 gas at a flow rate of 50 cc/min was passed through the reactor to remove traces of water. The temperature was lowered to 200° C. and HF and N_2 gas (.1, molar ratio) were passed through the reactor and the N_2 flow was decreased with time until neat HF was being passed through the reactor. At this point the temperature was gradually raised to 425° C. and maintained there for 15 to 300 minutes.

The temperature was then decreased to the indicated value and, thereafter, the other reactant flows were started. The flows were adjusted to give the indicated molar ratios and contact times in the Examples. The reactor effluent was sampled on-line with a Hewlett Packard HP 5890 gas chromatograph using a 20 foot long, one-eighth inch diameter, column containing Krytox® perfluorinated polyether on an inert support and a helium flow of 35 cc/min. Gas chromatographic conditions were 70° C. for three minutes followed by temperature programming to 180°C at a rate of 6° C./minute.

The general procedure of fluorination of recycle trichloroethylene with HCFC-133a was followed using 19.0 g (30 ml) of $CoCl_2/Al_2O_3$ (2% Co) as the initial catalyst charge. The HF/133a/trichloroethylene/O_2 molar ratio was 10/1/0.2/0.2. The product stream resulting from the reaction of HF with HCFC-133a, trichloroethylene and air over the prepared catalyst with a 20 second contact time at 390°C and after a reaction run of 53 hours showed the following: 17.4% CF_3CH_2F (HFC-134a), 80.8% CF_3CH_2Cl (HCFC-133a), 1.2% CF_2=CHCl (FC-1122), 0.3% C_2HCl_2F (FC-1121, two isomers), 0.1% trichloroethylene and 0.2% CH_2F_2. The conversion of trichloroethylene was greater than 99% and the ultimate selectivity to HFC-134a with recycle of the HCFC-133a, FC-1122 and FC-1121 was greater than 98% [51].

The general procedure for fluorination was followed using 21.1 g (30 ml) of ($CoCl_2$ + $CrCl_3$)/Al_2O_3 (1% Co+1% Cr) as the initial catalyst charge. The results of the reaction of HF with either HCFC-133a and trichloroethylene, or HCFC-133a, trichloroethylene, and air over the prepared catalyst with a 20 second contact time are give in Table 6.16. The HFC-133a/trichloroethylene/O_2 molar ratio was 10/1/0.2/0 for Examples 2-4 and 10/1/0.2/0.2 for Examples 5-7.

Rao and Walczak [52] provided a process for producing $CF_3CHFCHF_2$ (HFC-236ea) CF_2CF=CHF was reacted with HF at an elevated temperature over a catalyst selected from the group including AlF_3, fluorinated alumina, metals supported on AlF_3, metals supported on fluorinated alumina and trivalent chromium compounds. They presented a process to enrich CF_3CF=CHF from a starting mixture containing CF_3CF=CHF and CF_3CH=CF_2.

The process comprised of reacting the starting mixture with HF at an elevated temperature over a catalyst consisting of carbon, a catalyst consisting essentially of metal halides supported on carbon, a catalyst consisting of Cr_2O_3, or their mixtures (when Cr_2O_3 was present the temperature was ≤250°C) to produce a product mixture containing $CF_3CH_2CF_3$ (HFC-236fa) and unreacted CF_3CF=CHF in which the mole ratio of CF_3CF=CHF to CF_3CH=CF_2 was greater than the ratio in the starting mixture. Production of $CF_3CH_2CF_3$, (or both $CF_3CFHCHF_2$, and $CF_3CH_2CF_3$) took place by reacting a starting mixture containing both CF_3CF=CHF and CF_3CH=CF_2, with HF at an elevated temperature over a fluorination catalyst.

For example the reactor was charged with 19.5 g of chromium oxide that was activated according to the following procedure. The reactor was charged with chromium oxide and heated to 175°C for two hours in a flow of N_2 (25 ml/min). At the end flow of N_2 and HF in the molar ratio of 2:1 (total flow 100 ml/min) was started through the reactor. After passage of an initial exotherm in the range of 10-30°C the molar ratio of nitrogen to HF was adjusted to 1:3. The temperature was gradually increased over a three hour period to 400°C. The reactor and its contents were kept at this temperature for an additional 30 minutes and brought back to the desired operating temperature. The N_2 flow was stopped and the flow of HF and CF_3CF=CFH in a 4:1 molar ratio was started through the reactor. The contact time was 30 seconds. The organic feed contained 85.3% CF_3CF=CFH and 13.6% trifluoroethylene in addition to small amounts of other components. An analysis of the product at two different reactor temperatures is given in Table 6.17 [52].

Another invention of Rao and Walczak was the development of azeotrope or azeotrope-like compositions $CF_3CHFCHF_2$ and $CF_3CH_2CF_3$ with hydrogen fluoride. Binary

Table 6.17 Analysis of products of hydrofluorination of 1,1,1,2,3,3-hexafluoropropane (HFC-236ea) [52].

Reactor temperature, °C	CF_3CH_2F, mole%	$CF_3CF{=}CFH$, mole%	$CF_3CHFCHF_2$, mole%
300	15.1	35.1	46.9
325	14.9	50.8	32.3

compositions of hydrogen fluoride ranged from 31 to 60% mole of $CF_3CHFCHF_2$ and about 41 to 63% mole $CF_3CH_2CF_3$.

An azeotrope or an azeotrope-like composition is a mixture of two or more different components that, when in liquid form under given pressure, will boil at a substantially constant temperature. The temperature may be higher or lower than the boiling temperatures of the individual components. The vapor composition is essentially identical to the liquid composition undergoing boiling.

Rao and Walczak found that azeotropes of HF and HFC-236fa were formed at a variety of temperatures and pressures. At a pressure of 294 kPa and temperature 20°C, the azeotrope vapor composition was found to be about 50% mole HF and about 50% mole HFC-236fa. They calculated that an azeotropic composition of 37 mole percent HF and 63 mole percent HFC-236fa could be formed at 100°C and 2,900 kPa [53]. More generally, an azeotrope or azeotrope-like composition could be formed in the range 59 to 37 mole percent HF and from about 31 to 63 mole percent HFC-236fa. The boiling point was estimated at -25°C at 44 kPa to 100°C at 2,900 kPa. The HFC-236ea/HF and HFC-236fa/HF azeotropes were useful as feed to produce HFC-227ca ($CHF_2CF_2CF_3$) and/or HFC-227ea (CF_3CHFCF_3) [54].

Yamamoto *et al.* pursued two objectives in their research. The first objective was to provide a process to efficiently and economically manufacture 1,1,1,3,3-pentafluoropropane, especially a process of manufacturing 1,1,1,3,3-pentafluoropropane at a high yield rate in the reductive reaction of 2,2,3-trichloro-1,1,1,3,3-pentafluoropropane. The other objective was to develop a process to efficiently and economically manufacture 2,2,3-trichloro-1,1,1,3,3-pentafluoropropane and 2,3,3-trichloro-1,1,1-trifluoropropene as intermediates in the manufacture of 1,1,1,3,3-pentafluoropropane [55].

A stainless steel reactor tube had an inside diameter of 7 mm and a length of 150 mm. It was filled with 2.3 ml of a palladium catalyst supported by active carbon at a concentration of 3%. The reactor was heated to 250°C in an electric furnace while being purged with nitrogen gas. After the temperature reached 250°C, 2,2,3-trichloro-1,1,1,3,3-pentafluoropropane was introduced at the rate of 1.0 ml/min together with hydrogen at the rate of 10 ml/min. The product gas was washed with rinsed with water followed by analysis using gas chromatography, the results which are shown in Table 6.18.

Yamamoto *et al.* demonstrated both liquid and vapor phase catalytic reaction for the preparation 2,2,3-trichloro-1,1,1,3,3-pentafluoropropane and 2,3,3-trichloro-1,1,1-trifluoropropene as intermediates. Here is an example of manufacture of 1,1,1,3,3-pentafluoropropane from 2,2,3-trichloro-1,1,1,3,3-pentafluoropropane (CFC-215aa). CFC-215aa obtained through catalytic reactions was reduced with hydrogen; then 1,1,1,3,3-pentafluoropropane was obtained at the highest conversion yield and selectivity rates given in Table 6.18. At lower reactor temperatures the selectivity was reduced. This invention

Table 6.18 Conversion of 2,2,3-trichloro-1,1,1,3,3 pentafluoropropane to 1,1,1,3,3-pentafluoropropane [55].

Reactor temperature, °C	Conversion, %	Selectivity, %
250	100	96.8
200	100	95.9
200	100	93.5
150	100	79.7

reduces the manufacturing cost of HFC-245fa because of the relatively low cost of manufacturing and reducing 2,2,3-trichloro-1,1,1,3,3-pentafluoropropane at high yield rate [55].

Sievert *et al.* in 2001 [56] revealed a process for the manufacture of $CF_3CF{=}CF_2$, and optionally at least one of $CF_3CH_2CF_3$, (HFC-236fa) and $CF_3CHFCHF_2$ (HFC-236ea) compounds. The process comprised (a) of contacting a reactor feed consisted of a stream of at least one compound from halogenated propanes of the formula $CX_3CH_2CH_yX_{(3-y)}$, and halogenated propenes of the formula $CX_3CH = CH_yX_{(2-y)}$, in which each X is independently selected from Cl and F and y is 0, 1 or 2, provided that the average fluorine content of said precursor stream is no more than 5 fluorine substituents per molecule.

The feed stream was combined with HF and Cl_2 and brought into a chlorofluorination reactor containing a fluorination catalyst at a temperature between of 150°C to 400°C. Leaving the reactor was an effluent containing HF, HCl and a mixture of reaction products. At least one compound with the formula $C_3Cl_2F_6$ including $CC1F_2CC1FCF_3$ and at least one compound with the formula C_3HClF_6 including $CHF_2CC1FCF_3$ was present in the reaction products. The fluorine content of the products had on one fluorine substituent per molecule more than the average fluorine content of the feed stream.

Next the product stream was distilled to recover $CC1F_2CC1FCF_3$ and $CHF_2CC1FCF_3$. These two components were hydrogenated to produce a mixture of $CF_3CF{=}CF_2$ and $CF_3CHFCHF_2$. Next, $CF_3CF{=}CF$, was recovered from the product mixture and the under-fluorinated component were recycled to the chlorofluorination reactor.

In one example $CCl_3CH_2CH_2Cl$ was chlorofluorinated in presence of a chromium oxide catalyst obtained from the pyrolysis of ammonium dichromate [57]. Catalyst powder was placed an Inconel" nickel alloy tubular reactor that was heated in a fluidized sand bath. The catalyst was heated from 60°C to 175°C in a flow of nitrogen (50 ml/min) over the course of about one hour. HF was then entered in the reactor at a flow rate of 50 ml/min. After 35 minutes the nitrogen flow was decreased to 20 ml/min and the HF flow increased to 80 ml/min.

The reactor temperature was gradually increased to 400°C during a three-hour period and maintained at 400°C for an additional 55 minutes. At the end of this period, the HF flow was stopped and the reactor cooled to 250°C under 20 standard cm^3/min ($3.3x10^{-7}$ m^3/s) nitrogen flow. The results of the chlorofluorination of $CCl_3CH_2CH_2Cl$ are shown in Table 6.19 in which analytical data is given in units of % area under the curve of gas chromatograph (see Table 6.20 for the legend key). Total gas flow in the reactor was 120 standard cm^3/min ($2x10^{-6}$ m^3/s) for a 15 second contact time except for the data at 400°C that was conducted with a 30 second contact time.

Table 6.19 Analysis of reaction products of chlorofluuorination of 1,1,1,3 hydrochloropropane [56].

T °C.	Molar ratio HF:250fb:$C1_2$	CT sec.	% 1243zf	% 242	% 243db	% 244**	% 234bb	% 224aa	% 224ba
170	20:1:0	nc	97.5	0	0	0.1	0	0	0
200	30:1:6	nc	35.4	4.2	34.9	2.6	0	0	0
200	30:1:6	15	0.1	0	55.3	7.1	6.7*	4.2	0
225	30:1:6	15	0	0	10.6	3.1	19.0	20.0	0
250	30:1:6	15	0	0	0	0	0.8	31.7	25.3
275	30:1:6	15	0	0	0	0	0	3.1	4.6
300	30:1:6	15	0	0	0	0	0	0.5	0.7
325	30:1:6	15	0	0	0	0	0	0.3	0.3
350	30:1:6	15	0	0	0	0	0	0	0
375	30:1:6	15	0	0	0	0	0	0	0
400	30:1:6	30	0	0	0	0	0	0	0

C. T. = Contact Time, Sec., nc = no catalyst; reactants mixed in tube packed with Monel gauze (not exposed to the fluorination catalyst).
*Includes 1.7% of unidentified isomer.
**Sum of two isomers.

T °C.	% 1233zd	% 1233xf	% 233ab	% 223aa	% 1213xa	% 1214*	% 215aa**	% 225aa	% 225ba
170	0	0	0	0	0	0	0	0	0
200	4.6	4.0	3.2	2.5	0	0	0	0	0
200	5.8	1.5	10.7	0.5	2.4	0	0	0.1	0
225	1.3	0.4	32.0	2.1	6.8	0	0	0.4	0.5
250	0.1	0.1	0.3	0	6.5	4.3	2.8	7.1	18.8
275	0	0	0	0	0.5	0.2	7.0	37.3	18.7
300	0	0	0	0	0.1	0	7.4	26.3	3.3
325	0	0	0	0	0.1	0	8.9	5.2	1.2
350	0	0	0	0	0	0	0.8	1.1	0.2
375	0	0	0	0	0	0	0.9	0.6	0.2
400	0	0	0	0	0	0	0.2	0.2	0

*includes <1% of an unidentified isomer
*sum of two isomers
*includes <1% of an unidentified isomer

(Continued)

Table 6.19 Analysis of Reaction Products of chlorofluuorination of 1,1,1,3 hydrochloropropane [56]. (*Continued*)

T °C.	% 225ca	% 226ba	% 226ca	% 216ba	% 216aa	% 217ba	% 217ca	% 227ca	% 218
170	0	0	0	0	0	0	0	0	0
200	0	0	0	0	0	0	0	0	0
200	0	0	0	0	0	0	0	0	0
225	0	0	0	0	0	0	0	0	0
250	0	0.6	0	0.1	0.2	0	0	0	0
275	4.2	14.7	4.4	1.7	1.2	0.1	0	0.4	0
300	3.4	31.2	7.7	2.4	4.4	0.5	1.1	9.0	0
325	3.3	24.3	5.3	3.8	16.1	1.1	3.8	24.5	0
350	0.9	10.7	3.4	3.2	25.2	6.1	7.4	39.4	0.4
375	1.6	2.1	3.3	2.9	45.5	7.6	12.9	19.9	1.2
400	0.4	0.9	1.3	0.8	59.4	15.0	6.4	9.7	4.7

Other compounds observed at low levels include: 216cb, 226da, 232, 235cb, 245fa, 252, 1222, 1223, 1232, 1234ye, 1234ze.

This invention provided a process for producing $CF_3CHFCHF_2$ in which $CF_3CF{=}CHF$ was reacted with HF. The reaction took place at an elevated temperature over a catalyst chosen from the group that included aluminum fluoride, fluorinated alumina, metals supported on aluminum fluoride, metals supported on fluorinated alumina and trivalent chromium [58].

Another process enriched $CF_3CF{=}CHF$ from a starting mixture containing $CF_3CF{=}CHF$ and $CF_3CH{=}CF_2$. This process began with the reaction of the starting mixture with HF at elevated temperatures over a catalyst of carbon, metal halides supported on carbon, Cr_2O_3, or their mixtures. When Cr_2O_3 was present reaction temperature had to be ≤250°C. The product mixture contained $CF_3CH_2CF_3$ and unreacted $CF_3CF{=}CHF$ in which the mole ratio of $CF_3CF{=}CHF$ to $CF_3CH{=}CF_2$ was greater than the ratio in the starting mixture. This process allowed production of $CF_3CH_2CF_3$ (or both $CF_3CFHCHF_2$, and $CF_3CH_2CF_3$) by reacting a starting mixture containing both $CF_3CF{=}CHF$ and $CF_3CH{=}CF_2$ with HF at raised temperatures over a fluorination catalyst [54].

This invention specified compositions that consist essentially of hydrogen fluoride in combination with an effective amount of a compound selected from the group consisting of $CF_3CHFCHF_2$, and $CF_3CH_2CF_3$ to form an azeotrope or azeotrope-like composition with

Table 6.20 ASHRAE Legend key for halocarbons in Table 6.19 [56].

114a is CCl_2FCF_3	115 is $CClF_2CF_3$
214ab is $CCl_2FCCl_2CF_3$	215aa is $CClF_2CCl_2CF_3$
215ba is $CClF_2CClFCClF_2$	215bb is $CCl_2FCClFCF_3$
216aa-is-$CF_3CCl_2CF_3$	216ba is $CClF_2CClFCF_3$
216ca is $CClF_2CF_2CClF_2$	216cb is $CF_3CF_2CCl_2F$
217ba is $CF_3CClFCF_3$	217ca is $CClF_2CF_2CF_3$
218 is $CF_3CF_2CF_3$	223aa is $CF_3CCl_2CHCl_2$
224aa is CF_3CCl_2CHClF	224ba is $CF_3CClFCHCl_2$
225aa is $CHF_2Cl_2CF_3$	225ba is $CHClFCClFCF_3$
225ca is $CHCl_2CF_2CF_3$	226ba is $CF_3CClFCHF_2$
226ca is CF_3CF_2CHClF	226da is $CF_3CHClCF_3$
227ea is CF_3CHFCF_3	232 is $C_3H_2Cl_4F_2$
233ab is $CF_3CCl_2CH_2Cl$	233da is $CF_3CHClCHCl_2$
234 is $C_3H_2Cl_2F_4$	234bb is $CF_3CClFCH_2Cl$
234da is $CF_3CHClCHClF$	235cb is $CF_3CF_2CH_2Cl$
235da is $CF_3CHClCHF_2$	236fa is $CF_3CH_2CF_3$
242 is $C_3H_3Cl_3F_2$	243db is $CF_3CHClCH_2Cl$
244 is $C_3H_3ClF_4$	245fa is $CF_3CH_2CHF_2$
252 is $C_3H_4Cl_2F_2$	1213xa is $CCl_2{=}CClCF_3$
1214 is $C_3Cl_2F_4$	1215 is C_3ClF_5
1222 is C3 $HC1_3F_2$	1223 is C_3HCl_2F
1224 is C_3HClF_4	1231 is $C_3H_2Cl_3F$
1232 is $C_3H_2Cl_2F_2$	1233xf is $CH_2{=}CClCF_3$
1233zd is $CHCl{=}CHCF_3$	1234 is $C_3H_2F_4$
1234ye is $CHF{=}CFCHF_2$	1234ze is $CHF{=}CHCF_3$
1243 is $C_3H_3F_3$	1243zf is $CH_2{=}CHCF_3$

hydrogen fluoride. The composition contained 31-60% mole $CF_3CHFCHF_2$, or 41-63% mole $CF_3CH_2CF_3$ [58].

Azeotropic and azeotrope-like mixtures of 1-chloro-1,3,3,3-tetrafluoropropane and 1,2-dichloro-3,3,3-trifluoropropene were found to be useful intermediates in the production of HFC-245fa. To make HFC-245fa, approximately 182 kg of antimony pentachloride catalyst was charged into a 189-liter reactor. The reactor temperature was raised to about 95°C. About 11.4 kg/hr of HCC-240fa, 6.8 kg/hr of HF and 0.9 kg/hr of Cl_2 were fed to the reactor continuously, with the Cl_2 keeping the catalyst active. The reactor pressure was maintained at about 14.8 bar [59].

The product stream contained HFC-245fa, HF, HCl and small amounts of organic by-products, including 1,2-dichloro-3,3,3-trifluoropropene (HCFC-1223xd) and 1-chloro-1,3,3,3-tetrafluoropropane (HCFC-244fa). Next, unreacted HF and HCl were removed, along with most of the HFC-245fa, leaving a mixture of by-products. A distillation of these byproducts was performed, from which a lights cut was retained, containing primarily HFC-245fa, HCFC-244fa, and HCFC-1223xd. This fraction was further distilled to recover recyclable components. The HFC-245fa was removed by distillation at a pressure of 3.8 bar. Removal of the HCFC-244fa was accompanied throughout the entire cut by about 3.5% by weight HCFC-1223xd as an azeotrope, or azeotrope-like mixture. After removal of this cut, a heel of substantially pure HCFC-1223xd remained [59].

6.6 Preparation of Hydrofluoroolefins (HFOs)

With the global phase-out of chlorofluorocarbons (CFCs) and hydrochlorofluorocarbons (HCFCs) and planned phase-out of hydrofluorocarbons (HFCs) suitable replacements are required. A candidate agent must have no impact on the ozone layer and minimal or no impact on global warming in addition to meeting the basic requirements of flame resistance, low/no toxicity and useful boiling point and vapor pressure. Fluorinated olefins, especially those containing one or more hydrogen atoms in the molecule (referred to as hydrofluoroolefins) were considered for existing applications such as in refrigeration. Hydrofluoroolefins were found to be suitable replacements for the previous generations of fluorocarbons.

In general, fluoroalkenes are very useful compounds and have many applications. For example, fluorinated alkenes such as 3,3,3-trifluoropropene ($CF_3CH{=}CH_2$) are useful as monomers for the manufacture of polymers such as fluorosilicones. They are also starting materials for the production of fluorinated chemical intermediates such as trifluoropropene epoxide and 3,3,3-trifluoropropylbenzene. Fluorinated alkenes are also used as aerosol propellants and as refrigerants.

Fluorinated alkenes such as 3,3,3-trifluoropropene are typically made in multi-step processes. For example, carbon tetrachloride (CC_4) can be added to ethylene to produce 1-chloro-3,3,3-trichloropropane ($CCl_3CH_2CH_2Cl$). Henne *et al.* described fluorination of the latter using a $CCl_3CH_2CH_2Cl$ liquid phase fluorination process, such as HF in conjunction with an antimony catalyst, yield 1-chloro-3,3,3-trifluoropropane ($CF_3CH_2CH_2Cl$)

[60]. Subsequent dehydrochlorination of $CF_3CH_2CH_2Cl$ yields 3,3,3-trifluoropropene ($CF_3CH=CH_2$).

A number of patents [61, 62] described a multi-step process for making 3,3,3-trifluoropropene. In US 5,986,151 the first step of the process requires the dehydrofluorination of 1,1,1,3,3-pentafluoropropane ($CF_3CH_2CHF_2$; HFC-245fa) to 1,3,3,3-tetrafluoropropene ($CF_3CH=CHF$) (HFC 1234ze). In the next step, 1,3,3,3 tetrafluoropropene ($CF_3CH=CHF$) was reduced to 1,3,3,3 tetrafluoropropane ($CF_3CH_2CH_2F$) by reacting a stream of gaseous 1,3,3,3-tetrafluoropropene ($CF_3CH=CHF$) and hydrogen gas under catalytic conditions at suitable reaction temperatures. Finally, the 1,3,3,3-tetrafluoropropane ($CF_3CH_2CH_2F$) was dehydrofluorinated to yield 3,3,3-trifluoropropene.

Other methods for producing fluoroalkenes included the use of hazardous reagents. U.S. Pat. No. 6,211,421 [63], for example, describes the preparation of 1,1,1-trifluoropropene by liquid phase fluorination of 1,1,3-trichloroprop-1-ene with HF. U.S. Pat. No. 4,582,921 [64] described the reaction of vinyl bromide with *in situ* generated trifluoromethylorganometallic reagents to yield 3,3,3-trifluoropropene ($CF_3CH=CH_2$). U.S. Pat. No. 4,220,608 [65] described fluorination of $CCl_3CH_2CH_2Cl$ or $CCl_3CH=CCl_2$, with HF in the presence of an amine to yield 3,3,3-trifluoropropene ($CF_3CH=CH_2$). In another publication [66] 3,3,3-trifluoropropene ($CF_3CH=CH_2$) was obtained via thermal conversion of trifluoromethyl oxirane at 32% yield.

Many of the past processes resulted in the generation of one or more unwanted waste streams containing chlorinated products [67]. It is well known that chlorinated compounds are generally considered to be problematic from an environmental standpoint. For one thing, chlorinated hydrocarbons are identified with environmental problems such as ozone depletion and global warming. The waste streams typically were disposed of by incineration, which itself is frequently considered environmentally unfriendly as well as economically inefficient. A number of past processes for hydrofluoroalkenes preparation suffered from other drawbacks.

These drawbacks included the use of expensive or hazardous starting materials and/or the use of reagents and/or starting materials that need to be prepared in multiple steps. Another drawback of many past processes was that the conversion of starting materials to the desired products such as, for example, $CF_3CH=CH_2$, involved multiple steps that made such processes less cost effective than is desirable for many commercial purposes. Moreover, certain past processes required one or more dehydrohalogenation steps, and it was not uncommon that a significant amount of waste was produced as a result of those steps. It was clearly advantageous to use readily available starting materials, use more cost effective materials, and reduce the disposal need for environmentally unfriendly waste products.

Nair and Poss [67] devised methods of reducing the emission of environmentally detrimental compounds from the byproduct of process to produce hydrofluoroalkenes. The process produces a compound with the formula (I) R_f-$C(R_1)=C(R_2)H$ in which R_f is fluorine or a substituted or unsubstituted C_1-C_{20} fluorinated alkyl, and R_1 and R_2 are independently H, Cl, Br, or I as long as one the two groups is one of those halogens. In this process the hydrofluoroalkenes R_f-$C(R_1)=C(R_2)H$ reacts with a formate salt in the presence of apalladium catalyst. The product includes at least one fluoroalkene with a lower degree of chlorination, bromination or iodination than R_f-$C(R_1)=C(R_2)H$.

In this example CF_3—CCl=CHCl (HCFC-1223) was converted to CF_3CH=CH, (HFO-1243) using a palladium catalyst in a medium of tetrahydrofuran. It is abbreviated as $Pd(PPh_3)_4$ and called tetrakis(triphenylphosphine) palladium(0); its chemical structure can be seen in Figure 6.7.

A 250 ml Parr autoclave was charged with 30 ml of tetrahydrofuran to which 0.5 grams of $Pd(PPh_3)_4$, 10 grams of $HCONH_4$, and 10 grams of CF_3CCl=CHCl under a nitrogen blanket. The reactor was sealed immediately, cooled to -30 to -40°C, and partially evacuated. The contents in the reactor were brought to room temperature and gradually heated to 100°C while stirring. The reactants were held at 100°C for 24 hours. The reactor was cooled to 25°C and the volatile materials were collected for analysis. The schemes for the reaction that took place in the reactor can be seen in Figure 6.8.

Gas chromatography (GC) analysis indicated CF_3CH=CH_2 and CF_3CH=CHCl were present at a ratio of 57:42. Carbon dioxide was also present. The reaction mixture was cooled to 0°C and filtered under pressure, the filtrate was analyzed by GC which indicated about a 40% conversion of CF_3CCl=CHCl to CF_3CH=CHCl and CF_3CH=CH_2 [67].

The compounds 1-chloro-3,3,3-trifluorpropene and 1,2-dichloro-3,3,3-trifluorpropene are frequently present unwanted byproducts in waste streams of processes. In the absence of use they would have to be disposed of at additional cost or with the result of undesirable environmental consequences. The work by Nair and Poss found a ways to convert both chlorofluoropropenes into useful compounds such as, for example, 3,3,3-trifluoropropene (CF_3CH=CH) known as HFO-1243 with commercial applications.

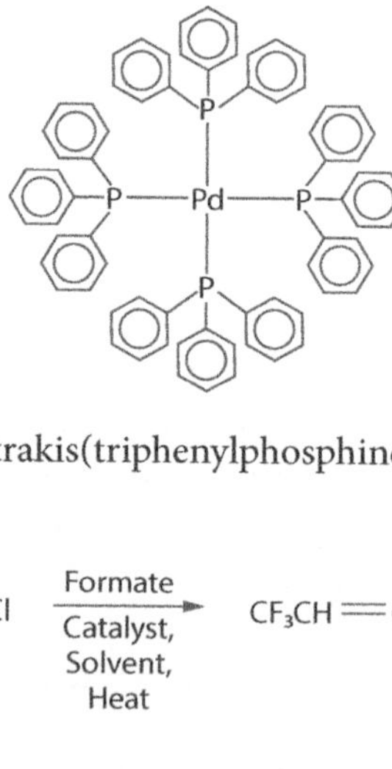

Figure 6.7 Chemical structure of tetrakis(triphenylphosphine) palladium(0) catalyst [$Pd(PPh_3)_4$].

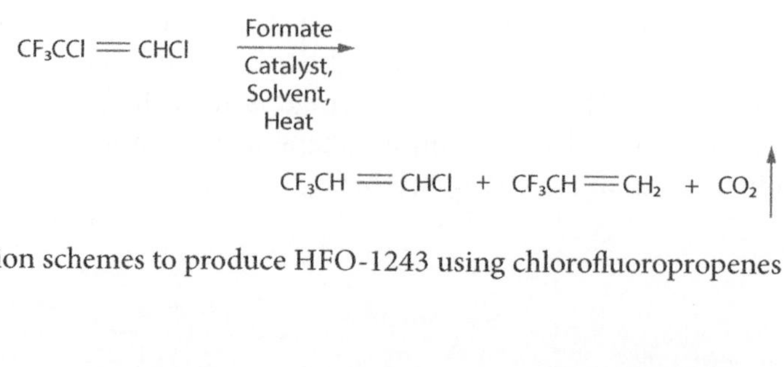

Figure 6.8 Reaction schemes to produce HFO-1243 using chlorofluoropropenes [67].

Another invention relates to formulated systems of *trans*- and/or *cis*-1-chloro-3,3,3-trifluoropropene (HCFO-1233zd) and/or 2-chloro-3,3,3-trifluoropropene (HCFO-1233xf) that are sufficiently thermally and chemically stable such that they can be effectively used without the need for additional stabilizers. The formulations of the present invention are particularly useful compositions for refrigeration, heat transfer, and foam pre-mixes. Stability results for select HCFCs are described in [68]. The stated findings were that R-123 is approximately ten times more stable than R-I l, and that R-141b and R-142b are more stable than R-123. The inventors discovered that HCFO-1233zd, without any additional stabilizer, was at least as stable as R-141b, and by extension it is also more stable than R-123 and R-I l [69].

A 2012 patent discusses a method to produce HFC-1225ye [70]. The process brought into contact, with chromium oxyfluoride, a hexafluoropropane selected from the group consisting of 1,1,1,2,2,3 hexafluoropropane (HFC-236cb) and HFC-236ea with a chromium oxyfluoride catalyst in a reactor. The product mixture contained 1,2,3,3,3-pentafluoropropene (HFC-1225ye, C_3HF_5), which consists of Z-HFC-1225ye and E-HFC-1225ye. In the final step HFC-1225ye was recovered from the product mixture.

A catalytic process was described to produce 1,3,3,3-tetrafluoropropene (HFO-1234ze) from 1,1,1,3,3-pentafluoropropane (HFC-245fa). Another process allowed preparation of 2,3,3,3-tetrafluoropropene (HFC-1234yf) from 1,1,1,2,2-pentafluoropropane (HFC-245cb) via a catalytic process using chromium oxyfluoride.

Miller *et al.* reported a process for production of hydrofluoroolefins by dehydrofluorinating them. Importantly they demonstrated a process for separation of a vast number of hydrofluoroolefins from hydrofluorocarbons and from hydrogen fluoride. The product mixture consisted of a hydrofluoroolefin, unreacted hydrofluorocarbon and hydrogen fluoride in which the hydrofluoroolefin and hydrofluorocarbon were both present in the mixture as an azeotrope composition with hydrogen fluoride [71].

A large number of hydrofluoroolefins selected from the group consisting of the following agents could be purified: $CF_2{=}C(CHF_2)_2$, $CHF{=}C(CHF_2)_2$, $CH_2{=}C(CH_2F)CF_3$, $CH_2{=}CFCF_2CF_2CF_3$, $CHF_2CF_2CF{=}CFCH_3$, $CF_3CF_2CF{=}CHCH_3$, $(CF_3)_2C{=}CFCHF_2$, $(CF_3)_2CFCF{=}CHCH_3$, $(CF_3)_2C{=}C(CH_3)_2$, $(CH_3)_2C{=}CFCF_2CF_3$, $C_2F_5CH{=}CHCF_2C_2F_5$, $CF_3CH{=}CHCF_2CF_2C_2F_5$, $CF_3CF_2CF_2CF_2CF{=}CHCH_3$, $CF_3CF_2CF_2CF{=}CHCH_2CH_3$, $(CH_3)_2C{=}CFCF_2CF_2CF_3$, and $CF_3CH{=}CFCH_2CF_3$.

The hydrofluoroolefin in the azeotrope mixtures contained one less fluorine atom and hydrogen atom than the corresponding hydrofluorocarbon indicating removal of one mole of HF. Purification took place via three consecutive distillation processes. Typically, a mixture of hydrofluoroolefin, hydrofluorocarbon and hydrogen fluoride was formed and distilled. In a distillation step the hydrofluoroolefin was removed thus forming an azeotrope or near-azeotrope mixture of hydrogen fluoride and the hydrofluoroolefin. The hydrofluoroolefin was separated from its azeotrope or near-azeotrope mixture with hydrogen fluoride by a two-step distillation process. The bottom fraction of the first distillation step was distilled in the second step. Similarly the bottom fraction of the second step was distilled, in a third step, at a different pressure from the second distillation leading to the recovery of virtually pure hydrofluoroolefin [71].

Other favorable findings included a process for producing $CF_3CH{=}CF_2$ in the absence of dehydrofluorination catalyst. In particular, $CF_3CH_2CF_3$ was pyrolyzed to yield $CF_3CH{=}CF_2$. Pyrolysis led to the thermal decomposition of the $CF_3CH_2CF_3$, at a temperature exceeding

700° C. The selective formation of $CF_3CH{=}CF_2$ had a number of benefits. First, it was surprising that the heat input of the pyrolysis process did not cause the $CF_3CH_2CF_3$ reactant to fragment to C-1, e.g., methanes, and C-2, e.g., ethane and ethylene, compounds. Second, the $CF_3CH{=}CF_2$ product was stable under pyrolysis conditions and did not undergo further conversion to rearranged products or to products containing fewer hydrogen and/or fluorine atoms. Third, it was desirable that the pyrolysis to form $CF_3CH{=}CF_2$ took place with high selectivity [71].

Other patents have reported formation of an azeotropic or azeotrope-like composition consisting essentially of 1,1,1,2,3,3-hexafluoropropane, hexafluoropropene and hydrogen fluoride. It is known that azeotrope and azeotrope-like compositions provide good mechanisms for product purification. The azeotrope-like mixture was then be separated into its component parts and HFC-236ea (1,1,1,2,3,3-hexafluoropropane) purified for further processing [72, 73].

A more recent invention disclosed a method for preparing HFO-1234ze [74]. One way to produce mixtures containing HFO-1234ze was by dehydrofluorination of 1,1,1,3,3-pentafluorpropane (HFC-245fa) by contact with a caustic solution. The reaction product contained cis-HFO-1234ze and trans-HFO-1234ze. It was brought to contact with one or more isomerization catalyst to convert the cis-HFO-1234ze to trans-HFO-1234ze as much as possible.

In an example 760 kg of HFC-245fa was fed into a reaction vessel containing a 20% molar excess of 45% KOH. A small amount of a phase transfer catalyst, Aliquat® 336 was added to aid the reaction. Phase-transfer catalyst or PTC is a catalyst that facilitates the migration of a reactant from one phase into another phase where reaction occurs. Aliquat® 336 is a quaternary ammonium chloride type agent with the chemical name tricaprylylmethylammonium chloride or trioctylmethylammonium chloride with the formula $CH_3(C_8H_{17})_3\text{-}N^+Cl^-$. The reaction was conducted at a temperature of 60°C and the pressure was allowed to increase to greater than 10.3 bar and the product was taken. Analysis of the product showed a 95.1% conversion of the HFC-245fa and a selectivity of 86.9% to trans-HFC-1234ze and 13.1% to cis-HFC-1234ze [74].

A more recent patent described preparation of 2,3,3,3-tetrafluoropropene (HFO-1234yf). More particularly, the invention relates to a two reaction step process wherein pentachloropropane, including 1,1,1,2,3-pentachloropropane (HCC-240db) and/or 1,1,2,2,3-pentachloropropane (HCC-240aa), is first contacted with hydrogen fluoride (HF), this first step preparing 2-chloro-3,3,3-trifluoropropene (HCFO-1233xf), which after separation is then converted into HFO-1234yf in a second reaction step [75].

6.7 Preparation Perfluorinated Alkanes

A number of perfluorinated alkanes were candidates for use as refrigerants (Table 6.21) because of being ozone safe (ODP=0). All these agents have massively large global warming potentials that render them unusable because their impacts on climate change.

There are four methods for preparation of perfluoroalkanes:

1. Direct fluorination of carbon.
2. Halogen exchange between chlorocarbons and hydrofluoric acid.

Table 6.21 Perfluorinated alkane compounds [76].

Type	ASHRAE number	IUPAC chemical name	Molecular Formula	CAS No./ Blend Name	Semi-empirical ODP	Net GWP-100 year
PFC	R-14	Tetrafluoromethane	CF_4	75-73-0	0	7,390
PFC	R-116	Hexafluoroethane	C_2F_6	76-16-4	0	12,200
PFC	R-218	Octafluoropropane	C_3F_8	76-19-7	0	8,830
PFC	R-C318	Octafluorocyclobutane (Perfluorocyclobutane	C_4F_8 or $-(CF_2)_4-$	115-25-3	0	10,300
PFC	R-3-1-10	Decafluorobutane (Perfluorobutane)	C_4F_{10}	355-25-9	0	8,860
PFC	R-4-1-12	Dodecafluoropentane (Perfluoropentane)	C_5F_{12}	678-26-2	0	9,160
PFC	R-5-1-14	Tetradecafluorohexane (Perfluorohexane)	C_6F_{14}	355-42-0	0	9,300

3. Fluorination of hydrocarbons using electrochemical and catalytic methods.
4. Fluorination techniques using metal fluorides such as cobalt trifluoride.

Of the agents listed in Table 6.21 the first three (CF_4, C_2F_6 and C_3F_8) are gaseous at the room temperature. Examples of preparation techniques for the three compounds are discussed here.

A process devised for the production of hexafluoroethane (CF_3-CF_3) is described. The first step was increasing the fluorine content of olefinic compounds of the formula $C_nH_mF_{2n-m}$, in which n was an integer from 2 to 6 and m was an integer from 1 to 2n. The process had three steps: 1. The olefinic compound was contacted with a metal fluoride composition of the formula $(AgF)(MF_2)_x$ in which M was selected from the group of Mn, Fe, Co, Ni, Cu, Zn or their mixtures. The x was a number from 0 to 1, at a temperature above 200°C sufficient to transfer F from the metal fluoride composition to the olefinic compound. Consequently a chemically reduced metal fluoride composition was produced that contained metallic silver. 2. The reduced metal fluoride composition from the Step 1 was oxidized in the presence of HF to regenerate the metal fluoride with the formula $(AgF)(MF_2)_x$. 3. Finally, the regenerated metal fluoride was recycled to Step 1. This invention described a novel composition of the formula $Ag_{10}F_8C_2$; and a process for producing hexafluoroethane that involved heating $Ag_{10}F_8C_2$ to a sufficiently high temperature for its decomposition [77].

When the ($Ag + Ag_{10}F_8C_2$) mixture was reacted excess ethylene at a temperature of least 350°C, then no vinyl fluoride was produced. Instead, the following products are formed: 1,1-difluoroethylene (CH_2=CF_2) a fluoromonomer, pentafluoroethane (CHF_2-CF_3) a refrigerant, 1,1,1-trifluoroethane (CH_3-CF_3) and hexafluoroethane (CF_3-CF_3). Subramanian believed that most of the silver was in the metallic state after this reaction. When $Ag_{10}F_8C_2$

was isolated and heated to 350°C in an inert atmosphere such as nitrogen, then the major fluorinated hydrocarbon product was hexafluoroethane, as shown in equation Eq.(6.33).

$$Ag_{10}F_8C_2 \rightarrow 8Ag + 2AgF + CF_3\text{-}CF_3 \quad (6.33)$$

A two-step process using a fairly available agent hexafluoropropene to manufacture octafluoropropane (C_3F_8) was disclosed by von Halasz [74]. In the first step hexafluoropropene was reacted with hydrogen fluoride in the presence of a chromium oxyfluoride catalyst. The 2H-heptafluoropropane (2HHF) obtained was reacted with fluorine in the presence of a catalyst that contained a subgroup element of the Periodic System as a free metal or as a fluoride. Octafluoropropane was isolated from the resulting gas mixture. Hydrogen fluoride obtained may be recycled to the first reaction step. The addition of hydrogen fluoride on hexafluoropropene with formation of 2H-heptafluoropropane according to equation Eq.(6.34) has been known.

$$CF_3\text{-}CF{=}CF_2 + HF \rightarrow CF_3\text{-}CHF\text{-}CF_3 \quad (6.34)$$

The reaction of hexafluoropropene, C_3F_6, with hydrogen fluoride, HF, was carried out in a vertically positioned copper tube having a length of 150 cm and a 25 inner diameter of 5.3 cm. The reactor tube was provided with a heating jacket on the outside, with thermocouples in the interior, and with a sieve at its lower end. The packing of the tube consists of chromium oxyfluoride grains prepared by fluorination of chromium oxyhydrate green with hydrogen fluoride. At the upper end of the reactor, there was an inlet orifice with tubular ducts for the feed of gaseous hydrogen fluoride and gaseous hexafluoropropene. To the lower end of the reactor, a further tubular duct was connected for the discharge of unreacted hydrogen fluoride into a receiver filled with water. The deacidified gaseous reaction product was dried in a calcium chloride tower and condensed in glass traps cooled by means of dry ice [78].

The starting hydrogen fluoride (HF) had 99% purity and hexafluoropropene ($CF_3CF{=}CF_2$) (HFP) purity was 98-99% both taken from a commercial steel cylinders. HFP was dried over phophorus pentoxide and dosed by means of a calibrated differential pressure gage. At reactor temperatures of 260-270°C., a total of 245 g of HF and 1470 g of C_3F_6, at a molar ratio of HF: C_3F_6 of 1.25:1, was fed in within 6.5 hours and passed over the chromium oxyfluoride catalyst contained in the reactor. Distillation of water washed and dried condensate yielded 1605 g 2H-heptafluoropropane (CF_3–CHF–CF_3) (2HHF) at a yield of 96.3% relative to the C_3F_6 used. The 2HHF obtained was identified by infrared spectroscopy.

2HHF was fluorinated with elementary fluorine, F_2, in a copper tube, Eq.(6.35). The packing of the tube was consisted of silver-plated copper chips. These silver-plated copper chips were preliminarily fluorinated with elementary fluorine at 100-150°C to form silver fluoride. The 2H-heptafluoropropane prepared was fed to the reactor via the opening at the tube's upper end.

$$CF_3\text{-}CHF\text{-}CF_3 + F_2 \rightarrow CF_3\text{-}CF_2\text{-}CF_3 + HF \quad (6.35)$$

A mixture of F_2 and N_2 was fed in via an additional stainless steel tube. Another tubular duct was connected to the lower end of the reactor and lead to a trap. From there it went to HF absorption towers connected in parallel and packed with sodium fluoride granules. With alternate switching, HF was absorbed from the crude gas mixture at 90-120°C in one of these HF absorption towers, while simultaneously hydrogen fluoride previously absorbed was desorbed at 250-290°C from the other HF absorber. The hydrogen fluoride thus formed was nearly completely recovered, collected in a reservoir and reused for addition on hexafluoropropene. After further separation and drying in towers containing $CaCl_2$ or P_2O_5, the octafluoropropane formed was condensed in traps cooled by dry ice [78].

In a two-step process, similar to von Halasz's, 2H-heptafluoropropane obtained in the first step was directly fluorinated [79]. In this step 2HHF was reacted with fluorine, diluted with another gas, in the gas phase at a temperature in the range of 250 to 500° C. In spite of the absence of catalyst in the reaction octafluoropropane was produced with good yield. This step seen in Eq.(6.35) had the following three beneficial features:

(a) In the case of reacting hydrofluorocarbon with fluorine gas to produce perfluorocarbon, a large heat of reaction was generated. The heat of reaction was proportional to the number of fluorine moles reacted per one molecule. As the amount of fluorine becomes larger, the heat of reaction became larger, the probability of cleavage of a carbon-carbon bond, polymerization, cyclization addition even explosion grew higher. In that case the yield decreases which was a problem in a commercial operation.

An illustrative example for heat generation is fluorination of methane and ethane by fluorine gas. Methane and ethane contain four and six hydrogen atoms, respectively. Each hydrogen atom combines with a fluorine atom generating a mole of hydrofluoric acid as seen in Eq.(6.36) and Eq.(6.37). Large quantities of heat are produced rapidly because of the extreme reactivity of fluorine

$$CH_4 + 4F_2 \rightarrow CF_4 + 4HF \qquad (6.36)$$

$$(\Delta H = -479 \text{ kcal/mol})$$

$$C_2H_6 + 6F_2 \rightarrow C_2F_6 + 6HF \qquad (6.37)$$

$$(\Delta H = -690 \text{ kcal/mol})$$

To prevent sudden generation of the heat of reaction fluorine could be diluted with another gas. An inert gas such as nitrogen and helium as a diluent worked but it was unhelpful because of the cost, of separation and purification of the perfluorocarbon. Instead a diluting gas selected from the group consisting of hydrogen fluoride, tetrafluoromethane, hexafluoroethane and octafluoropropane solved the problem of costly purification [79].

(b) In the direct fluorination reaction process there was a serious probability of combustion or explosion because of the high reactivity of fluorine. The probability was enhanced in case of compounds containing hydrogen 2HHF like 2HHF. It was important to keep the mixed gas composition out of the explosive range. Ohno and Ohi studied the explosion range between 2HHF and fluorine gas and found the lower limit of the explosive range was 8% by mole. Thus, the concentration of 2H-heptafluoropropane at the reaction inlet was set within the safe range.

(c) If excess 2H-heptafluoropropane relative to fluorine gas was used all fluorine was consumed. A fluorine separation was, therefore, not required but great difficulties were incurred in the later separation and purification. The reaction efficiency grew when fluorine gas was used in excess but the reaction product contained the excess fluorine in addition to perfluorocarbon and hydrogen fluoride. Using as inorganic oxide such as alumina or soda lime resulted in the production water that caused corrosion of the equipment. To avoid that problem, the excess fluorine gas was removed by contacting the gas with a hydrofluorocarbon at a molar ratio of 1.1 to F_2 [79].

6.8 Summary

This chapter presents the preparation methods of fluorocarbons gases from their inception to the present time. The first compounds had high ODP and GWP characteristics. After the discovery of ozone depletion due to the impact of CFC's alternatives were developed (HFC) with zero ODP. The drawback to HFCs, however, was large GWP values. The industry developed a family of hydrofluoroolefins (HFO) which have zero or very low global warming potential values.

References

1. *Chlorofluorocarbons, Encyclopedia of Environmental Science*, D.E. Alexander and R.W. Fairbridge (Eds.), pp. 78–80, Chapman & Hall, Kluwer Academic, Boston, MA, 1999.
2. IHS Chemical Consulting, *The Economic Benefits of Chlorine Chemistry in Fluorocarbons in the United States and Canada*, American Chemistry Council, May 2016.
3. Vetra, E., *Refrigeration System*, Badak LNG, www.badaklng.co.id, December 2016.
4. American Society of Heating, *Refrigerating and Air-Conditioning Engineers (ASHRAE)*, www.Ashrae.org, May 2020.
5. T. Midgely, Jr., A.L. Henne, R.R. McNary, US Patent 1,833,847, Nov. 24, 1931.
6. T. Midgely, Jr., US Patent 1,926,395, Sept. 12, 1933.
7. T. Midgely, Jr. and A.L. Henne, US Patent 1,926,396, Sept. 12, 1933.
8. T. Midgely, Jr., A.L. Henne, R.R. McNary, US Patent 1,930,129, Oct.10, 1933.
9. T. Midgely, Jr., A.L. Henne, R.R. McNary, US Patent 1,968,049, assigned to General Motors Corp, July 31, 1934.
10. T. Midgely, Jr., A.L. Henne, R.R. McNary, US Patent 2,024,008, assigned to General Motors Corp, Dec.10, 1935.
11. A.L. Henne, US Patent 1,973,069, assignments, to General Motors Corp, Sep. 11, 1934.
12. T. Midgely, Jr. and A.L. Henne, US Patent 2,013,062, assigned to Frigidaire Corp, Sep 3, 1935.
13. H.W. Daudt and M.A. Youker, US Patent 2,005,705, assigned to Kinetic Chemicals Co., June 18, 1935.
14. H.W. Daudt and M.A. Youker, US Patent 2,005,706, assigned to Kinetic Chemicals Co., June 18, 1935.
15. H.W. Daudt and M.A. Youker, US Patent 2,005,707, assigned to Kinetic Chemicals Co., June 18, 1935.
16. H.W. Daudt and M.A. Youker, US Patent 2,005,708, assigned to Kinetic Chemicals Co., June 18, 1935.

17. H.W. Daudt and M.A. Youker, US Patent 2,005,709, assigned to Kinetic Chemicals Co., June 18, 1935.
18. H.W. Daudt and M.A. Youker, US Patent 2,005,710, assigned to Kinetic Chemicals Co., June 18, 1935.
19. H.W. Daudt and M.A. Youker, US Patent 2,062,743, assigned to Kinetic Chemicals Co., Dec 1, 1936.
20. Hexachloroethane, Immediately Dangerous to Life or Health Concentrations (IDLH). Centers for Disease Control, Dec 2016.
21. Banks, R.E., Smart, B.E., Tatlow, J.C., Organofluorine Chemistry—Principles and Commercial Applications, Springer Science, 1994.
22. A.L. Henne, US Patent 1,970,562, assigned to General Motors Corp, Aug. 21, 1934.
23. T. Midgely, Jr., A.L. Henne, R.R. McNary, US Patent 2,007,208, assigned to Frigidaire Corp, July 9, 1935.
24. A.L. Henne, US Patent 2,013,050, assigned to General Motors Corp, Sep.3, 1935.
25. A.L. Henne, US Patent 1,978,840, assigned to General Motors Corp, Oct 30, 1934.
26. M. Hauptschein and A.H. Fainberg, US Patent 3,087,974, assigned to Pennsalt Chemicals Corp, Apr. 30, 1963.
27. M. Vecchio, I. Canmarata, L. Lodi, US Patent 3,294,852, assigned to Edison Corp, Dec. 27, 1966.
28. M. Vecchio, I. Canmarata, L. Lodi, US Patent 3,442,962, assigned to Edison Corp, May 6, 1969.
29. G. Fernschild, W. Rudolph, J. Massonne, US Patent 4,005,176, assigned to Kali-Chemie, AG, Jan 25, 1977.
30. S.K. Lee, US Patent 4,411,874, assigned to SCA Services, Oct 25, 1983.
31. V. Kalcevic and J.F. McGahan, US Patent 4,751,063, assigned to International Tech Corp, June 14, 1988.
32. R.F. Ruh, R.A. Davis, M.R. Broadworth, US Patent 2,885,427, assigned to Dow Chemical Co, May 5, 1959.
33. F.W. Swamer and B.W. Howk, US Patent 3,258,500, assigned to Du Pont Co, June 28, 1966.
34. J.F. Knaak, US Patent 3,978,145, assigned to Du Pont Co, Aug 31, 1976.
35. E.J. Carlson, J.N. Amor, W.J. Cunningham, A.M. Smith, US Patent 4,828,818, assigned to Allied Signal Co, May 9, 1989.
36. F.B. Downing, A.F. Benning, R.C. Mcharness, US Patent 2,413,696, assigned to Kinetic Chemicals, January 7, 1947.
37. E.R. Larsen, US Patent 2,920,112, assigned to The Dow Chemical Company, Jan 5, 1960.
38. D.G.H. Ballard, J. Farrar, D.A. Laidler, US Patent 4,745,237, Imperial Chemical Industries, May 17, 1988.
39. L.E. Manzer and V.M.N. Rao, US Patent 4,766,260, to DuPont Co, Aug. 23, 1988.
40. W.H. Gumprecht, L.E. Manzer, V.N.M. Rao, US Patent 4,843,181, assigned to DuPont Co, Jun. 27, 1989.
41. W.H. Manogue, V.N.M. Rao, F.J. Weigert, US Patent 5,030,372, assigned to DuPont Co, Jul. 9, 1991.
42. S. Okazaki, M. Ogura, Ichikawa, Y. Mochizuki, US Patent 5,118,887, assigned to DuPont-Mitsui Fluorochemicals, June 2, 1992.
43. A.C. Sievert, C.G. Krespan, F.J. Weigert, US Patent 5,157,171, assigned to DuPont Co., Oct. 20, 1992.
44. R.M. Joyce, Jr., US Patent No. 2,462,402, assigned to Du Pont Co., Feb 22, 1949.
45. Coffman, D.D., Cramer, R., Rigby, G.W., *J. Am. Chem. Soc.*, 71, 979–980, 1949.
46. W.H. Manogue, V.N.M. Rao, F.J. Weigert, US Patent 5,406,009, assigned to Du Pont Co., Apr 11, 1995.

47. L.E. Manzer and V.M.N. Rao, US Patent 4,902,838, assigned to DuPont Co, Feb. 20, 1990.
48. Benson, S.W. *et al.*, *Chem. Rev.*, 69, 279, 1969.
49. G.J. Moore and H.M. Massey, US Patent 4,950,815, assigned to Imperial Chemical Industries, Aug. 21, 1990.
50. G.J. Moore and H.M. Massey, US Patent 5,091,600, assigned to Imperial Chemical Industries, Feb. 25, 1992.
51. L.E. Manzer, US Patent 5,345,016, assigned to DuPont Co, Sep. 6, 1994.
52. V.M.N. Rao and Walczak, US Patent 5,563,304, assigned to DuPont Co, Oct. 8, 1996.
53. Schotte, W., *Ind. Eng. Chem. Process. Des. Dev.*, 19, 3, 432–439, 1980.
54. D.W. Bruhnke, J.J. Lerou, V.N.M. Rao, W.C. Seidel, F.J. Weigert, US Patent 5,177,273, assigned to DuPont Co, Jan. 5, 1993.
55. A. Yamamoto, E. Seki, H. Aoyama, S. Takubo, T. Nakada, US Patent 5,714,655, assigned to Daikin Industries, Feb. 3, 1998.
56. A.C. Sievert, V.N.M. Rao, F.J. Walczak, US Patent 6,329,559, assigned to DuPont Co, Dec 11, 2001.
57. J.J. Lerou, US Patent 5,036,036, assigned to DuPont Co, July 30, 1991.
58. V.M.N. Rao and F.J. Walczak, US Patent 6,388,147, assigned to DuPont Co, May 14, 2002.
59. R.C. Johnson, D.C. Merkel, H. Tung, US Patent 6,759,381, assigned to Honeywell International, July 6, 2004.
60. Henne, A.L. and Kaye, S., *J. Am. Chem. Soc.*, 72, 3369, 1950.
61. M. Van Der Puy, U.S. Patent No. 5,986,151, assigned to Honeywell International, Nov 16, 1999.
62. S.A. Cottrell, H.S. Tung, H. Wang, Y. Chiu, G. Cerri, U.S. Patent No. 9,255,046, assigned to Honeywell International, Feb 9, 2016.
63. M.Y. Elsheikh and B. Chen, U.S. Patent 6,211,421, assigned to Arkema Corp, Apr 3, 2001.
64. D.J. Burton, D.M. Wiemers, J.C. Easdon, US Patent 4,582,921, assigned to Iowa Research Foundation, Apr 15, 1986.
65. A. Feiring, US Patent 4,220,608, assigned to DuPont Co, Sep 2, 1980.
66. Gable, K.P. and Brown, E.C., *Organometallics*, 19, 944–946, 2000.
67. H.K. Nair and A.J. Poss, US Patent 6,958,424, assigned to Honeywell International, Oct. 25, 200s5.
68. Huttenlocher, D.F., *Chemical and Thermal Stability of Refrigerant-Lubricant Mixtures with Metals*, DOE/CE/23810-5, Spauschus Inc., 1992.
69. B.L. Van Horn, M.Y. Elsheikh, P. Bonnett, WO2009/114397A2, assigned to Arkema, Inc, Sep 17, 2009.
70. M.J. Nappa and V.M.N. Rao, US Patent US 8,148,586, assigned to DuPont Co, Apr. 3, 2012.
71. R.N. Miller, M.J. Nappa, V.N.M. Rao, A.C. Sievert, US Patent 8,399,723, assigned to DuPont Co, Mar 19, 2013.
72. R. Hulse, H.T. Pham, R.R. Singh, H.S. Tung, D.C. Merkel, K.A. Pokrovski, US Patent 8,436,218, assigned to Honeywell International, May. 7, 2013.
73. R. Hulse, H.T. Pham, R.R. Singh, H.S. Tung, D.C. Merkel, K.A. Pokrovski, US Patent 8,653,311, assigned to Honeywell International, Feb. 18, 2014.
74. S.A. Cottrell, H.S. Tung, H. Wang, Y. Chiu, G. Cerri, US Patent 9,255,046, assigned to Honeywell International, Feb. 9, 2016.
75. D. Deur-Bert, B. Collier, L. Wendlinger, US Patent 10,329,227, assigned to, Arkema, June 25, 2019.
76. Forster, P., Ramaswamy, V., *et al.*, Chapter 2: Changes in Atmospheric Constituents and in Radiative Forcing, in: *Climate Change 2007: The Physical Science Basis. Contribution of Working Group I to the Fourth Assessment Report of the Intergovernmental Panel on Climate Change*,

S. Solomon and H.L. Miller (Eds.), Cambridge University Press, Cambridge, UK, October 9, 2016, 2007.
77. M.A. Subramanian, US Patent 6,096,932, assigned to DuPont Co, Aug 1, 2000.
78. S.P. von Halasz, US Patent 4,158.023, assigned to Solvay Corp, June 12, 1979.
79. H. Ohno and T. Ohi, US Patent 7,102, 039, assigned to Showa Denko KK, Sep 5, 2006.

7

Properties of Fluorocarbons

The necessity to phase out chlorofluorocarbons (CFCs) and replace them with less environmentally harmful compounds has resulted in proliferation of the transition materials hydrochlorofluorocarbons (HCFC). The large number of HCFCs and hydrofluorocarbons (HFC) relative to the number of CFCs also had to do with the inability of the manufacturers to find exact replacements for the CFCs. In the end blends of HCFCs had to be made in order to match the essential properties of CFCs.

The sheer number of all CFCs, HCFCs, fluorocarbons (HFCs) and hydrofluoroolefins (HFO) renders the task of listing all their properties impractical. The properties of the less common compounds have not been broadly published. The most commonly used agents have been fully characterized and their properties published. Properties of the most common and largest consumption volume CFCs, HCFCs, HFCs, HFOs and perfluorocarbons have been listed (Tables 7.1–7.4). References have been provided for readers who wish to search for the properties of less common products.

The desirable characteristics for a widely used refrigerant include:

- No Ozone Depletion Potential;
- Low or no Global Warming Potential;
- High volumetric cooling capacity;
- Chemical stability at all temperatures and environments;
- Compatibility with other materials which are used in the system (metals, elastomers);
- Solubility with lubricants;
- Low or no toxicity;
- Low or no flammability;
- Commercial availability;
- Low cost (per kg or per system charge).

There is no single refrigerant that combines all of the above properties. Because different refrigerants possess different properties, the suitability of a refrigerant varies depending on the application.

Fluorocarbons have had a vital role as blowing agents for polyurethanes (PU). The important characteristics of blowing agents include:

- Boiling Point at or near room temperature
- Expand with heat

Sina Ebnesajjad, PhD. Concise Handbook of Fluorocarbon Gases: Applications in Refrigeration and Other Industries, (155–216) © 2021 Scrivener Publishing LLC

Table 7.1 Highest consumption chlorofluorocarbons (CFCs).

ASHRAE code	CFC name	Chemical formula
11	Trichlorofluoromethane	CCl_3F
12	Dichlorodifluoromethane	CCl_2F_2
13	Chlorotrifluoromethane	$CClF_3$
113	1,1,2 Chloro,trifluoroethane	$CClF_3$
114	1,2-Dichloro-1,1,2,2-tetrafluoro-ethane	CF_4
115	1-Chloro-1,1,2,2,2-pentafluoroethane	C_2ClF_5

*When CFCs were in use.

Table 7.2 Highest consumption hydrochlorofluorocarbons (HCFC).

ASHRAE code	HCFC name	Chemical formula
123	2,2-Dichloro-1,1,1-trifluoroethane	CF_3CHCl_2
124	1-Chloro-1,2,2,2-tetrafluoroethane	$CHClFCF_3$
141b	1,1-Dichloro-1-fluoroethane	CCl_2FCH_3
142b	1-Chloro-1,1-difluoroethane	$CClF_2CH_3$
21	Dichlorofluoromethane	CCl_2FH
22	Difluorochloromethane	$CClF_2H$
225ca	1,1-Dichloro-2,2,3,3,3-pentafluoropropane	$CCl_2HCF_2CF_3$
225cb	1,3-Dichloro-1,1,2,2,3-pentafluoropropane	$CClF_2CF_2CHClF$

- Solubility in PU ingredients
- No solubility in PU polymer
- Low to no Flammability
- Good Thermal Properties

Fire extinguishing agent characteristics:

- High fire fighting efficiency
- Low residue
- Low electrical conductivity

Table 7.3 Highest consumption hydrofluorocarbons (HFC).

ASHRAE code	HFC name	Chemical formula
23	Trifluoromethane	CHF_3
32	Difluoromethane	CH_2F_2
HFC-43-10mee	1,1,1,2,2,3,4,5,5,5-decafluoropentane	CH_3F
125	Pentafluoroethane	CF_3CHF_2
134a	1,1,1,2-Tetrafluoroethane	CF_3CH_2F
143a	1,1,1-Trifluoroethane	CF_3CH_3
152a	1,1-Difluoroethane	CHF_2CH_3
227ea	1,1,1,2,3,3,3-Heptafluoropropane	CF_3CHFCF_3
236fa	1,1,1,3,3,3-Hexafluoropropane	$CF_3CH_2CF_3$
245fa	1,1,1,3,3-Pentafluoropropane	$CF_3CH_2CHF_2$
365mfc	1,1,1,3,3-Pentafluorobutane	$CF_3CH_2CF_2CH_3$
43-10mee	1,1,1,2,3,4,4,5,5,5-Decafluoropentane	$CF_3CHFCHFCF_2CF_3$

Table 7.4 Highest consumption hydrofluoroolefins (HFO).

ASHRAE code	HFO name	Chemical formula
1132a	1,1-Difluoroethylene	CF_2=CH_2
1233zdE	*trans* 1-Chloro-3,3,3-trifluoroprop-1-ene	*trans*-CHCl=CH-CF_3
1234yf	2,3,3,3-Tetrafluoropropene	CH_2=CF-CF_3
1234ze	1,3,3,3-Tetrafluoropropene	CHF=CH-CF_3
1243zf	3,3,3-Trifluoropropene	CF_3-CH=CH_2
1327	1,1,3,3,4,4,4-Heptafluorobutene	CF_2=CH-CF_2-CF_3
1336mzz-Z	*cis*-1,1,1,4,4,4-Hexafluoro-2-butene	CF_3-CH=CH-CF_3
1345zfc	3,3,4,4,4-Pentafluoro-1-butene	CH_2=CH-CF_2-CF_3
1354	2,4,4,4-Tetrafluorobutene	CF_3CH=CF-CH_3
HBFO-1233xfB*	2-bromo-3,3,3-trifluoropropene	CF_3CBr=CH_2

*Hydrobromofluoroolefin.

- Low metals corrosion
- High materials compatibility
- Long-term stability
- Low toxicity of the agent and its combustion products

Tables 7.5 through 7.35 provide some of the important properties of a number of fluorocarbons.

Table 7.5 Properties of R-11 - Trichloromonofluoromethane.

Fluocarbon: Trichloromonofluoromethane				
Property	**Value**	**Unit**	**Temperature, °C**	**Pressure, bar**
Code	R-11			
CAS Number	75-69-4			
UN Number	1029			
Group	CFC			
Chemical Formula	CCl_3F			
Appearance/Color	Colorless liquid			
Odor	Slightly Ethereal			
*GWP (CO_2=1), 100 years	4750			
**ODP (R11=1)	1			
Molecular Weight	137.4	g/mol		
Melting point	-111	°C		1.013
Latent heat of fusion	50.2			
Boiling point	23.7	°C		1.013
Latent heat of vaporization@	181.4	kJ/kg		1.013
Liquid density	1479.3	kg/m^3	23.7	1.013
Vapor pressure	1.065	bar	25	
Critical temperature	197.96	°C		
Critical pressure	44.17	bar		
Critical density	619.435	Kg/m^3		

(Continued)

Table 7.5 Properties of R-11 - Trichloromonofluoromethane. (*Continued*)

Fluocarbon: Trichloromonofluoromethane				
Property	**Value**	**Unit**	**Temperature, °C**	**Pressure, bar**
Triple point temperature	-111.11	°C		5.856×10^{-5}
Vapor density	5.8531	Kg/m^3	23.7	1.013
Specific gravity (air=1)	4.9			
Specific volume	0.1717	m^3/kg	25	1.013
Compressibility Factor (Z)	0.96		25	1.013
Heat capacity (C_p)	0.0835	kJ/(mol.K)	25	1.013
Heat capacity (C_v)	0.0733	kJ/(mol.K)	25	1.013
Solubility in water	0.179	vol/vol	25	1.013
Dynamic Viscosity (liquid)	4.3×10^{-4}	Poise	20	1.013
Thermal Conductivity	8.4546	mW/(m.K)	25	1.013
Flammability	Not flammable			
Autoignition Temperature	593	°C		
Explosive limits	NA			
Upper flammability limit	NA			
Compatibility	Aluminum, Brass, Ferritic Steel, Stainless Steel			
Incompatibility	Alkali or alkaline earth metals and metal powders Al, Zn, Be, etc			
Chemical stability	Stable under normal conditions			

*GWP = Global Warming Potential.
**ODP = Ozone Depletion Potential.

Table 7.6 Properties of R-12 - Dichlorodifluoromethane.

Fluocarbon: Dichloromonofluoromethane				
Property	**Value**	**Unit**	**Temperature, °C**	**Pressure, bar**
Code	R-12			
CAS Number	75-71-8			
UN Number	1028			
Group	CFC			
Chemical Formula	CCl_2F_2			
Appearance/ Color	Colorless gas			
Odor	Faint ethereal odor			
*GWP (CO_2=1), 100 years	10900			
**ODP (R11=1)	1			
Molecular Weight	120.9	g/mol		
Melting point	-158	°C		1.013
Latent heat of fusion	34.2			
Boiling point	-29.75	°C		1.013
Latent heat of vaporization@	166.2	kJ/kg		1.013
Liquid density	1487	Kg/m^3	23.7	1.013
Critical temperature	112	°C		
Critical pressure	41.4	bar		
Critical density	565	Kg/m^3		
Triple point temperature	-157	°C		2.449 x 10^{-6}
Vapor density	6.289	Kg/m^3	29.75	1.013
Vapor density	5.2252	Kg/m^3	15	1.013

(Continued)

Table 7.6 Properties of R-12 - Dichlorodifluoromethane. (*Continued*)

Fluocarbon: Dichloromonofluoromethane				
Property	**Value**	**Unit**	**Temperature, °C**	**Pressure, bar**
Specific gravity (air=1)	4.3			
Specific volume	0.1985	m^3/kg	25	1.013
Compressibility Factor (Z)	0.98		15	1.013
Heat capacity (C_p)	0.74	kJ/(mol.K)	25	1.013
Heat capacity (C_v)	0.065	kJ/(mol.K)	25	1.013
Dynamic Viscosity (gas)	2.6×10^{-3}	Poise	21.1	1.013
Thermal Conductivity	8.7331	mW(m.K)	25	1.013
Flammability	NA			
Autoignition Temperature	NA	°C		
Explosive limits	NA			
Upper flammability limit	NA			
Compatibility	Aluminum, Brass, Copper, Ferritic Steel, Stainless Steel. Polytetrafluoroethylene, Chloropene, Polyamide, Polypropylene			
Incompatibility	Polychlorotrifluoroethylene, Butyl rubber, Nitril rubber, Viton®, EPDM, Silicone, Hydrocarbon and Fluorocarbon lubricants			
Chemical stability	Stable under normal conditions			

*GWP = Global Warming Potential.
**ODP = Ozone Depletion Potential.

Table 7.7 Properties of R-13 - Chlorotrifluoromethane.

Fluorocarbon: Chlorotrifluoromethane				
Property	**Value**	**Unit**	**Temperature, °C**	**Pressure, bar**
Code	R-13			
CAS Number	75-72-9			
UN Number	1022			
Group	CFC			
Chemical Formula	$CClF_3$			
Appearance/ Color	Colorless gas			
Odor	Faint ethereal odor			
*GWP (CO_2=1), 100 years	14,400			
**ODP (R11=1)	1			
Molecular Weight	120.9	g/mol		
Melting point	-158	°C		0.013
Latent heat of fusion	34.2			
Boiling point	-29.75	°C		1.013
Latent heat of vaporization @	166.2	kJ/kg		1.013
Liquid density	1487	Kg/m^3	23.7	1.013
Vapor pressure	0.374	bar	20	
Vapor density	6.5	air=1		
Critical temperature	28.7	°C		
Critical pressure	41.4	bar		
Critical density	565	Kg/m^3		
Triple point temperature	-157	°C		2.449×10^{-6}

(Continued)

Table 7.7 Properties of R-13 - Chlorotrifluoromethane. (*Continued*)

Fluorocarbon: Chlorotrifluoromethane				
Property	**Value**	**Unit**	**Temperature, °C**	**Pressure, bar**
Vapor density	6.289	Kg/m^3	29.75	1.013
Vapor density	5.2252	Kg/m^3	15	1.013
Specific gravity (air=1)	4.3			
Specific volume	0.1985	m^3/kg	25	1.013
Compressibility Factor (Z)	0.98		15	1.013
Heat capacity (C_p)	0.074	kJ/(mol.K)	25	1.013
Heat capacity (C_v)	0.065	kJ/(mol.K)	25	1.013
Dynamic Viscosity (gas)	1.0764×10^{-4}	Poise	25	1.013
Thermal Conductivity	8.7331	mW/(m.K)	25	1.013
Flammability	NA			
Autoignition Temperature	NA	°C		
Explosive limits	NA			
Upper flammability limit	NA			
Compatibility	Aluminum, Brass, Copper, Ferriic Steel, Stainless Steel. Polytetrafluoroethylene, Polychlorotrifluoroethylene, Polyvinylidene fluoride, Polyamide, polypropylene			
Incompatibility	Alkali or alkaline earth metals and metal powders Al, Zn, Be, etc, Butyl rubber, Nitril rubber, Vitron®, EPDM, Silicone, Hydrocarbon and Fluorocarbon lubricants			
Chemical stability	Stable under normal conditions			

*GWP = Global Warming Potential.
**ODP = Ozone Depletion Potential.

Table 7.8 Properties of R - 113 - Trichloromonofluoromethane.

Fluorocarbon: 1,1,2 chloro, trifluoroethane				
Property	**Value**	**Unit**	**Temperature, °C**	**Pressure, bar**
Code	R-113			
CAS Number	76-13-1			
UN Number	3082			
Group	CFC			
Chemical Formula	CCl_2FCClF_2			
Appearance/Color	Colorless volatile liquid			
Odor	Faint solvent odor			
*GWP (CO_2=1), 100 years	6,130			
**ODP (R11=1)	1			
Molecular Weight	187.37	g/mol		
Melting point	-36.23	°C		1.013
Latent heat of fusion	13.166	kJ/kg		
Boiling point	47.6	°C		1.013
Latent heat of vaporization @bp	144.319	kJ/kg		1.013
Liquid density	1508.18	kg/m^3	23.7	1.013
Vapor pressure	0.484	bar	25	
Vapor pressure	0.29	bar	15	
Critical temperature	214.1	°C		
Critical pressure	33.9	bar		
Critical density	576.0	kg/m^3		
Triple point temperature	-36.22	°C		2.449×10^{-6}
Triple point pressure	0.0152	kg/m^3		

(*Continued*)

Table 7.8 Properties of R - 113 - Trichloromonofluoromethane. (*Continued*)

Fluorocarbon: 1,1,2 chloro, trifluoroethane				
Property	**Value**	**Unit**	**Temperature, °C**	**Pressure, bar**
Vapor density	7.425	kg/m^3	15	1.013
Specific gravity (air=1)	1.56		21.1	
Specific volume vapor	0.13	m^3/kg	15	1.013
Compressibility Factor (Z)	0.89			1.013
Heat capacity (C_p)	0.17231	kJ/ (mol.K)	27	1.013
Heat capacity (C_v)	0.12576	kJ/ (mol.K)	27	1.013
Solubility in water	0.17	Weight%	25	1.013
Dynamic Viscosity (liquid)	0.0149	Poise	25	1.013
Thermal Conductivity	73.16	mW/ (m.K)	27	1.013
Autoignition Temperature	680	°C		
Explosive limits	NA			
Incompatibility	Decomposes on contact with hot surfaces or flames. This produces toxic and corrosive gases of hydrogen chloride, phosgene, hydrogen fluoride and carbonyl fluoride. Attacks some forms of plastics, rubber, and coatings.			
Chemical stability	Reacts violently with powdered metals. This generates fire and explosion hazard. Attacks magnesium and its alloys.			

*GWP = Global Warming Potential.
**ODP = Ozone Depletion Potential.
Source: PubChem Open Chemistry Database https://pubchem.ncbi.nlm.nih.gov/compound/Dichloro tetrafluoroethane#section=Top.
National Center for Biotechnology Information, National Institutes of Health, January 2017.

Table 7.9 Properties of R-114 – 1,2-dichloro-1,1,2,2-tetrafluoro-ethane.

Fluorocarbon: 1,2-dichloro-1,1,2,2-tetrafluoro-ethane				
Property	**Value**	**Unit**	**Temperature, °C**	**Pressure, bar**
Code	R-114			
CAS Number	76-14-2			
UN Number	1958			
Group	CFC			
Chemical Formula	$CClF_2CClF_2$			
Appearance/Color	Colorless gas			
Odor	Faint ethereal			
*GWP (CO_2=1), 100 years	10000			
**ODP (R11=1)	1			
Molecular Weight	170.9	g/mol		
Melting point	-92.53	°C		1.013
Latent heat of fusion	8.834	kJ/kg		
Boiling point	3.6	°C		1.013
Latent heat of vaporization (@bp)	136	kJ/kg		1.013
Liquid density	1518	Kg/m^3	23.7	1.013
Vapor pressure	1.88	bar	21.1	
Vapor pressure	1.536	bar	15	
Critical temperature	145.68	°C		
Critical pressure	32.57	bar		
Critical density	580.00	Kg/m^3		
Triple point temperature	-92.52	°C		2.449×10^{-6}
Triple point pressure	0.002363	bar		
Vapor density	7.83	Kg/m^3	15	1.013

(Continued)

Table 7.9 Properties of R-114 – 1,2-dichloro-1,1,2,2-tetrafluoro-ethane. (*Continued*)

Fluorocarbon: 1,2-dichloro-1,1,2,2-tetrafluoro-ethane				
Property	**Value**	**Unit**	**Temperature, °C**	**Pressure, bar**
Specific gravity (air=1)	6.1		15	
Specific volume	0.1336	m^3/kg	15	1.013
Compressibility Factor (Z)	0.97		15	1.013
Heat capacity (C_p)	0.68956	kJ/(mol.K)	15	1.013
Heat capacity (C_v)	0.63349	kJ/(mol.K)	15	1.013
Solubility in water	0.13	Weight%	25	1.013
Dynamic Viscosity (gas)	$1.2x10^{-4}$	Poise	60	1.013
Thermal Conductivity	9.702	mW/(m.K)	15	1.013
Flammability	Not flammable			
Autoignition Temperature	NA	°C		
Explosive limits	NA			
Upper flammability limit	NA			
Incompatibility	will attack some forms of plastics, rubber, & coatings. hemically-active metals such as sodium, potassium, calcium, powdered aluminum, zinc and magnesium; acids; acid fumes.			
Chemical stability	chemically inert in many situations			

*GWP = Global Warming Potential.

**ODP = Ozone Depletion Potential.

Source: PubChem Open Chemistry Database https://pubchem.ncbi.nlm.nih.gov/compound/Dichlorotetrafluoroethane#section=Top.

National Center for Biotechnology Information, National Institutes of Health, January 2017.

Table 7.10 Properties of R-115 – 1-chloro-1,1,2,2,2-pentafluoroethane.

Fluorocarbon: 1-chloro-1,1,2,2,2-pentafluoroethane				
Property	**Value**	**Unit**	**Temperature, °C**	**Pressure, bar**
Code	R-115			
CAS Number	76-15-1			
UN Number	1020			
Group	CFC			
Chemical Formula	$CClF_2CF_3$			
Appearance/ Color	Colorless gas			
Odor	Odorless			
*GWP (CO_2=1), 100 years	7370			
**ODP (R11=1)	0.44			
Molecular Weight	154.5	g/mol		
Melting point	-99.44	°C		
Latent heat of fusion	12.16	kJ/kg		
Boiling point	-39.22	°C		
Latent heat of vaporization @bp	125.4	kJ/kg		
Liquid density	1546.6	kg/m^3		
Vapor pressure	9.081	bar		
Vapor pressure	6.923	bar		
Critical temperature	79.95	°C		
Critical pressure	31.29	bar		
Critical density	614.8	kg/m^3		
Triple point temperature	-99.4	°C		2.449×10^{-6}

(Continued)

Table 7.10 Properties of R-115 – 1-chloro-1,1,2,2,2-pentafluoroethane. (*Continued*)

Fluorocarbon: 1-chloro-1,1,2,2,2-pentafluoroethane				
Property	**Value**	**Unit**	**Temperature, °C**	**Pressure, bar**
Triple point pressure	0.02189	kg/m³		1.013
Vapor density	8.406	kg/m³	15	1.013
Specific gravity (air=1)	5.4		15	
Specific volume	0.1499	m³/kg	15	1.013
Compressibility Factor (Z)	0.97927		15	1.013
Heat capacity (C_p)	0.70739	kJ/(mol.K)	15	1.013
Heat capacity (C_v)	0.64853	kJ/(mol.K)	15	1.013
Solubility in water	0.58	Weight%	25	1.013
Dynamic Viscosity (liquid)	0.93×10^{-3}	Poise	25	1.013
Thermal Conductivity	11.658	mW/(m.K)		1.013
Flammability	NA			
Autoignition Temperature	NA	°C		
Explosive limits	NA			
Flammability limits	NA			
Incompatibility	Strong oxidizing reagent			
Chemical stability	When heated to decomposition it emits toxic vapors/hydrogen fluoride and hydrogen chloride			

*GWP = Global Warming Potential.
**ODP = Ozone Depletion Potential.

Table 7.11 Properties of R-123 - 2,2-dichloro-1,1,1-trifluoroethane.

Fluorocarbon: 2,2-dichloro-1,1,1-trifluoroethane				
Property	**Value**	**Unit**	**Temperature, °C**	**Pressure, bar**
Code	R-123			
CAS Number	306-83-2			
UN Number	1078			
Group	HCFC			
Chemical Formula	$CHCl_2CF_3$			
Appearance/ Color	Colorless			
Odor	Either-like			
*GWP (CO_2=1), 100 years	93			
**ODP (R11=1)	0.02			
Molecular Weight	152.93	g/mol		
Melting point	-107	°C		1.013
Boiling point	28	°C		1.013
Latent heat of vaporization @b	170	kJ/kg		1.013
Vapor Pressure	0.91	bar	25	
Critical temperature	183.568	°C		
Critical pressure	36.88	bar		
Critical density	550.00	kg/m^3		
Critical volume	0.001818	m^3/kg		
Liquid density	1463	kg/m^3	25	1.013
Vapor density	6.47	kg/m^3	28	1.013
Special gravity (air=1)				

(*Continued*)

Table 7.11 Properties of R-123 - 2,2-dichloro-1,1,1-trifluoroethane. (*Continued*)

Fluorocarbon: 2,2-dichloro-1,1,1-trifluoroethane				
Property	**Value**	**Unit**	**Temperature, °C**	**Pressure, bar**
Specific Volume	0.154560	m^3/kg	25	1.013
Compressibility Factor (Z)				
Heat capacity liquid @bp	0.2392	kJ/(mol.K)	28	1.013
Heat capacity vapor @bp	0.1719	kJ/(mol.K)	28	1.013
Solubility in water	0.39	weight%		1.013
Dynamic Viscosity (liquid)	0.00456	Poise	25	1.013
Dynamic Viscosity (vapor)	0.00011	Poise	25	1.013
Thermal Conductivity Liquid	0.081	W/(m.K)	25	1.013
Thermal Conductivity Vapor	0.0112	W/(m.K)	25	1.013
Autoignition Temperature	NA	°C		
Explosive limits	NA			
Upper flammability limit	Non-flammable			
Incompatibility	Alkali metals. Alkaline earth metals. Powdered metals. Powdered metal salts			
Chemical Stability				

*GWP = Global Warming Potential.
**ODP = Ozone Depletion Potential.

Table 7.12 Properties of R-124 - 1-chloro-1,2,2,2-tetrafluoroethane.

Fluorocarbon: 1-chloro-1,2,2,2-tetrafluorethane				
Property	**Value**	**Unit**	**Temperature, °C**	**Pressure, bar**
Code	R-124			
CAS Number	2837-89-0			
UN Number	1021			
Group	HCFC			
Chemical Formula	$CHFClClF_3$			
Appearance/ Color	Clear, colorless			
Odor	Faint ethereal odor			
*GWP (CO_2=1), 100 years	609			
**ODP (R11=1)	0.02			
Molecular Weight	136.48	g/mol		
Melting point	-199	°C		1.013
Boiling point	-12.1	°C		1.013
Latent heat of vaporization @bp	194	kJ/kg		1.013
Vapor Pressure	0.00386	Bar	25	
Critical temperature	122.2	°C		
Critical pressure	0.0036	Bar		
Critical density	565.00	kg/m^3		
Critical volume	0.001770	m^3/kg		
Liquid density	1370	kg/m^3	25	1.013
Vapor density	4.7	kg/m^3	28	1.013
Special gravity (air=1)	1.34		30	

(*Continued*)

Table 7.12 Properties of R-124 - 1-chloro-1,2,2,2-tetrafluoroethane. (*Continued*)

Fluorocarbon: 1-chloro-1,2,2,2-tetrafluorethane				
Property	**Value**	**Unit**	**Temperature, °C**	**Pressure, bar**
Specific Volume vapor	0.21	m^3/kg	28	1.013
Heat capacity liquid	1.13	kJ/(mol.K)	25	1.013
Heat capacity vapor	0.742	kJ/(mol.K)	25	1.013
Solubility in water	0.44	weight%	35	1.013
Dynamic Viscosity (liquid)	0.00299	Poise	25	1.013
Dynamic Viscosity (vapor)	0.00012	Poise	25	1.013
Thermal Conductivity Liquid	0.0722	W/(m.K)	25	1.013
Thermal Conductivity Vapor	0.0106	W/(m.K)	25	1.013
Autoignition Temperature	NA	°C		
Explosive limits	NA			
Upper flammability limit	NA			
Incompatibility	potassium, calcium, powdered Aluminum, magnesium, and zinc			
Chemical Stability	Decomposition products: Halogens, halogen acids and Possibly carbonyl halides			

*GWP = Global Warming Potential.
**ODP = Ozone Depletion Potential.

Table 7.13 Properties of R-125 – 1,1,1,2,2-pentafluoroethane.

Fluorocarbon: 1,1,1,2,2-pentafluoroethane				
Property	**Value**	**Unit**	**Temperature, °C**	**Pressure, bar**
Code	R-125			
CAS Number	354-33-6			
UN Number	3220			
Group	HFC			
Chemical Formula	CHF_2ClF_3			
Appearance/Color	Colorless			
Odor	Faint ethereal odor			
*GWP (CO_2=1), 100 years	3500			
**ODP (R11=1)	0			
Molecular Weight	120	g/mol		
Melting point	-100.6	°C		1.013
Boiling point	-48.1	°C		1.013
Latent heat of vaporization @bp	164	kJ/kg		1.013
Vapor Pressure	13.8	bar	25	
Critical temperature	66	°C		
Critical pressure	36.2	bar		
Critical density	571.90	kg/m^3		
Critical volume	0.00175	m^3/kg		
Liquid density	1189.7	kg/m^3	25	1.013
Vapor density	4.983	kg/m^3	15	1.013
Special gravity (air=1)	4.2		25	
Compressibility Factor (Z)				

(Continued)

Table 7.13 Properties of R-125 – 1,1,1,2,2-pentafluoroethane. (*Continued*)

Fluorocarbon: 1,1,1,2,2-pentafluoroethane				
Property	**Value**	**Unit**	**Temperature, °C**	**Pressure, bar**
Heat capacity liquid	1.37	kJ/(kg.K)	25	1.013
Heat capacity vapor	0.809	kJ/(kg.K)	25	1.013
Ratio C_p/C_V				
Solubility in water	0.0923	weight%	25	1.013
Dynamic Viscosity (liquid)	0.00137	Poise	25	1.013
Dynamic Viscosity (vapor)	0.00013	Poise	25	1.013
Thermal Conductivity Liquid	52	mW/(m.K)	25	1.013
Thermal Conductivity Vapor	13	mW/(m.K)	25	1.013
Autoignition Temperature	>750	°C		
Explosive limits	NA	%		
Upper flammability limit	NA	%		
Incompatibility	Freshly abraded aluminum Surfaces potassium, calcium, powdered aluminum, magnesium, and zinc			
Chemical Stability	Exposure to flames, hot spots or Welding may yield toxic and/or corrosive decomposition products.			

*GWP = Global Warming Potential.
**ODP = Ozone Depletion Potential.

Table 7.14 Properties of R-142b - 1-Chloro-1,1-difluoroethane.

Fluorocarbon: 1-Chloro-1,1-dif luoroethane				
Property	**Value**	**Unit**	**Temperature, °C**	**Pressure, bar**
Code	R-142b			
CAS Number	75-68-3			
UN Number	2517			
Group	HCFC			
Chemical Formula	CH_3CClF_2			
Appearance/Color	Colorless gas			
Odor	Nearly odorless			
*GWP(CO_2=1), 100 years	2270			
**ODP(R11=1)	0.07			
Molecular Weight	100.49	g/mol		
Melting point	-131	°C		1.013
Boiling point	-9.8	°C		1.013
Latent heat of vaporization @bp	223.15	kJ/kg		1.013
Vapor pressure	3.37	bar	25	
Vapor pressure	3.03	bar	21	
Critical temperature	137.1	°C		
Critical pressure	41.23	bar		
Critical density	435.01	kg/m^3		
Critical volume	0.00230	°C		
Liquid density		kg/m^3		
Vapor density	1.107	kg/m^3	25	1.013
Specific gravity (air=1)	3.5		25	
Specific volume vapor	0.90334	m^3/kg		1.013
Heat capacity liquid	1.292	kJ/(kg.K)	25	1.013
Heat capacity vapor	0.8529	kJ/(kg.K)	25	1.013

(Continued)

Table 7.14 Properties of R-142b - 1-Chloro-1,1-difluoroethane. (*Continued*)

Fluorocarbon: 1-Chloro-1,1-dif luoroethane				
Property	**Value**	**Unit**	**Temperature, °C**	**Pressure, bar**
Ratio C_p/C_v	1.126		25	
Solubility in water	0.14	weight%	25	1.013
Dynamic Viscosity (liquid)	0.453	Poise	-20.9	1.013
Dynamic Viscosity (vapor)	0.0001099	Poise	25	1.013
Thermal Conductivity Liquid		W/(m.K)	25	1.013
Thermal Conductivity Vapor	12.4	mW/(m.K)	10	1.013
Autoignition Temperature	632	°C		
Explosive limits	6.2-18	%		
Upper flammability limit		%		
Incompatibility	Oxidizers such as perchlorates, peroxides, permanganates, chloreas, and nitrates since vigorous reactions occur.			
Chemical stability:	Decomposes in heat to form phosgene and hydrofluoric and hydrochloric acides.			

*GWP = Global Warming Potential
**ODP = Ozone Depletion Potential

Table 7.15 Properties of R-21 - dichlorofluoromethane.

Fluorocarbon: Dichlorofluoromethane				
Property	**Value**	**Unit**	**Temperature, °C**	**Pressure, bar**
Code	R-142b			
CAS Number	75-68-3			
UN Number	2517			
Group	HCFC			
Chemical Formula	CH_3CClF_2			
Appearance/Color	Colorless gas			
Odor	Nearly odorless			
*GWP (CO_2=1), 100 years	2270			
**ODP (R11=1)	0.07			
Molecular Weight	100.49	g/mol		
Melting point	-131	°C		1.013
Boiling point	-9.8	°C		1.013
Latent heat of vaporization @bp	223.15	kJ/kg		1.013
Vapor Pressure	3.37	bar	25	
Vapor Pressure	3.03	bar	21	
Critical temperature	137.1	°C		
Critical pressure	41.23	bar		
Critical density	435.01	kg/m^3		
Critical volume	0.00230	°C		
Liquid density		kg/m^3		
Vapor density	1.107	kg/m^3	25	1.013
Special gravity (air=1)	3.5		25	
Specific volume vapor	0.90334	m^3/kg		1.013

(Continued)

Table 7.15 Properties of R-21 - dichlorofluoromethane. (*Continued*)

Fluorocarbon: Dichlorofluoromethane				
Property	**Value**	**Unit**	**Temperature, °C**	**Pressure, bar**
Heat capacity liquid	1.292	kJ/(kg.K)	25	1.013
Heat capacity vapor	0.8529	kJ/(kg.K)	25	1.013
Ratio C_P/C_V	1.126		25	
Solubility in water	0.14	weight%	25	1.013
Dynamic Viscosity (liquid)	0.453	Poise	-20.9	1.013
Dynamic Viscosity (vapor)	0.0001099	Poise	25	1.013
Thermal Conductivity Liquid		W/(m.K)	25	1.013
Thermal Conductivity Vapor	12.4	mW/(m.K)	10	1.013
Autoignition Temperature	632	°C		
Explosive limits	6.2-18	%		
Upper flammability limit		%		
Incompatibility	Oxidizers such as perchlorates, peroxides, permanganates, chloreas, and nitrates since vigorous reactions occur.			
Chemical Stability	Decomposes in heat to form phosgene and hydrofluoric and hydrochloric acides.			

*GWP = Global Warming Potential.
**ODP = Ozone Depletion Potential.

Table 7.16 Properties of R-22 - chlorodifluoromethane.

Fluorocarbon: Chlorodifluoromethane				
Property	**Value**	**Unit**	**Temperature, °C**	**Pressure, bar**
Code	R-22			
CAS Number	75-45-6			
UN Number	1018			
Group	HCFC			
Chemical Formula	$CHClF_2$			
Appearance/Color	Colorless			
Odor	Sweetish			
*GWP (CO_2=1), 100 years	1810			
**ODP (R11=1)	0.055			
Molecular Weight	86.47	g/mol		
Melting point	-157.42	°C		1.013
Boiling point	-40.7	°C		1.013
Latent heat of vaporization @bp	233.3	kJ/kg		1.013
Vapor pressure	10.4	bar	25	
Critical temperature	96.15	°C		
Critical pressure	50	bar		
Critical density	523.80	kg/m^3		
Critical volume	0.0019	m^3/kg		
Liquid density	1194	kg/m^3	25	1.013
Vapor density	3.66	kg/m^3	15	1.013
Specific gravity (air=1)	3.11		25	
Specific volume	0.2788	m^3/kg	25	1.013
Compressibility Factor (Z)	0.98537		25	1.013

(Continued)

Table 7.16 Properties of R-22 - chlorodifluoromethane. (*Continued*)

Fluorocarbon: Chlorodifluoromethane				
Property	**Value**	**Unit**	**Temperature, °C**	**Pressure, bar**
Heat capacity liquid	1.25	kJ/(kg.K)	25	1.013
Heat capacity vapor	0.66	kJ/(kg.K)	25	1.013
Ratio C_p/C_v	1.1847		25	
Solubility in water	0.3628	Weight %	25	1.013
Dynamic Viscosity (liquid)	0.0023	Poise	25	1.013
Dynamic Viscosity (vapor)	0.00013	Poise	25	1.013
Thermal Conductivity Liquid	0.0879	W/(m.K)	25	1.013
Thermal Conductivity Vapor	10.58	m.W/(m.K)	25	1.013
Autoignition Temperature	632	°C		
Explosive limits	NA	%		
Upper flammability limit	Non-flammable	%		
Incompatibility	Alkalis, alkaline earth metals (e.g., powdered aluminum, sodium, potassium, zinc)			
Chemical stability				

*GWP = Global Warming Potential.
**ODP = Ozone Depletion Potential.

Table 7.17 Properties of R-225ca - 1,1-dichloro-2,2,3,3,3-pentafluoropropane.

Fluorocarbon: 1, 1-dichloro-2,2,3,3,3-pentafluoropropane				
Property	**Value**	**Unit**	**Temperature, °C**	**Pressure, bar**
Code	R-225ca			
CAS Number	422-56-0			
UN Number	3082			
Group	HCFC			
Chemical Formula	$CF_3CF_2CHCl_2$			
Appearance/Color	Colorless			
Odor	Odorless			
*GWP (CO_2=1), 100 years	525			
**ODP (R11=1)	0.03			
Molecular Weight	202.933	g/mol		
Melting point	-94	°C		1.013
Boiling point	54	°C		1.013
Latent heat of vaporization @bp	147	kJ/kg		1.013
Vapor Pressure	0.32	bar	25	
Critical temperature	205	°C		
Critical pressure	29.74	bar		
Critical density	578.00	kg/m^3		
Critical volume	0.001730	m^3/kg		
Liquid density	1550	kg/m^3	25	1.013
Vapor density		kg/m^3	15	1.013
Special gravity (air=1)				
Compressibility Factor (Z)				
Health Capacity liquid		kJ/(kg.K)		
Heat Capacity vapor		kJ/(kg.K)		
Ratio C_p/C_V				

(Continued)

Table 7.17 Properties of R-225ca - 1,1-dichloro-2,2,3,3,3-pentafluoropropane. (*Continued*)

Fluorocarbon: 1, 1-dichloro-2,2,3,3,3-pentafluoropropane				
Property	**Value**	**Unit**	**Temperature, °C**	**Pressure, bar**
Solubility in water		Weight %		
Dynamic Viscosity (liquid)		Poise		
Dynamic Viscosity (vapor)		Poise		
Thermal Conductivity Liquid		W/(m.K)		
Thermal Conductivity Vapor		W/(m.K)		
Autoignition Temperature		°C		
Explosive limits		%		
Upper flammability limit		%		
Incompatibility	Alkali metals, Finely divided Metals (Al, Mg, Zn). Strong bases, Strong oxidizing agents			
Chemical Stability				

*GWP = Global Warming Potential.
**ODP = Ozone Depletion Potential.

Table 7.18 Properties of R-225cb - 1,3-dichloro-1,1,2,2,3-pentafluoropropane.

Fluorocarbon: 1,3-dichloro-1,1,2,2,3-pentafluoropropane				
Property	**Value**	**Unit**	**Temperature, °C**	**Pressure, bar**
Code	R-225cb			
CAS Number	507-55-1			
UN Number	3082			
Group	HCFC			
Chemical Formula	$CClF_2CF_2CHClF$			

(*Continued*)

Table 7.18 Properties of R-225cb - 1,3-dichloro-1,1,2,2,3-pentafluoropropane. (*Continued*)

Fluorocarbon: 1,3-dichloro-1,1,2,2,3-pentafluoropropane				
Property	**Value**	**Unit**	**Temperature, °C**	**Pressure, bar**
Appearance/Color	Colorless			
Odor	Slight ethereal			
*GWP (CO_2=1), 100 years	595			
**ODP (R11=1)	0.033			
Molecular Weight	201.94	g/mol		
Melting point	-130.55	°C		1.013
Boiling point	54	°C		1.013
Latent heat of vaporization @bp		kJ/kg		1.013
Vapor Pressure	0.38	bar	25	
Critical temperature	211	°C		
Critical pressure	30.12	bar		
Critical density	557.00	kg/m^3		
Critical volume	0.00180	m^3/kg		
Liquid density	1565	kg/m^3	25	1.013
Vapor density		kg/m^3	15	
Special gravity (air=1)	7		25	
Compressibility Factor (Z)				
Health Capacity liquid		kJ/(kg.K)	25	
Heat Capacity vapor		kJ/(kg.K)	25	
Ratio C_P/C_V				
Solubility in water	0.033	Weight %	25	1.013
Dynamic Viscosity (liquid)		Poise		
Dynamic Viscosity (vapor)		Poise		

(*Continued*)

Table 7.18 Properties of R-225cb - 1,3-dichloro-1,1,2,2,3-pentafluoropropane. (*Continued*)

Fluorocarbon: 1,3-dichloro-1,1,2,2,3-pentafluoropropane				
Property	**Value**	**Unit**	**Temperature, °C**	**Pressure, bar**
Thermal Conductivity Liquid		W/(m.K)		
Thermal Conductivity Vapor		W/(m.K)		
Autoignition Temperature		°C		
Explosive limits		%		
Upper flammability limit		%		
Incompatibility				
Chemical Stability				

*GWP = Global Warming Potential.
**ODP = Ozone Depletion Potential.

Table 7.19 Properties of R-23 – Trifluoromethane.

Fluorocarbon: Trifluoromethane				
Property	**Value**	**Unit**	**Temperature, °C**	**Pressure, bar**
Code	R-23			
CAS Number	507-55-1			
UN Number	3136			
Group	CFC			
Chemical Formula	CHF_3			
Appearance/Color	Colorless			
Odor	Slight ether like			
*GWP (CO_2=1), 100 years	14800			
**ODP (R11=1)	0			
Molecular Weight	70.01	g/mol		
Melting point	-155	°C		1.013

(*Continued*)

Table 7.19 Properties of R-23 – Trifluoromethane. (*Continued*)

Fluorocarbon: Trifluoromethane				
Property	**Value**	**Unit**	**Temperature, °C**	**Pressure, bar**
Boiling point	-82	°C		1.013
Latent heat of vaporization @bp	239	kJ/kg		1.013
Vapor Pressure	47.3	bar	25	
Vapor Pressure	40	bar	20	
Critical temperature	25.9	°C		
Critical pressure	48.2	bar		
Critical density	525.02	kg/m3		
Critical volume	0.0019	m3/kg		
Liquid density	673	kg/m^3	25	1.013
Vapor density	2.976	kg/m^3	15	1.013
Special gravity (air=1)	2.4			
Compressibility Factor (Z)				
Health Capacity liquid		kJ/(kg.K)	25	
Heat Capacity vapor		kJ/(kg.K)	25	
Ratio C_P/C_V				
Solubility in water	0.1	weight %	25	1.013
Dynamic Viscosity (liquid)		Poise	25	
Dynamic Viscosity (vapor)		Poise	25	
Thermal Conductivity Liquid	55	mW/(m.K)	25	1.013
Thermal Conductivity Vapor	14	W/(m.K)	25	1.013
Autoignition Temperature		°C		
Explosive limits		%		
Upper flammability limit		%		
Incompatibility	Strong oxidizing agents			
Chemical Stability				

*GWP = Global Warming Potential.
**ODP = Ozone Depletion Potential.

Table 7.20 Properties of R-32 – Difluoromethane.

Fluorocarbon: Difluoromethane				
Property	**Value**	**Unit**	**Temperature, °C**	**Pressure, bar**
Code	R-32			
CAS Number	75-10-5			
UN Number	3252			
Group	HFC			
Chemical Formula	CH_2F_2			
Appearance/Color	Colorless			
Odor	Odorless			
*GWP (CO_2=1), 100 years	675			
**ODP (R11=1)	0			
Molecular Weight	52.02	g/mol		
Melting point	-136	°C		1.013
Boiling point	-52	°C		1.013
Latent heat of vaporization @bp	382	kJ/kg		1.013
Vapor Pressure	15.2	bar	21.1	
Vapor Pressure	16.8	bar	25	
Critical temperature	78.4	°C		
Critical pressure	858.14	bar		
Critical density	430.00	kg/m^3		
Critical volume	0.002326	m^3/kg		
Liquid density	961	kg/m^3	25	1.013
Vapor density	81	kg/m^3	43	1.013
Special gravity (air=1)	1.7		25	
Compressibility Factor (Z)				1.013
Health Capacity liquid	2.24	kJ/(kg.K)	43	1.013
Heat Capacity vapor	2.131	kJ/(kg.K)	43	1.013
Ratio C_P/C_V				

(Continued)

Table 7.20 Properties of R-32 – Difluoromethane. (*Continued*)

Fluorocarbon: Difluoromethane				
Property	**Value**	**Unit**	**Temperature, °C**	**Pressure, bar**
Solubility in water	0.44	weight %	25	1.013
Dynamic Viscosity (liquid)	0.0009078	Poise	43	1.013
Dynamic Viscosity (vapor)	0.0001339	Poise	43	1.013
Thermal Conductivity Liquid	135	mW/(m.K)	25	1.013
Thermal Conductivity Vapor	12	W/(m.K)	25	1.013
Autoignition Temperature	648	°C		
Explosive limits	13.3-29.3	%		
Upper flammability limit	Highly flammable	%		
Incompatibility	Strong oxidizing reducing			
Agents, many amines, aikali				
Metals, and epoxides.				
Chemical Stability	Highly flammable			

*GWP = Global Warming Potential.
**ODP = Ozone Depletion Potential.

Table 7.21 Properties of R-134a – 1,1,1,2-tetrafluoroethane.

Fluorocarbon: 1,1,1,2-tetrafluoroethane				
Property	**Value**	**Unit**	**Temperature, °C**	**Pressure, bar**
Code	R134a			
CAS Number	811-97-2			
UN Number	UN3159			
REACH Registration No.	01-2119459374-33-0002			
Group	HFC			
Formula	CH_2FCF_3			
Appearance/Color	Colourless gas			
Odor	Ethereal			

(*Continued*)

Table 7.21 Properties of R-134a – 1,1,1,2-tetrafluoroethane. (*Continued*)

Fluorocarbon: 1,1,1,2-tetrafluoroethane				
Property	**Value**	**Unit**	**Temperature, °C**	**Pressure, bar**
GWP (CO_2=1)*	1370			
ODP (R11=1)**	0			
Molecular Weight	102.03	g/mol		
Melting point	-101	°C		1.013
Boiling point	-26.55	°C		1.013
Latent heat of vaporization	215.9	kJ/kg		1.013
Liquid density	1206	kg/m^3	25	1.013
Vapor Pressure	3.5	bar	5	
Vapor Pressure	4.9	bar	15	
Vapor Pressure	5.7	bar	20	
Vapor Pressure	13.2	bar	50	
Critical temperature	100.95	°C		
Critical pressure	40.6	bar		
Critical density	512	kg/m^3		
Triple point temperature	-103.3	°C		
Gas density	5.28	kg/m^3	-26.55	1.013
Gas density	4.25	kg/m^3	15	1.013
Specific gravity	3.25			
Specific volume	0.235	m^3/kg	15	1.013
Compressibility Factor (Z)	1		15	1.013
Health Capacity (C_p)	0.08754	kJ/(mol.K)	25	1.013
Solubility in water	0.21	vol/vol	25	1.013
Flammability	Non-flammable			
Ignition Temperature	NA			
Explosive limits	NA			
Upper flammability limit	NA			

(*Continued*)

Table 7.21 Properties of R-134a – 1,1,1,2-tetrafluoroethane. (*Continued*)

Fluorocarbon: 1,1,1,2-tetrafluoroethane				
Property	**Value**	**Unit**	**Temperature, °C**	**Pressure, bar**
Incompatibility	Akali metals, alkaline earth			
Metals, powdered metals				
Powdered metal salts				
Chemical Stability	Stable under normal conditions			

*GWP = Global Warming Potential.
**ODP = Ozone Depletion Potential.

Table 7.22 Properties of R-143a – 1,1,1-trifluoroethane.

Fluorocarbon: 1,1,1,trifluoroethane				
Property	**Value**	**Unit**	**Temperature, °C**	**Pressure, bar**
Code	R143a			
CAS Number	420-46-2			
UN Number	2035			
Group	HFC			
Chemical Formula	CF_3CH_3			
Appearance/Color	Colorless			
Odor	Slight ethereal			
*GWP (CO_2=1),100 years	0			
**ODP (R11=1)	4470			
Molecular Weight	84.04	g/mol		
Melting point	-111	°C		
Boiling point	-47	°C		
Latent heat of vaporization @bp	227	kJ/kg		
Vapor Pressure	12.6	bar	25	
Critical temperature	72.6	°C		
Critical pressure	37.6	bar		

(*Continued*)

Table 7.22 Properties of R-143a – 1,1,1-trifluoroethane. (*Continued*)

Fluorocarbon: 1,1,1,trifluoroethane				
Property	**Value**	**Unit**	**Temperature, °C**	**Pressure, bar**
Critical density	431	kg/m^3		1.013
Critical volume		m^3/kg		
Liquid density	930	kg/m^3	25	1.013
Vapor density		kg/m^3		
Specific gravity (air=1)	3.1			
Compressibility Factor (Z)				
Health Capacity liquid		kJ/(kg.K)		
Health Capacity vapor		kJ/(kg.K)		
Ratio C_p/C_v				
Solubility in water	0.00548	weight %		
Dynamic Viscosity (liquid)		Poise		
Dynamic Viscosity (vapor)		Poise		
Thermal Conductivity Liquid	67	mW/(m.K)	25	1.013
Thermal Conductivity Vapor	14	mW/(m.K)	25	
Autognition Temperature	750	°C		
Explosive limits	7.1- 16.1	%		
Upper flammability limit		%		
Incompatibility	Strong oxidizing and reducing agents, many amines, nitrides, azo/diazo compounds, alkali metals, and epoxides			
Chemical Stability				

*GWP = Global Warming Potential.
**ODP = Ozone Depletion Potential.

Table 7.23 Properties of R-152a – 1,1-difluoroethane.

Fluorocarbon: 1,1-difluoroethane				
Property	**Value**	**Unit**	**Temperature, °C**	**Pressure, bar**
Code	R152a			
CAS Number	75-37-6			
UN Number	1030			
Group	HFC			
Chemical Formula	CHF_2CH_3			
Appearance/Color	Colorless			
Odor	Odorless			
*GWP (CO_2=1), 100 years	124			
**ODP (R11=1)	0			
Molecular Weight	66.05	g/mol		
Melting point	-118.6	°C		
Boiling point	-24.7	°C		
Heat of vaporization	288.9	kJ/kg	25	
Latent heat of vaporization @bp	335	kJ/kg		
Vapor pressure	5.16	bar	20	
Vapor pressure	6.07	bar	25	
Critical temperature	113.4	°C		
Critical pressure	45.2	bar		
Critical density	368	kg/m^3		1.013
Critical volume		m^3/kg		
Liquid density	860	kg/m^3	40	1.013
Vapor density	19	kg/m^3	27	1.013
Specific gravity (air=1)	2.3			
Specific volume		m^3/kg		
Compressibility Factor (Z)				
Health capacity liquid		kJ/(kg.K)		

(Continued)

Table 7.23 Properties of R-152a – 1,1-difluoroethane. (*Continued*)

Fluorocarbon: 1,1-difluoroethane				
Property	**Value**	**Unit**	**Temperature, °C**	**Pressure, bar**
Health capacity vapor		kJ/(kg.K)		
Ratio C_p/C_v				
Solubility in water	0.2	weight %		
Dynamic Viscosity (liquid)	0.00135	Poise	40	
Dynamic Viscosity (vapor)	0.000104	Poise	27	1.013
Thermal Conductivity Liquid	105	mW/(m.K)	25	1.013
Thermal Conductivity Vapor	13	mW/(m.K)	25	
Flash point	<-50	°C		
Autognition Temperature	455	°C		
Explosive limits	3.7-18	%		
Flammability limit	4-19.6	%		
Incompatibility	Alkali metals, Alkaline earth			
Metals, Powdered metals,				
Powdered metal salts				
Chemical Stability				

*GWP = Global Warming Potential.
**ODP = Ozone Depletion Potential.

Table 7.24 Properties of R-227ea – 1,1,1,2,3,3,3-heptafluoropropane.

Fluorocarbon: 1,1,1,2,3,3,3-heptafluoropropane				
Property	**Value**	**Unit**	**Temperature, °C**	**Pressure, bar**
Code	R227ea			
CAS Number	431-89-0			
UN Number	3296			
Group	HFC			
Chemical Formula	CF_3CHFCF_3			
Appearance/Color	Colorless			
Odor	Light ethereal			
*GWP (CO2=1), 100 years	3220			
**ODP (R11=1)	0			
Molecular Weight	170.03	g/mol		
Melting point	-126.8	°C		
Boiling point	-16.3	°C		
Latent heat of vaporization @bp	131.8	kJ/kg		
Vapor pressure	4.55	bar	25	
Vapor pressure		bar		
Critical temperature	139.4	°C		
Critical pressure	34.2	bar		
Critical density	594.25	kg/m³		1.013
Critical volume		m³/kg		
Liquid density	1394	kg/m³	25	1.013
Vapor density	7.15	kg/m³	25	1.013
Specific gravity (air=1)	4.2		20	
Specific volume		m³/kg		
Compressibility Factor (Z)				
Health capacity liquid	1.18	kJ/(kg.K)		
Health capacity vapor	0.813	kJ/(kg.K)		
Ratio C_p/C_v				

(Continued)

Table 7.24 Properties of R-227ea – 1,1,1,2,3,3,3-heptafluoropropane. (*Continued*)

Fluorocarbon: 1,1,1,2,3,3,3-heptafluoropropane				
Property	**Value**	**Unit**	**Temperature, °C**	**Pressure, bar**
Solubility in water	0.0198	weight %	25	
Dynamic Viscosity (liquid)	0.00184	Poise	25	
Dynamic Viscosity (vapor)	0.000132	Poise	25	1.013
Thermal Conductivity Liquid	0.0605	W/(m.K)		1.013
Thermal Conductivity Vapor	0.0133	W/(m.K)		
Flash point		°C		
Autognition Temperature		°C		
Explosive limits		%		
Flammability limits		%		
Incompatibility				
Chemical Stability				

*GWP = Global Warming Potential.
**ODP = Ozone Depletion Potential.

Table 7.25 Properties of R-245fa – 1,1,1,3,3-pentafluoropropane.

Fluorocarbon: 1,1,1,3,3-pentafluoropropane				
Property	**Value**	**Unit**	**Temperature, °C**	**Pressure, bar**
Code	R245fa			
CAS Number	460-73-1			
UN Number	3163			
Group	HFC			
Chemical Formula	$CF_3CH_2CHF_3$			
Appearance/Color	Colorless			
Odor	Odorless			
*GWP (CO_2=1), 100 years	950			
**ODP (R11=1)	0			
Molecular Weight	134	g/mol		

(*Continued*)

Table 7.25 Properties of R-245fa – 1,1,1,3,3-pentafluoropropane. (*Continued*)

Fluorocarbon: 1,1,1,3,3-pentafluoropropane				
Property	**Value**	**Unit**	**Temperature, °C**	**Pressure, bar**
Melting point	<-107	°C		
Boiling point	15.3	°C		
Latent heat of vaporization @bp	196.7	kJ/kg		
Vapor pressure	1.23	bar	20	
Vapor pressure		bar		
Critical temperature	154.05	°C		
Critical pressure	36.4	bar		
Critical density	517	kg/m^3		1.013
Critical volume		m^3/kg		1.013
Liquid density	1339	kg/m^3	25	1.013
Vapor density	5.92	kg/m^3	15.1	1.013
Specific gravity (air=1)	4.6		20	
Specific volume		m^3/kg		
Compressibility Factor (Z)				
Health capacity liquid	1.36	kJ/(kg.K)	25	
Health capacity vapor	0.8931	kJ/(kg.K)	25	
Ratio C_p/C_v				
Solubility in water	0.718	weight %	25	
Dynamic Viscosity (liquid)	4.027	Poise	25	
Dynamic Viscosity (vapor)	0.103	Poise	25	1.013
Thermal Conductivity Liquid	0.081	W/(m.K)		1.013
Thermal Conductivity Vapor	0.0125	W/(m.K)		
Flash point	-25	°C		

(*Continued*)

Table 7.25 Properties of R-245fa – 1,1,1,3,3-pentafluoropropane. (*Continued*)

Fluorocarbon: 1,1,1,3,3-pentafluoropropane				
Property	**Value**	**Unit**	**Temperature, °C**	**Pressure, bar**
Autognition Temperature	412	°C		
Explosive limits	None	%		
Flammability limits	None	%		
Incompatibility	Oxidizing agents, May react Violently with alkaline-earth and alkali metals			
Chemical Stability				

*GWP = Global Warming Potential.
**ODP = Ozone Depletion Potential.

Table 7.26 Properties of R-365mfc – 1,1,1,3,3-pentafluorobutane.

Fluorocarbon: 1,1,1,3,3-pentafluorobutane				
Property	**Value**	**Unit**	**Temperature, °C**	**Pressure, bar**
Code	R365mfc			
CAS Number	460-58-6			
UN Number	1993			
Group	HFC			
Chemical Formula	$CF_3CH_2CF_2CH_3$			
Appearance/Color	Colorless			
Odor	Ether-like			
*GWP (CO_2=1), 100 years	794			
**ODP (R11=1)	0			
Molecular Weight	148.076	g/mol		
Melting point	-20	°C		
Boiling point	40	°C		
Latent heat of vaporization @bp	181	kJ/kg		

(*Continued*)

Table 7.26 Properties of R-365mfc – 1,1,1,3,3-pentafluorobutane. (*Continued*)

Fluorocarbon: 1,1,1,3,3-pentafluorobutane				
Property	**Value**	**Unit**	**Temperature, °C**	**Pressure, bar**
Vapor pressure	0.435	bar	20	
Critical temperature	187.7	°C		
Critical pressure	27.54	bar		
Critical density		kg/m^3		
Critical volume		m^3/kg		
Liquid density	1270	kg/m^3	20	1.013
Vapor density		kg/m^3		
Specific gravity (air=1)	5.11		20	
Compressibility Factor (Z)				
Heat capacity liquid		kJ/(kg.K)		
Heat capacity vapor		kJ/(kg.K)		
Ratio C_P/C_V				
Solubility in water	0.718	weight %	25	
Dynamic Viscosity (liquid)	0.004	Poise	25	
Dynamic Viscosity (vapor)		Poise		
Thermal Conductivity Liquid	12	mW/(m.K)	25	1.013
Thermal Conductivity Vapor		W/(m.K)		
Flash Point	-27	°C		
Autoignition Temperature	580	°C		
Explosive limits	3.6 - 13.3	%		
Flammability limits	Highly flammable	%		
Incompatibility	Strong alkali or alkaline earth metals. Finely powdered metals such as Al, Be, Zn, Na, Mg, etc. Strong bases such as NaOH, KOH			
Chemical stability				

*GWP = Global Warming Potential.
**ODP = Ozone Depletion Potential.

Table 7.27 Properties of R-43-10mee – 1,1,1,2,2,3,4,5,5,5-decafluoropentane.

Fluorocarbon: 1,1,1,2,2,3,4,5,5,5-decafluoropentane (liquid, not a refrigerant)				
Property	**Value**	**Unit**	**Temperature, °C**	**Pressure, bar**
Code	R43-10mee			
CAS Number	138495-42-8			
UN Number	Not classified as dangerous for transportation			
Group	HFC			
Chemical Formula	$CF_3CF_2CHFCHFCF_3$			
Appearance/Color	Colorless			
Odor	Ether-like			
*GWP (CO_2=1), 100 years	1700			
**ODP (R11=1)	0			
Molecular Weight	252.05	g/mol		
Melting point	-84	°C		
Boiling point	55	°C		
Latent heat of vaporization @bp	129.7	kJ/kg		
Vapor pressure	0.298	bar	20	
Vapor pressure	0.854	bar	50	
Critical temperature	181	°C		
Critical pressure	22.88	bar		
Critical density	581.99	kg/m^3		
Critical volume	0.001718	m^3/kg		
Liquid density	1600	kg/m^3	20	1.013
Vapor density		kg/m^3		
Specific gravity (air=1)				
Compressibility Factor (Z)				

(Continued)

Table 7.27 Properties of R-43-10mee – 1,1,1,2,2,3,4,5,5,5-decafluoropentane. (*Continued*)

Fluorocarbon: 1,1,1,2,2,3,4,5,5,5-decafluoropentane (liquid, not a refrigerant)				
Property	**Value**	**Unit**	**Temperature, °C**	**Pressure, bar**
Heat capacity liquid	1.58	kJ/(kg.K)	25	
Heat capacity vapor		kJ/(kg.K)		
Ratio C_P/C_V				
Solubility in water	0.010-0.014	weight %	20	
Dynamic Viscosity (liquid)		Poise		
Dynamic Viscosity (vapor)		Poise		
Thermal Conductivity Liquid		mW/(m.K)		
Thermal Conductivity Vapor		W/(m.K)		
Flash Point	125	°C		
Autoignition Temperature		°C		
Explosive limits		%		
Flammability limits	Non-flammable	%		
Incompatibility	Strong oxidizing agents, alkaline earth metals, strong bases			
Chemical stability				

*GWP = Global Warming Potential.
**ODP = Ozone Depletion Potential.

Table 7.28 Properties of R-1132a – 1,1-difluoroethylene.

Fluorocarbon: 1,1-difluoroethylene				
Property	**Value**	**Unit**	**Temperature, °C**	**Pressure, bar**
Code	R1132a			
CAS Number	75-38-7			
UN Number	1959			
Group	HFO			
Chemical Formula	CF_2=CH_2			
Appearance/Color	Colorless			
Odor	Faint ethereal			
*GWP (CO_2=1), 100 years	<1			
**ODP (R11=1)	0			
Molecular Weight	64.03	g/mol		
Melting point	-144	°C		
Boiling point	-84	°C		
Latent heat of vaporization @bp	276.4	kJ/kg		
Vapor pressure	35.7	bar	25	
Critical temperature	29.75	°C		
Critical pressure	44.6	bar		
Critical density		kg/m^3		
Critical volume		m^3/kg		
Liquid density	0.617	kg/m^3	24	1.013
Vapor density		kg/m^3		
Specific gravity (air=1)	2.2		20	
Specific volume		m^3/kg		
Compressibility Factor (Z)				

(Continued)

Table 7.28 Properties of R-1132a – 1,1-difluoroethylene. (*Continued*)

Fluorocarbon: 1,1-difluoroethylene				
Property	**Value**	**Unit**	**Temperature, °C**	**Pressure, bar**
Heat capacity liquid		kJ/(kg.K)		
Heat capacity vapor		kJ/(kg.K)		
Ratio CP/Cv				
Solubility in water	0.018	weight %	25	
Dynamic Viscosity (liquid)	0.776	Poise		
Dynamic Viscosity (vapor)		Poise		
Thermal Conductivity Liquid		W/(m.K)		
Thermal Conductivity Vapor		W/(m.K)		
Flash Point	Flammable	°C		
Autoignition Temperature	640	°C		
Explosive limits	5.5-21.3	%		
Flammability limits	Flammable	%		
Incompatibility	Oxidizers, react violently with HCl. Alkyl boron and alkyl hyponitrite compounds initiate polymerization. Forms peroxides on exposure to pure O_2			
Chemical stability				

*GWP = Global Warming Potential.
**ODP = Ozone Depletion Potential.

Table 7.29 Properties of R-1234yf –2,3,3,3-tetrafluoropropene.

Fluorocarbon: 2,3,3,3-tetrafluoropropone				
Property	**Value**	**Unit**	**Temperature, °C**	**Pressure, bar**
Code	R1234yf			
CAS Number	754-12-1			
UN Number	3161			
Group	HFO			
Chemical Formula	CH_2=CF-CF_3			
Appearance/ Color	Colorless			
Odor	Slight			
*GWP (CO_2=1), 100 years	1			
**ODP (R11=1)	0			
Molecular Weight	114.04	g/mol		
Melting point	-152.2	°C		
Boiling point	-29.4	°C		
Heat of vaporization	170	kJ/kg	-20	
Heat of vaporization	145.3	kJ/kg	25	
Vapor pressure	6.1	bar	21.1	
Vapor pressure	14.2	bar	54.4	
Critical temperature	94.85	°C		
Critical pressure	33.82	bar		
Critical density	478	kg/m³		
Critical volume	0.00209	m³/kg		
Liquid density	1100	kg/m³	25	1.013
Vapor density	37.92	kg/m³	25	1.013

(Continued)

Table 7.29 Properties of R-1234yf –2,3,3,3-tetrafluoropropene. (*Continued*)

Fluorocarbon: 2,3,3,3-tetrafluoropropone				
Property	**Value**	**Unit**	**Temperature, °C**	**Pressure, bar**
Specific gravity (air=1)	4		20	
Specific volume vapor	0.0264	m^3/kg		
Compressibility Factor (Z)-estimate	0.867			
Heat capacity liquid	1.514	kJ/(kg.K)	47.15	15
Heat capacity vapor		kJ/(kg.K)	25	
Ratio C_P/C_V				
Solubility in water	0.02	weight %	25	
Dynamic Viscosity (liquid)	1.55	Poise	25	
Dynamic Viscosity (vapor)	0.114	Poise	25	1.013
Thermal Conductivity Liquid	0.064	W/(m.K)	25	1.013
Thermal Conductivity Vapor		W/(m.K)		
Flash Point		°C		
Autoignition Temperature	405	°C		
Explosive limits	6.2-12.3	%		
Flammability	Slightly flammable			

(*Continued*)

Table 7.29 Properties of R-1234yf –2,3,3,3-tetrafluoropropene. (*Continued*)

Fluorocarbon: 2,3,3,3-tetrafluoropropone				
Property	**Value**	**Unit**	**Temperature, °C**	**Pressure, bar**
Flammability limits	6.5-12.3	%		
Incompatibility	Oxidizing materials and metals			
Chemical stability				

*GWP = Global Warming Potential.
**ODP = Ozone Depletion Potential.

Table 7.30 Properties of R-1233zd – trans-1-chloro-3,3,3-trifluoropropene.

Fluorocarbon: Trans-1-chloro-3,3,3-trifluoropropene				
Property	**Value**	**Unit**	**Temperature, °C**	**Pressure, bar**
Code	R1233zd			
CAS Number	102687-65-0			
UN Number	3163			
Group	HCFO*			
Chemical Formula	(E)-CHCl=CH-CF_3			
Appearance/Color	Low boiling liquid			
Odor	Slight			
*GWP (CO_2=1), 100 years	1			
**ODP (R11=1)	0			
Molecular Weight	130	g/mol		
Melting point	<-90	°C		
Boiling point	18.32	°C		
Latent heat of vaporization @bp	194	kJ/kg		
Vapor pressure	1.52	bar	30	
Vapor pressure	1.06	bar	20	

(*Continued*)

Table 7.30 Properties of R-1233zd – Trans-1-chloro-3,3,3-trifluoropropene. (*Continued*)

Fluorocarbon: Trans-1-chloro-3,3,3-trifluoropropene				
Property	**Value**	**Unit**	**Temperature, °C**	**Pressure, bar**
Critical temperature	165.6	°C		
Critical pressure	35.72	bar		
Critical density	478.92	kg/m^3		
Critical volume	0.00209	m^3/kg		
Liquid density	1296	kg/m^3	20	1.013
Vapor density	6.1	kg/m^3	20	1.013
Specific gravity (air=1)	5.3			
Specific volume vapor	0.164	m^3/kg	20	
Compressibility Factor (Z)-estimate	0.859			
Heat capacity liquid		kJ/(kg.K)		
Heat capacity vapor		kJ/(kg.K)		
Ratio C_P/C_V				
Solubility in water	0.19	weight %	25	
Dynamic Viscosity (liquid)	0.00489	Poise	20	
Dynamic Viscosity (vapor)		Poise		
Thermal Conductivity Liquid		W/(m.K)		
Thermal Conductivity Vapor	10.2	mW/(m.K)	20	
Autoignition Temperature	380	°C		
Explosive limits		%		
Flammability limits	Not flammable	%		
Incompatibility	Strong oxidizing agents, magnesium and aluminum			
Chemical stability				

*HCFO = Hydrochloro fluoroolefin.
***ODP = Ozone Depletion Potential.
**GWP = Global Warming Potential.

Table 7.31 Properties of R-1234ze – 1,3,3,3-tetrafluoropropene.

Fluorocarbon: 1,3,3,3-tetrafluoropropane				
Property	**Value**	**Unit**	**Temperature, °C**	**Pressure, bar**
Code	R1234ze			
CAS Number	29118-24-9			
UN Number	3163			
Group	HFO			
Chemical Formula	$CHF{=}CH{-}CF_3$			
Appearance/Color	Colorless			
Odor	Slight ether-like			
*GWP (CO_2=1), 100 years	<1			
**ODP (R11=1)	0			
Molecular Weight	114.04	g/mol		
Melting point	-156	°C		
Boiling point	-19	°C		
Latent heat of vaporization		kJ/kg		
Vapor pressure	4.3	bar	20	
Vapor pressure	11.1	bar	54.4	
Critical temperature	109.52	°C		
Critical pressure	36.4	bar		
Critical density	489.2	kg/m^3		
Critical volume	0.00204	m^3/kg		
Liquid density	1170	kg/m^3	21.1	1.013
Vapor density	5.71	kg/m^3	-19	1.013
Specific gravity (air=1)	4		20	
Specific volume vapor	0.175	m^3/kg	-19	
Compressibility Factor (Z)-estimate	0.859			
Heat capacity liquid	1.383	kJ/(kg.K)	25	15

(Continued)

Table 7.31 Properties of R-1234ze – 1,3,3,3-tetrafluoropropene. (*Continued*)

Fluorocarbon: 1,3,3,3-tetrafluoropropane				
Property	**Value**	**Unit**	**Temperature, °C**	**Pressure, bar**
Heat capacity vapor	0.9822	kJ/(kg.K)	25	
Ratio C_p/C_V				
Solubility in water	0.0373	weight %	25	
Dynamic Viscosity (liquid)	1.99	Poise	25	
Dynamic Viscosity (vapor)	0.122	Poise	25	1.013
Thermal Conductivity Liquid	0.0781	W/(m.K)	25	1.013
Thermal Conductivity Vapor	0.0136	W/(m.K)		
Flash point	NA	°C		
Autoignition Temperature	368	°C		
Explosive limits		%		
Flammability limits	7-12	%	100	
Incompatibility	Alkali metals and finely divided metals (Al, Mg, Zn). Oxidant gases and other oxidants.			
1.013				
Chemical stability				

*GWP = Global Warming Potential.
**ODP = Ozone Depletion Potential.

Table 7.32 Properties of R-1243zf – 3,3,3-trifluoropropene.

Fluorocarbon: 3,3,3-trifluoropropene				
Property	**Value**	**Unit**	**Temperature, °C**	**Pressure, bar**
Code	R1243zf			
CAS Number	677-21-4			
UN Number	NA			
Group	HFO			
Chemical Formula	CF_3-CH=CH_2			
Appearance/Color	Colorless			
Odor	NA			
*GWP (CO_2=1), 100 years	9			
**ODP (R11=1)	0			
Molecular Weight	96.05	g/mol		
Melting point	-145	°C		
Boiling point	-21.5	°C		
Latent heat of vaporization @bp	229.05	kJ/kg		
Vapor pressure	6.4	bar	29	
Critical temperature	105.44	°C		
Critical pressure	36.1	bar		
Critical density	422.62	kg/m^3		
Critical volume	0.00237	m^3/kg		
Liquid density	940	kg/m^3	22.5	1.013
Specific gravity (air=1)	3.3			
Compressibility Factor (Z)				
Heat capacity liquid		kJ/(kg.K)		

(Continued)

Table 7.32 Properties of R-1243zf – 3,3,3-trifluoropropene. (*Continued*)

Fluorocarbon: 3,3,3-trifluoropropene				
Property	**Value**	**Unit**	**Temperature, °C**	**Pressure, bar**
Heat capacity vapor		kJ/(kg.K)		
Ratio C_P/C_V				
Solubility in water	Very soluble	weight %	25	
Dynamic Viscosity (liquid)		Poise		
Dynamic Viscosity (vapor)		Poise		
Thermal Conductivity Liquid		W/(m.K)		
Thermal Conductivity Vapor		W/(m.K)		
Flash point	-91	°C		
Autoignition Temperature		°C		
Explosive limits	4.7-13.5	%		
Flammability limits		%		
Incompatibility	Polymerization can occur in the presence of a strong initiator			
Chemical stability	Reaction with the lighter divalent metals may give much more reactive materials analogous to Grignard reagents. Polymerization can occur in the presence of a strong initiator			

*GWP = Global Warming Potential.
**ODP = Ozone Depletion Potential.

Table 7.33 Properties of R-1327 – 1,1,3,3,4,4,4-heptafluorobutene.

Fluorocarbon: 1,1,3,3,4,4,4-heptafluorobutene				
Property	**Value**	**Unit**	**Temperature, °C**	**Pressure, bar**
Code	R1327			
CAS Number	677-21-4			
UN Number	NA			
Group	HFO			
Chemical Formula	CF_3-CH=CH_2			
Appearance/Color	Colorless			
Odor	NA			
*GWP (CO_2=1), 100 years	9			
**ODP (R11=1)	0			
Molecular Weight	96.05	g/mol		
Melting point	-145	°C		
Boiling point	-22.3	°C		
Latent heat of vaporization @bp	229.05	kJ/kg		
Vapor pressure	6.4	bar	29	
Critical temperature	105.44	°C		
Critical pressure	36.1	bar		
Critical density	422.62	kg/m^3		1.013
Critical volume	0.00237	m^3/kg		1.013
Liquid density	940	kg/m^3	22.5	1.013
Vapor density		kg/m^3		1.013
Specific gravity (air=1)	3.3			
Specific volume liquid	0.001064	m^3/kg	22.5	

(Continued)

Table 7.33 Properties of R-1327 – 1,1,3,3,4,4,4-heptafluorobutene. (*Continued*)

Fluorocarbon: 1,1,3,3,4,4,4-heptafluorobutene				
Property	**Value**	**Unit**	**Temperature, °C**	**Pressure, bar**
Compressibility Factor (Z)				
Heat capacity liquid		kJ/(kg.K)		
Heat capacity vapor		kJ/(kg.K)		
Ratio C_p/C_v				
Solubility in water	very soluble	weight %	25	
Dynamic Viscosity (liquid)		Poise		
Dynamic Viscosity (vapor)		Poise		1.013
Thermal Conductivity Liquid		W/(m.K)		1.013
Thermal Conductivity Vapor		W/(m.K)		
Flash point	-91	°C		
Autoignition Temperature		°C		1.013
Explosive limits	4.7-13.5	%		1.013
Flammability limits		%		1.013
Incompatibility	Polymerization can occur in the presence of a strong initiator			1.013
Chemical stability	Reaction with the lighter divalent metals may give much more reactive materials analogous to Grignard reagents. Polymerization can occur in the presence of a strong initiator			

*GWP = Global Warming Potential.
**ODP = Ozone Depletion Potential.

Table 7.34 Properties of R-1336mzz-Z – cis-1,1,1,4,4,4–hexafluoro-2-butene.

Fluorocarbon: cis-1,1,1,4,4,4-hexafluoro-2-butene				
Property	**Value**	**Unit**	**Temperature, °C**	**Pressure, bar**
Code	R1336mzz-Z			
CAS Number	692-49-9			
UN Number	1993			
Group	HFO			
Chemical Formula	CF_3-CH=CH-CF_3			
Appearance/Color	Colorless			
Odor	Odorless			
*GWP (CO_2=1), 100 years	2			
**ODP (R11=1)	0			
Molecular Weight	164.05	g/mol		
Melting point	-90.5	°C		
Boiling point	33	°C		
Latent heat of vaporization @bp		kJ/kg		
Vapor pressure	0.41	bar	20	
Critical temperature	171.3	°C		
Critical pressure	29	bar		
Critical density	476-528	kg/m^3		
Critical volume	0.00189-0.0021	m^3/kg		
Liquid density	1380	kg/m^3	20	1.013
Vapor density	6.03	kg/m^3	30	1.013
Specific gravity (air=1)				
Compressibility Factor (Z)				

(Continued)

Table 7.34 Properties of R-1336mzz-Z – cis-1,1,1,4,4,4–hexafluoro-2-butene. (*Continued*)

Fluorocarbon: cis-1,1,1,4,4,4-hexafluoro-2-butene				
Property	**Value**	**Unit**	**Temperature, °C**	**Pressure, bar**
Heat capacity liquid		kJ/(kg.K)		
Heat capacity vapor		kJ/(kg.K)		
Ratio C_p/C_v				
Solubility in water	0.3834	weight %	25	
Dynamic Viscosity (liquid)		Poise		
Dynamic Viscosity (vapor)		Poise		1.013
Thermal Conductivity Liquid		W/(m.K)		1.013
Thermal Conductivity Vapor		W/(m.K)		
Flash point	-21	°C		
Autoignition Temperature		°C		
Explosive limits		%		
Flammability limits	Not flammable	%		
Incompatibility	Strong acids and strong bases			
Applications				
	Foam blowing agent for polyurethane and polyisocyanurate			

*GWP = Global Warming Potential.
**ODP = Ozone Depletion Potential.

Table 7.35 Properties of R-1345zfc Pentafluoro-1-butene.

Pentafluoro-1-butene				
Property	**Value**	**Unit**	**Temperature, °C**	**Pressure, bar**
Code	R1345zfc			
CAS Number	374-27-6			
UN Number	3161			
Group	HFO			
Chemical Formula	CH_2=CH-CF_2-CF_3			
Appearance/Color				
Odor				
*GWP (CO_2=1), 100 years	0.09			
**ODP (R11=1)	0			
Molecular Weight	146.06	g/mol		
Melting point		°C		
Boiling point	5	°C		
Latent heat of vaporization @bp	155.28	kJ/kg		
Latent heat of vaporization		kJ/kg		
Vapor pressure	4	bar	25	
Vapor pressure		bar		
Critical temperature		°C		
Critical pressure		bar		
Critical density		kg/m^3		
Critical volume		m^3/kg		
Liquid density	1248	kg/m^3	20	1.013
Vapor density		kg/m^3		
Specific gravity (air=1)				
Specific volume		m^3/kg		
Compressibility Factor (Z)				
Heat capacity liquid		kJ/(kg.K)		

(*Continued*)

Table 7.35 Properties of R-1345zfc Pentafluoro-1-butene. (*Continued*)

Pentafluoro-1-butene				
Property	**Value**	**Unit**	**Temperature, °C**	**Pressure, bar**
Heat capacity vapor		kJ/(kg.K)		
Ratio C_p/C_v				
Solubility in water		weight %	25	
Dynamic Viscosity (liquid)		Poise		
Dynamic Viscosity (vapor)		Poise		
Thermal Conductivity Liquid		W/(m.K)		
Thermal Conductivity Vapor		W/(m.K)		
Flash point		°C		
Autoignition Temperature		°C		
Explosive limits		%		
Flammability limits	Flammable	%		
Incompatibility				
Chemical stability				

*GWP = Global Warming Potential.
**ODP = Ozone Depletion Potential.

8

Environmental, Safety, Health and Sustainability

It is not possible to speak about fluorocarbons without bringing up the issues of the impact of these compounds on the Earth's Ozone Layer and the Montreal Protocol (described later in this section) that banned some of those fluorocarbons. Examples of the three main commercial classes, based on their interaction with ozone, of fluorocarbons are summarized in Table 8.1. CFC's have been completely phased out and hydrochlorofluorocarbons (HCFC's) and hydrofluorocarbons (HFCs) are at the different stages of being phased out.

HFCs and perfluorocarbons (PFCs), have served as alternatives to ozone depleting substances phased out under the Montreal Protocol. Table 8.2 summarizes the applications of HFCs and PFCs.

There are two sets of issues with commercial fluorocarbons, ozone depletion and global warming. Chlorofluorocarbons were found to contribute to destruction of the ozone layer over the earth. Most man-made ODS are also very potent greenhouse gases. Some of them are up to 14,000 times stronger than carbon dioxide (CO_2), the main greenhouse gas. Eliminating these substances therefore also contributes significantly to the fight against climate change. Both topics are discussed in the next two sections.

8.1 Montreal Protocol

Countries of the world met in the 1980's and reached an agreement in 1987, called *Montreal Protocol on Substances that Deplete the Ozone Layer*, to reduce/eliminate harmful fluorocarbons. It was agreed to phase out the CFCs and replace them with harmless chemicals as quickly as possible. Montreal Protocol has survived the test of times. Montreal Protocol has been revised periodically since its inception in 1987 followed by implementation beginning January 1, 1989. Appendix I is the Section 1 of Montreal Protocol which the reader should consult for better understanding the impact and status and evolution of fluorocarbons.

The General Obligations of Montreal Protocol are quoted here:

1. The Parties shall take appropriate measures in accordance with the provisions of this Convention and of those protocols in force to which they are party to protect human health and the environment against adverse effects resulting or likely to result from human activities, which modify or are likely to modify the ozone layer.
2. To this end the Parties shall, in accordance with the means at their disposal and their capabilities:
   a. Co-operate by means of systematic observations, research and information exchange in order to better understand and assess the effects of

Sina Ebnesajjad, PhD. Concise Handbook of Fluorocarbon Gases: Applications in Refrigeration and Other Industries, (217–240) © 2021 Scrivener Publishing LLC

Table 8.1 Main commercial classes of fluorocarbons.

Fluorocarbon type	Description
CFC (chlorofluorocarbon)	Substance that contains chlorine and has strong characteristics of destroying the ozone layer. Examples: CFC-11, 12, 113 etc.
HCFC (hydrochlorofluorocarbons)	It contains chlorine but it also contains hydrogen, therefore propensity to destroy the ozone layer is considerably weakened. Examples: HCFC-22, 123, 141b, 142b etc.
HFC (hydrofluorocarbons)	It does not contain chlorine and has hydrogen that does not destroy ozone layer. Examples: HFC-23, 32, 125, 134a, 143a, 152a etc. HFC blend refrigerants R-410A, 407C, 404A
HFO (hydrofluoro-olefins)	It contains carbon, hydrogen and fluorine and carbon-carbon double bond and characterized by very low GWP of less than one

human activities on the ozone layer and the effects on human health and the environment from modification of the ozone layer;

b. Adopt appropriate legislative or administrative measures and co-operate in harmonizing appropriate policies to control, limit, reduce or prevent human activities under their jurisdiction or control should it be found that these activities have or are likely to have adverse effects resulting from modification or likely modification of the ozone layer;
c. Co-operate in the formulation of agreed measures, procedures and standards for the implementation of this Convention, with a view to the adoption of protocols and annexes;
d. Co-operate with competent international bodies to implement effectively this Convention and protocols to which they are party.

3. The provisions of this Convention shall in no way affect the right of Parties to adopt, in accordance with international law, domestic measures additional to those referred to in paragraphs 1 and 2 above, nor shall they affect additional domestic measures already taken by a Party, provided that these measures are not incompatible with their obligations under this Convention.
4. The application of this article shall be based on relevant scientific and technical considerations.

Examples of worst source compounds of chlorine and their impacts are shown in Figure 8.1. The sources of nearly 95% of all chlorine in the stratosphere are eight chemicals; all other compounds account for 5% of the Cl. Montreal Protocol devised a variety of approaches including the recovery, phasing out and replacement of ozone depleting fluorocarbons. Considering the applications of fluorocarbons gases and liquids, it would be virtually impossible to maintain the present living standards without them thus ruling out elimination. Those applications include refrigerants, fire retardant, explosion prevention, aerosols (asthma inhalers), solvents, etchant (semicon chips), foam blowing and a host of other uses.

Table 8.2 Main applications for HFC and PFC substitutes for ODS [1].

Chemical	Refrigeration and air conditioning	Fire suppression and explosion protection	Aerosols		Solvent cleaning	Foam blowing	Other applications[2]
			Propellants	Solvents			
HFC-23	X	X					
HFC-32	X						
HFC-125	X	X					
HFC-134a	X	X	X			X	X
HFC-143a	X						
HFC-152a	X		X			X	
HFC-227ea	X	X	X			X	X
HFC-236fa	X	X					
HFC-245fa				X		X	
HFC-365mfe				X	X	X	

(Continued)

Table 8.2 Main applications for HFC and PFC substitutes for ODS [1]. (*Continued*)

Chemical	Refrigeration and air conditioning	Fire suppression and explosion protection	Aerosols		Solvent cleaning	Foam blowing	Other applications[2]
			Propellants	Solvents			
HFC-43-10mee				X	X		
PFC-14[3] (CF_4)		X					
PFC-116 (C_2F_6)							X
PFC-218 (C_3F_8)							
PFC-31-10 (C_4F_{10})		X					
PFC-51-14[4] (C_6F_{14})					X		

[1] Several applications use HFCs and PFCs as components of blends. The other components of these blends are sometimes ODSs and/or non-greenhouse gases. Several HFCs, PFCs and blends are sold under various trade names.

[2] Other applications include sterilization equipment, tobacco expansion applications, plasma etching of electronic chips (PFC-116) and as solvents in the manufacture of adhesive coatings and inks.

[3] PFC-14 (chemically CF_4) is used as a minor component of a proprietary blend. Its main use is for semiconductor etching.

[4] PFC-51-14 is an inert material, which has little or nil ability to dissolve soils. It can be used as a carrier for other solvents or to dissolve and deposit disk drive lubricants. PFCs are also used to test that sealed components are hermetically sealed.

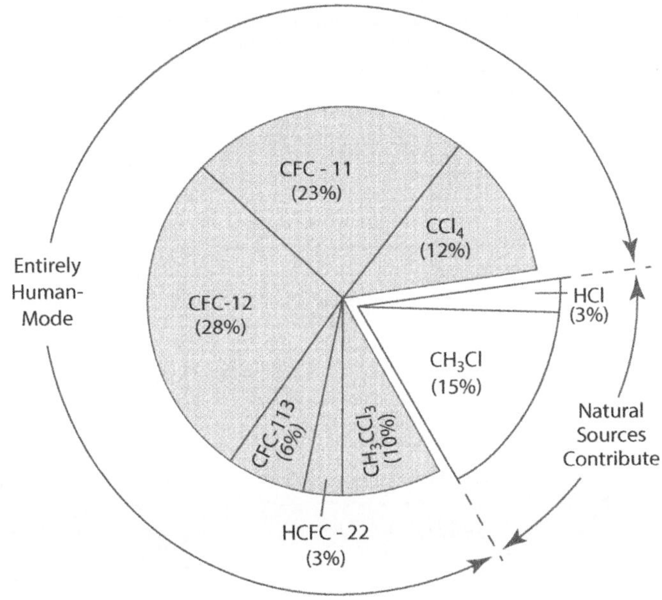

Figure 8.1 Primary sources of stratospheric chlorine in early 1990's [2].

Manufacturers of fluorocarbons went to work quickly to implement the Montreal Protocol. The task to replace CFCs was immense and thus divided into intermediate and final phases. Just imagine the magnitude of work to exchange the gas in every new and old refrigerator and in home and auto air conditioning systems in the world! In the intermediate phase CFCs were replaced by hydrofluorocarbons (HCFCs) to be replaced later by hydrofluorocarbons (HFCs) (Table 8.3). HCFCs are significantly less harmful to ozone because of having hydrogen in their chemical structures. The values of ODP and GWP of commercial fluorocarbons are in the following order: ODP_{CFC}> ODP_{HCFC} >ODP_{HFC}> ODP_{HFO} and GWP_{CFC}> GWP_{HCFC} >GWP_{HFC}> GWP_{HFO}.

Halons were completely banned by Montreal Protocol as of January 1, 2010 because of their large ozone depleting (ODP) potential values (see *Group II* in Table 2.6) [4]. ODP values of Halon's are several times larger than even the worst CFC's. Halon's were the fire extinguishing of choice for cargo compartment of aircrafts because of their extraordinary effectiveness not leaving any residues behind after use.

The worst CFCs (CFC-11, 12, 113, 114 and 115) were phased out by 1996 (see *Group I* in Table 8.4). Phase-out of a few CFCs and Halons took until 2010. CFCs were replaced by the far less active HCFCs in an intermediate step. HCFCs are scheduled for phase-out by 2020 in the developed countries and by 2040 in the developing countries.

In September of 2007, the Parties to the Montreal Protocol agreed to accelerate the phase-out schedule for HCFC's in developed countries. The Parties agreed to reduce HCFC consumption and production between 2010 and 2015 from 65% to 75%, which restricts by approximately 20% the amount of HCFCs that would have been available between 2010 and 2015.

HCFC's were to be phased out by 2020 in the developed countries (non-Article 5) and by 2040 in the developing countries (Article 5). They include R22, R401A, R401B, R402A,

Table 8.3 ODPs and GWPs of the main commercial classes of fluorocarbons [3].

Refrigerant type	Refrigerant name	ODP[1]	GWP[2]
CFC	R11	1.00	4,750
	R12	0.82	10,900
HCFC	R22	0.06	1,810
	R124	0.03	610
HFC	R134a	0	1,430
	R404A	0	3,920
	R407C	0	1,770
	R410A	0	2,090
	R422D	0	2,730
HFO	R1234yf	0	4
Natural refrigerants	R717	0	0
	R744	0	1
	Hydrocarbons e.g. R290	0	<20

[1] Ozone depletion potential, UNEP (2006). R11 = 1
[2] Global warming potential (100year), IPCC 2005, 2007. $CO_2 = 1$

Table 8.4 Ozone depleting potential values of CFC's and Halon's classified according to Montreal Protocol [5].

Group	Substance	Ozone-depleting potential
Group I		
$CFCl_3$	(CFC-11)	1.0
CF_2Cl_2	(CFC-12)	1.0
$C_2F_3Cl_3$	(CFC-113)	0.8
$C_2F_4Cl_2$	(CFC-114)	1.0
C_2F_5Cl	(CFC-115)	0.6
Group II		
CF_2BrCl	(halon-1211)	3.0
CF_3Br	(halon-1301)	10.0
$C_2F_4Br_2$	(halon-2402)	6.0

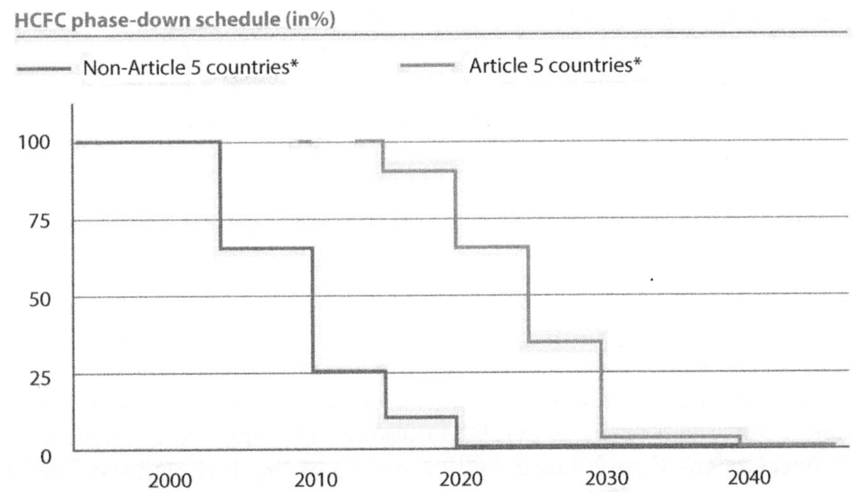

Figure 8.2 Phase-down plan of HCFC for developed countries (non-Article 5) and developing countries (Article 5) [6].

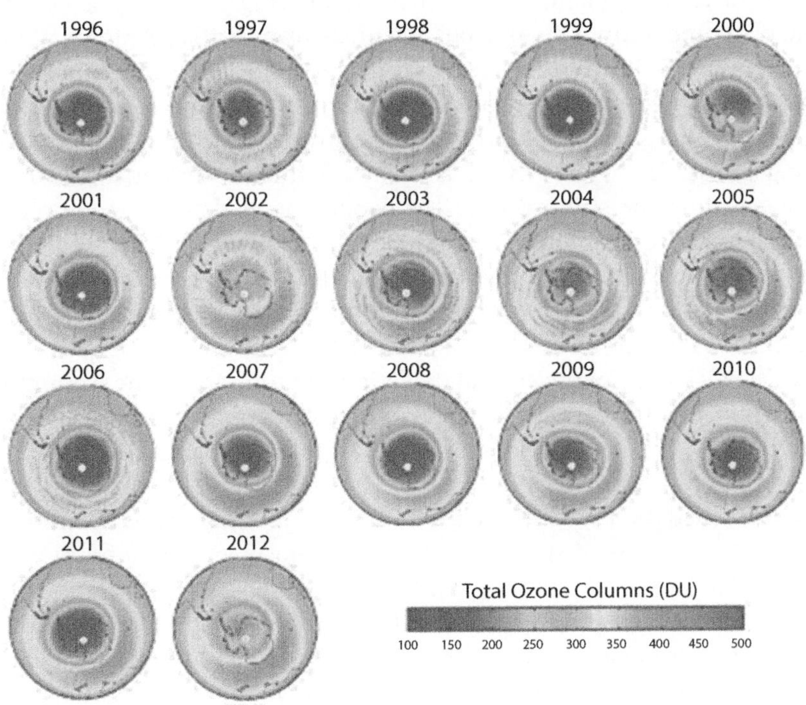

Figure 8.3 Size of the ozone hole over Antarctica during 1996-2012 [7].

R402B, R408A and R409A but R123 has been excluded and its use allowed until 2030 in developed countries. The reduction in consumption (phase-down) is based on a stepwise plan depicted in Figure 8.2. EU regulations reduced use of HCFC's by 15% starting in 2002, use was reduced by 55% in 2003, 70% in 2004, 75% in 2008, and 100% (phase-out) in 2010. HFCs contain hydrogen but no chlorine; thus do not destroy the ozone layer at all.

Ozone depletion is more extreme in Antarctica than at the North Pole because high wind speeds cause a fast-rotating vortex of cold air, leading to extremely low temperatures. Long-term observations reveal that the Ozone Layer has been healing following the international agreements to protect this vital layer of the atmosphere. According to the ozone sensor on Europe's MetOp weather satellites, the hole over Antarctica in 2012 was the smallest in the prior decade as seen in Figure 8.3 [7].

The Ozone Layer is far from full recovery. There are still ozone depleting chemicals, albeit non-fluorocarbons, in commercial use and emitted to the atmosphere. Those emissions have slowed the pace of recovery of the ozone layer. The United States EPA (www.EPA.gov) states: continued declines in ODS emissions are expected to result in an almost complete recovery of the Ozone Layer by the middle of the 21st century. The long time scale for this recovery is due to the slow rate of ODS removal from the atmosphere by natural processes.

8.2 Ozone Depletion

The Ozone Layer is the Earth's natural sunscreen that protects humans from melanoma (an aggressive form of cancer), plants and animals by filtering out harmful UV-B radiation. Its role in the plant and animal life is vital. Scientists from British Antarctic Survey began monitoring ozone in 1957. In 1985 scientists discovered that since mid 1970's ozone values, over Halley and Faraday research stations, in Antarctica, had been steadily dropping when the sun reappeared each spring. Something in the stratosphere (about 20km above Earth) was destroying ozone. The ozone "hole" is really a reduction in concentrations of ozone high above the earth in the stratosphere. The ozone hole is defined geographically as the area wherein the total ozone amount is less than 220 Dobson Units.

Paul Crutzen, Mario Molina and Sherwood Rowland raised concern about the effect of man-made chemicals, especially chlorofluorocarbons, on the Ozone Layer as early as in the 1970's. Their pioneering work was recognized in 1995 by the awarding of the Nobel Prize in Chemistry. The discovery of the annual depletion of ozone above the Antarctica was first announced in a paper by Joe Farman, Brian Gardiner and Jonathan Shanklin, in the journal *Nature* in May 1985 [8]. Later, NASA scientists re-analyzed their satellite data and found that the entire Antarctica was affected.

NASA scientists used their satellite data to confirm that not only was the hole over British research stations but it covered the entire Antarctic continent (Figure 8.4). International efforts by scientists and politicians then led to steps to control the production and use of CFC's and other ozone depleting chemicals. The international efforts culminated in the Montreal Protocol. In comparison (Figure 8.5) the ozone hold reached its smallest recorded size in September 2019 as a result of reduction of ozone depleting compounds in the atmosphere. A contributing factor to the unusually small size was uncommon weather patterns during Sep-Oct 2019.

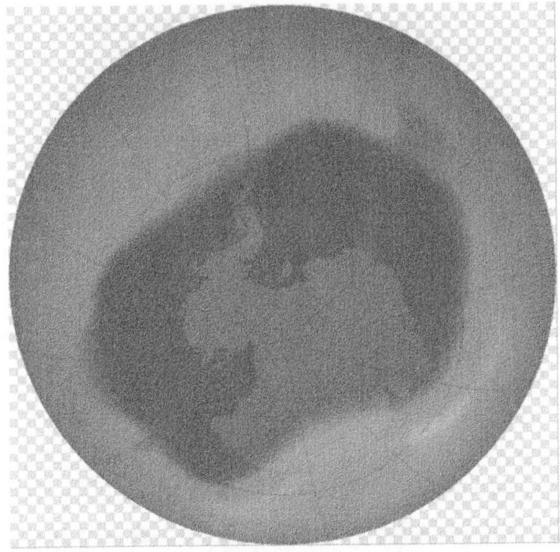

Figure 8.4 Image from the Antarctic Ozone Hole (blue hue) - the average area of the ozone hole was the largest ever observed, at 27.5 million square kilometers between September 21 and 30, 2006 [9].

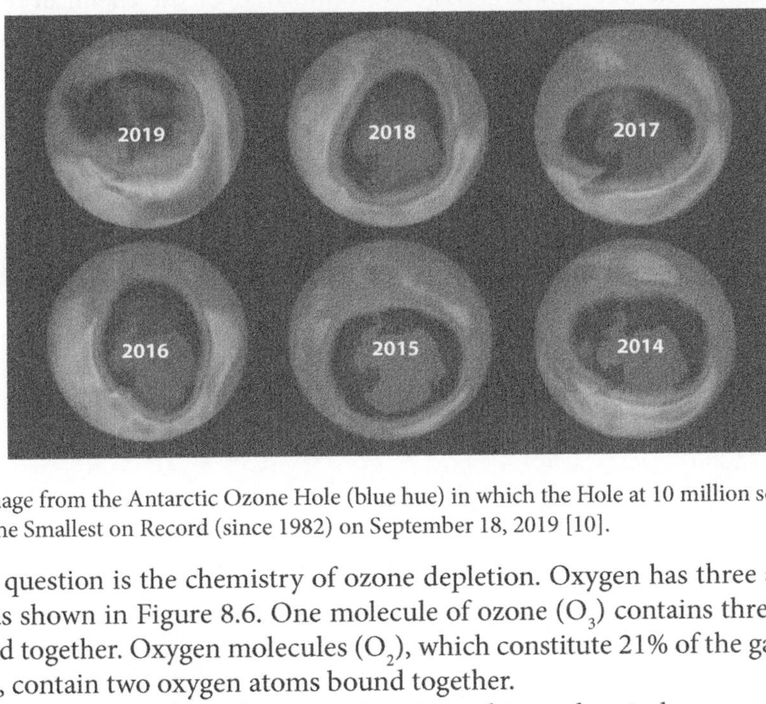

Figure 8.5 Image from the Antarctic Ozone Hole (blue hue) in which the Hole at 10 million square kilometers is the Smallest on Record (since 1982) on September 18, 2019 [10].

The next question is the chemistry of ozone depletion. Oxygen has three allotropes: O, O_2 and O_3 as shown in Figure 8.6. One molecule of ozone (O_3) contains three oxygen (O) atoms bound together. Oxygen molecules (O_2), which constitute 21% of the gases in Earth's atmosphere, contain two oxygen atoms bound together.

Ozone is formed throughout the atmosphere in multistep chemical processes that require sunlight. In the stratosphere, the process begins with an oxygen molecule (O_2) being broken

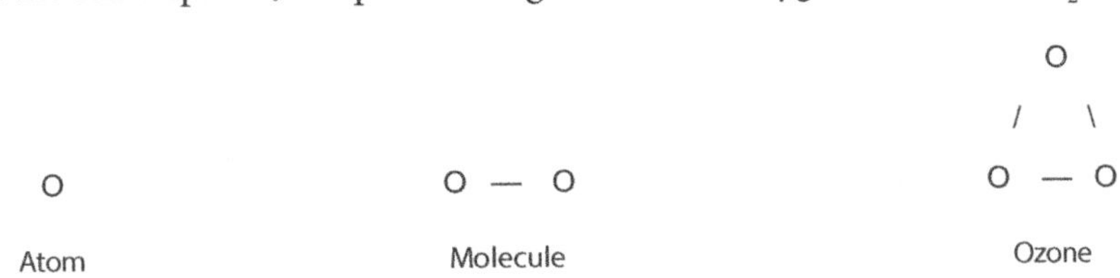

Figure 8.6 Chemical structures of three allotropes of Oxygen.

apart by ultraviolet radiation from the Sun. The combination of an atom (O) and a molecule (O_2) of oxygen produce O_3. In the lower atmosphere (*troposphere*), ozone is formed by different chemical reactions that involve naturally occurring gases and those from pollution sources. The tropospheric ozone does not significantly contribute to the concentration of stratospheric ozone [1, 11]. Actually direct contact with ozone is harmful to plants, animals and humans. Ground level, "bad," ozone forms when nitrogen oxide gases from vehicles and industrial emissions react with volatile organic chemicals (like volatile paint thinners) [12].

In the first step, solar ultraviolet radiation breaks apart one oxygen molecule (O_2) to produce two oxygen atoms (2O) (see Figure 8.7). In the second step, each of these highly reactive atoms combines with an oxygen molecule to produce an ozone molecule (O_3). These reactions occur continually whenever solar ultraviolet radiation is present in the stratosphere. Reactive gases that destroy the ozone are hydrogen and nitrogen oxides and any containing chlorine and bromine.

Ozone is measured in the atmosphere using instruments on the ground, aircraft, high-altitude balloons, and satellites. Some devices measure ozone in sampled air and others measure ozone remotely. Instruments use optical techniques, with the Sun and lasers as light sources; detect the microwave emissions from ozone; or use chemical reactions that are unique to ozone. Regular measurements are made around the globe to monitor total ozone amounts and their variations over time.

Ozone degrades through several steps (Figure 8.8) in the stratosphere starting with the emission of chlorine and bromine source gases emanating primarily from human activities.

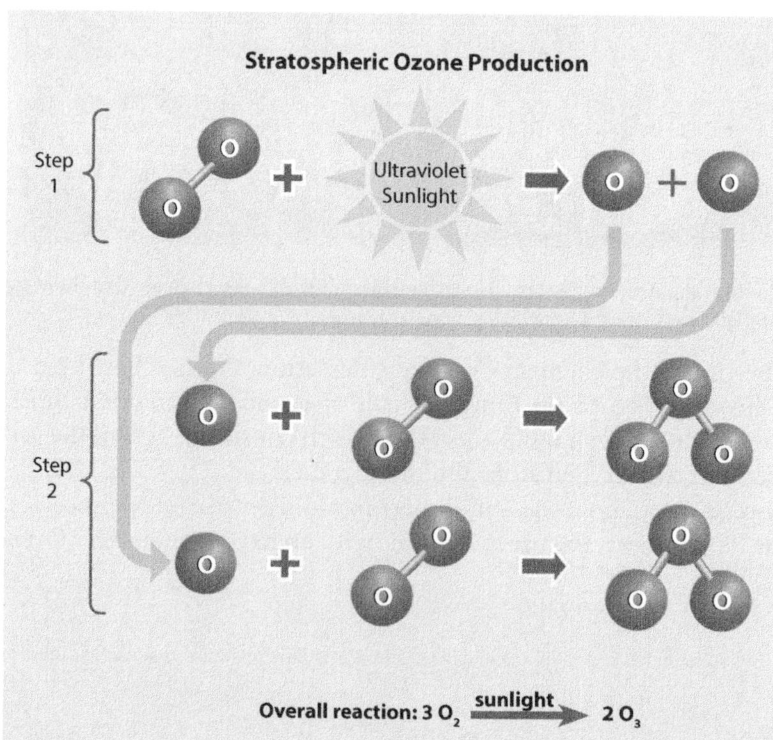

Figure 8.7 O_3 is naturally produced in the stratosphere by a two-step reactive process [11].

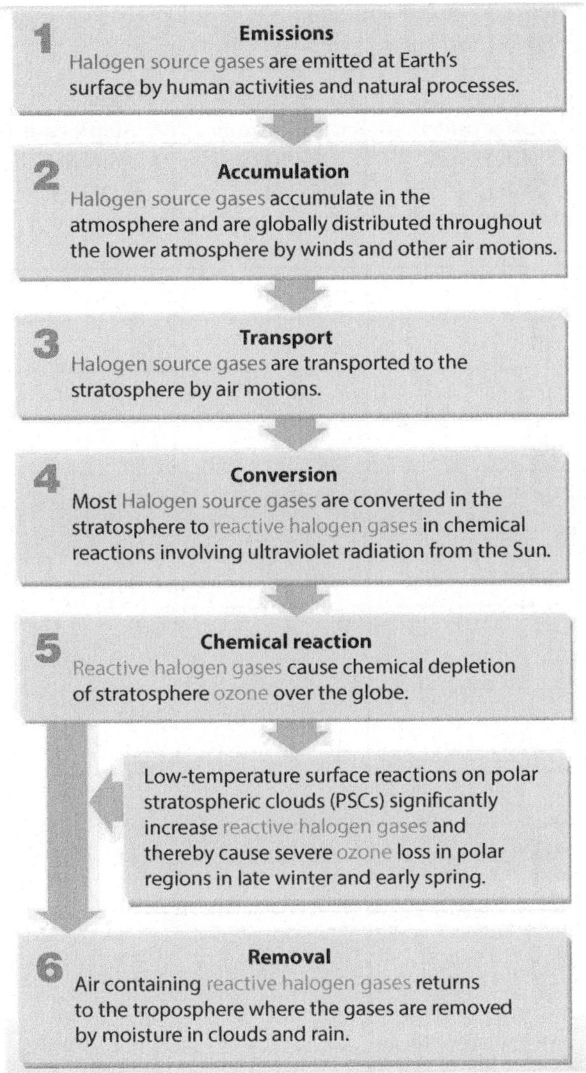

Figure 8.8 Main steps in stratospheric depletion of O_3 [11].

Those gases are called ozone-depleting substances (ODSs). The other steps in the action of ODS include their accumulation, transport, conversion, chemical reaction in stratosphere, and final removal. Ozone depletion only stops when reactive halogen gases are removed by rain and snow in the troposphere and land on the ground [11].

Halogen source gases are compared in their effectiveness to destroy the stratospheric ozone using a factor called Ozone Destruction Potential (ODP) [13]. ODP is calculated relative to CFC-11 as reference substance, which has been assumed to have an ODP value of 1. A gas with a larger ODP destroys more ozone over its atmospheric lifetime than one with a lower ODP value (Table 8.5).

Chlorofluorocarbons (CFC) compounds were proven to be the culprits. They rise over time and reach the stratosphere. During the rise ultraviolet light (UV) those containing chlorine or bromine degrade releasing atomic chlorine or bromine. For example, CFC 11

Table 8.5 Atmospheric lifetimes, global emissions, ODP, and global warming potentials of some halogen source gases and HFC substitutes [11].

Gas	Atmospheric lifetime (years)	Global emissions in 2008 (Kt/yr)[a]	Ozone depletion potential (ODP)[c]	Global warming potential (GWP)[c]
Halogen source gases				
Chlorine gases				
CFC-11	45	52–91	1	4750
CFC-12	100	41–99	0.82	10900
CFC-113	85	3–8	0.85	6130
Carbon tetrachloride (CCl_4)	26	40–80	0.82	1400
HCFCs	1–17	385–481	0.01–0.12	77–2220
Methyl chloroform (CH_3CCl_3)	5	Less than 10	0.16	146
Methyl chloride (CH_3Cl)	1	3600–4600	0.02	13
Bromine gases				
Halon-1301	65	1-3	15.9	7140
Halon-1211	16	4–7	7.9	1890
Methyl bromide (CH_3Br)	0.8	110–150	0.66	5
Very short-lived gases (e.g., $CHBr_3$)	Less than 0.5	[b]	[b] very low	[b] very low
Hydrofluorocarbons (HFCs)				
HFC-134a	13.4	149 ± 27	0	1370
HFC-23	222	12	0	14200
HFC-143a	47.1	17	0	4180
HFC-125	28.2	22	0	3420
HFC-152a	1.5	50	0	133
HFC-32	5.2	8.9	0	716

[a] Includes both human activities (production and banks) and natural sources. Emissions are in units of kilo tons per year (1 kilo ton = 1000 metric tons = 1 giga gram = 109 grams).

[b] Estimates are very uncertain for most species.

[c] 100-year GWPs.

($CFCl_3$) degrades by UV light exposure yielding Cl atoms that participate in the cycle shown in Figure 8.9. The reaction paths for bromine is similar to chlorine except that bromine is 50 times more potent than chlorine as an ODS [14].

In *Cycle 1* chlorine reacts with ozone and produces ClO, which is quite reactive. ClO group combines with O to form Cl atom (Figure 8.9). Then Cl reacts with ozone and re-forms ClO leading to consumption of ozone in the process; chlorine is produced and the cycle starts all over. The chlorine atom acts as a catalyst destroying ozone. Until the chlorine atom reacts with another atom, or is removed in another way, a single atom of Cl can continue to consume ozone. Cycle 1 is most important in the stratosphere at tropical and middle latitudes where solar ultraviolet radiation is most intense [16].

Major destruction of ozone takes place in the Polar Regions because ClO concentration is quite high (Figure 8.10). The cycles initiated by the reaction of ClO with another ClO (*Cycle 2*) or the reaction of ClO with BrO (*Cycle 3*) efficiently destroy ozone. The net reaction in

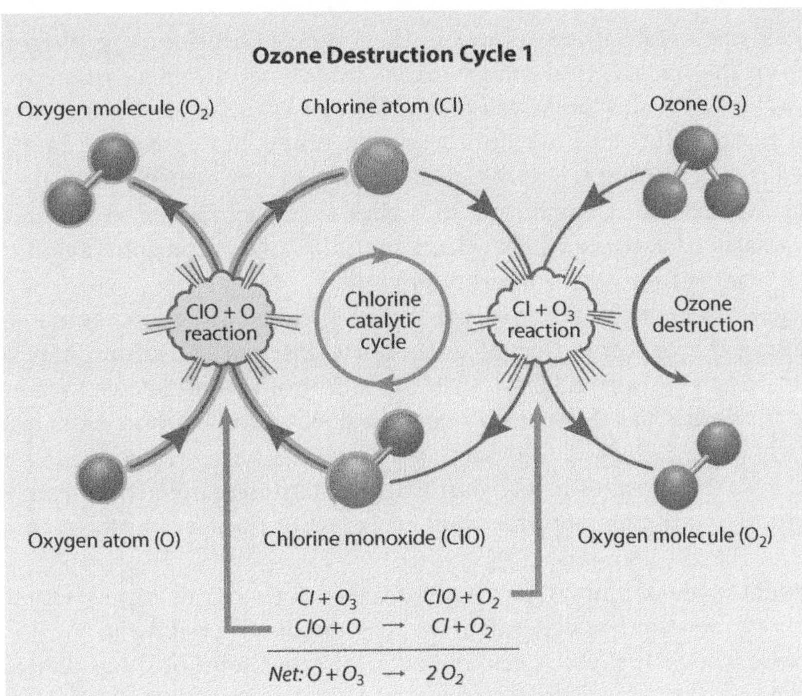

Figure 8.9 Ozone destruction in tropical and middle latitude (*Cycle 1*) [15].

Ozone Destruction Cycles in Polar Regions

Cycle 2

$$ClO + ClO \rightarrow (ClO)_2$$
$$(ClO)_2 + \text{sunlight} \rightarrow ClOO + Cl$$
$$ClOO \rightarrow Cl + O_2$$
$$2(Cl + O_3 \rightarrow ClO + O_2)$$
$$\text{Net: } 2O_3 \rightarrow 3O_2$$

Cycle 3

$$ClO + BrO \rightarrow Cl + Br + O_2$$
or
$$\left(\begin{array}{l} ClO + BrO \rightarrow BrCl + O_2 \\ BrCl + \text{sunlight} \rightarrow Cl + Br \end{array} \right)$$
$$Cl + O_3 \rightarrow ClO + O_2$$
$$Br + O_3 \rightarrow BrO + O_2$$
$$\text{Net: } 2O_3 \rightarrow 3O_2$$

Figure 8.10 Ozone destruction cycles in polar regions (*Cycles 2* and *3*) [11].

both cases is two ozone molecules forming three oxygen molecules. The reaction of ClO with BrO has two pathways to form the Cl and Br product gases. Ozone destruction in Cycles 2 and 3 are catalytic, as in *Cycle 1*, because chlorine and bromine gases react and are re-formed each time the reaction cycle is completed. Visible sunlight is required to complete each cycle and to help form and maintain ClO concentrations. During polar night and other periods of darkness, ozone cannot be destroyed by these reactions [11].

8.3 Global Warming

Global warming is the unusually rapid increase in Earth's average surface temperature over the past century primarily due to the greenhouse gases released as people burn fossil fuels. The global average surface temperature rose 0.6 to 0.9°C between 1906 and 2005, and the rate of temperature increase has nearly doubled in the last 50 years. Temperatures are certain to go up further.

As the rocks, the air, and the seas warm, they radiate "heat" energy (thermal infrared radiation). From the surface, this energy travels into the atmosphere where much of it is absorbed by water vapor and long-lived greenhouse gases including carbon dioxide, methane and other gaseous chemicals like fluorocarbons. When they absorb the energy radiating from Earth's surface greenhouse gas molecules turn into tiny heaters— like the bricks in a fireplace, they radiate heat even after the fire goes out. They radiate in all directions. The energy that radiates back toward Earth heats both the lower atmosphere and the surface, enhancing the heating they get from direct sunlight.

What has scientists concerned now is that over the past 250 years, humans have been artificially raising the concentration of greenhouse gases in the atmosphere at an ever-increasing rate, mostly by burning fossil fuels, but also from cutting down carbon-absorbing forests. Since the Industrial Revolution began in about 1750, carbon dioxide levels have increased nearly 38 percent as of 2009 and methane levels have increased 148 percent [17].

To fully address the threat of global warming, we must demand action from our elected leaders to support and implement a comprehensive set of climate solutions [18]:

1. Expand the use of renewable energy and transform our energy system to one that is cleaner and less dependent on coal and other fossil fuels.
2. Increase vehicle fuel efficiency and support other solutions that reduce U.S. oil use.
3. Place limits on the amount of carbon that polluters are allowed to emit.
4. Build a clean energy economy by investing in efficient energy technologies, industries, and approaches.
5. Reduce tropical deforestation and its associated global warming emissions.

The Global Warming Potential (GWP) was developed as a factor to allow comparisons of the global warming impacts of different gases. Specifically, it is a measure of how much energy the emissions of one ton of a gas will absorb over a given period of time, relative to the emissions of one ton of carbon dioxide. Higher values of GWP indicate that a given gas warms the Earth compared to carbon dioxide over that time period. The time period usually used for GWPs is 100 years. GWP provides a uniform measure that allows analysts to add up emissions estimates of different gases (e.g., to compile a national greenhouse gas inventory), and allows policy-makers to compare emissions-reductions opportunities across sectors and gases.

1. Carbon dioxide (CO_2), by definition, has a GWP of 1 regardless of the time period used, because it is the gas being used as the reference. Carbon dioxide emissions cause increases in atmospheric concentrations of carbon dioxide that will last thousands of years.
2. Methane (CH_4) is estimated to have a GWP of 28-36 over 100 years. CH_4 emitted today lasts about a decade on average, significantly less time than CO_2. But CH_4 also absorbs much more energy than CO_2. The net effect of the shorter lifetime and higher energy absorption is reflected in the GWP.
3. Nitrous Oxide (N_2O) has a GWP 265-298 times that of CO_2 for a 100-year timescale.
4. Chlorofluorocarbons (CFCs), hydrofluorocarbons (HFCs), hydrochlorofluorocarbons (HCFCs), perfluorocarbons (PFCs), and sulfur hexafluoride (SF_6) are, in addition to ODS, also called high-GWP gases because. The GWPs for these gases and others can be seen in Table 8.3.

Table 8.6 lists the GWP and atmospheric lifetime of fluorocarbons and several other gases. Figure 8.11 shows the ozone depletion and global warming potential of some of the

Table 8.6 Global warming potentials (GWP) for some representative compounds [19–24].

Compound	GWP (CO_2 equivalents)	Residence time (years)
CO_2	1 (defined)	
CH_4	23	10
HCF_3 (HFC-23)	12,000–13,000	243
CH_2CF_2 (HFC-32)	550–710	5.6
CHF_2CF_3 (HFC-125)	3400–3600	32.6
CH_2FCF_3 (HFC-134a)	1300	13.6
$CF_3CH_2CF_3$ (HFC-236fa)	6300–8400	226
CF_4	5700	50,000
C_2F_6	9200	10,000
C_6F_{14}	9000	3200
CFC-11	4500	45
CFC-12	10,600	100
NF_3	10,800	740
SF_6	23,900	3200

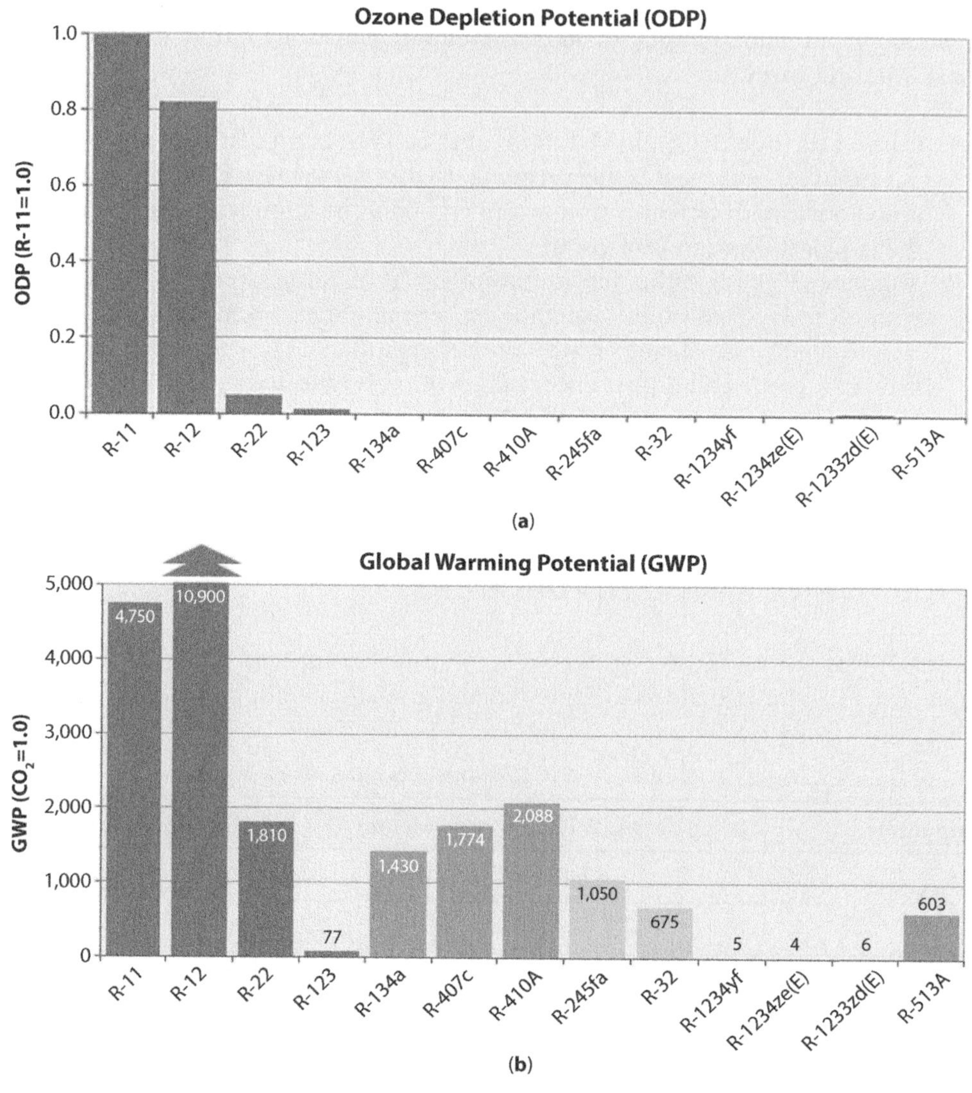

Figure 8.11 Comparison of (a) ODP and (b) GWP of various fluorocarbon refrigerants [25].

fluorocarbon refrigerants clearly indicating there are candidate gases with no impact on either ODP and GWP. A range of values is given where sources differ. Some of the gases last for thousands of years, most of which have large global warming impacts. Sulfur hexafluoride, used for blanketing during magnesium processing, has an astounding GWP value of 23,900.

An important factor in the selection of refrigerants is the lifetime warming impact of a system. The term *total equivalent warming impact* (TEWI) is defined to describe the overall impact. It includes the impacts of refrigerant leakage, refrigerant recovery losses and energy consumption. TEWI should be calculated when comparing system design options for specific applications [26]. TEWI may be calculated from the following formulas:

TEWI = Total equivalent warming impact
= (Direct effect) + (Indirect effect)
= (Leakage + Recovery losses) + (Power generation emissions)
$= \{(GWP \times L \times n) + [GWP \times m(1 - a)]\} + (n \times E \times b]$
GWP = Global warming potential
L = Leakage rate per year (kg)
n = Number of years of operation
m = Refrigerant charge (kg)
a = Recycling factor
E = Energy consumption per year (kWh)
b = CO_2 Emission factor, kg CO_2 emitted per kWh of electricity supplied

Comprehensive methods details with calculation examples are discussed in the Methods of calculating Total Equivalent Warming Impact and other references [26, 27].

There are regulatory deadlines that have already taken effect for using lower GWP of fluorocarbons. For instance, European Union's F-gas regulations [28, 29] specified that starting in 2011 new models and in 2017 new vehicles must have air conditioning must contain gases having GWPs < 150. Among possible candidates that meet this criterion are HFC-152a (Table 8.3), CO_2, and HFO-1234yf. HFC-152a, if used, would likely be implemented in an indirect system (secondary loop) because of its flammability. CO_2, if used, would be implemented in a transcritical cycle and would require significant modifications to automotive air-conditioning systems currently in use. HFO-1234yf has a GWP of 4, [30] and among the fluorinated propene isomers that has been commercialized [31].

8.3.1 Paris Agreement

On 12 December 2015, Parties to the United Nations Framework Convention on Climate Change (UNFCCC) reached a landmark agreement to combat climate change and to accelerate and intensify the actions and investments needed for a sustainable low carbon future. The Paris Agreement builds upon the UNFCCC and – for the first time – brings all nations into a common cause to undertake ambitious efforts to combat climate change and adapt to its effects, with enhanced support to assist developing countries to do so. As such, it charts a new course in the global climate effort. Article 2 of the Agreement, reprinted below, sets forth the objectives of this pact in specific terms [32]. Readers can find the entire text of Paris Agreement in Appendix III.

"Article 2

1. This Agreement, in enhancing the implementation of the Convention, including its objective, aims to strengthen the global response to the threat of climate change, in the context of sustainable development and efforts to eradicate poverty, including by:
   (a) Holding the increase in the global average temperature to well below 2°C above pre-industrial levels and pursuing efforts to limit the temperature increase to 1.5°C above pre-industrial levels, recognizing that this would significantly reduce the risks and impacts of climate change;

(b) Increasing the ability to adapt to the adverse impacts of climate change and foster climate resilience and low greenhouse gas emissions development, in a manner that does not threaten food production; and
(c) Making finance flows consistent with a pathway towards low greenhouse gas emissions and climate-resilient development.

2. This Agreement will be implemented to reflect equity and the principle of common but differentiated responsibilities and respective capabilities, in the light of different national circumstances."

The Paris Agreement requires all Parties to put forward their best efforts through nationally determined contributions (NDCs) and to strengthen these efforts in the years ahead. This includes requirements that all Parties report regularly on their emissions and on their implementation efforts. The Paris Agreement entered into force on 4 November 2016, thirty days after the date on which at least 55 Parties to the Convention accounting in total for at least an estimated 55 % of the total global greenhouse gas emissions have deposited their instruments of ratification, acceptance, approval or accession with the Depositary.

On May 31, 2017 the United States Government formally withdrew from the Paris Agreement. The impact of the Government withdraw from the Agreement is mitigated by the reaction of the private companies organizations which are among the main consumers of energy and emitters of global warming chemicals. Hundreds of U.S. companies – large and small – have declared their support for continuing U.S. leadership on climate change and the Paris Agreement. Those companies recognize that proactively addressing climate change will be a driver of innovation, economic growth and prosperity.

Indeed, the renewable energy sector in the United States now employs more workers than the fossil fuel industry, and the rate of job growth in the renewables sector is four times that of the rest of the U.S. economy. Maintaining U.S. leadership on climate is good for jobs, growth and competitiveness. State and local governments increasingly recognize these benefits and continue to support and invest in a clean energy future. Withdrawing from the Paris Agreement sends precisely the wrong signals to U.S. companies and places the United States at a competitive disadvantage [33].

8.4 Phase Out of Old Fluorocarbon Gases

A breakthrough took place among the signatories of the Montreal Protocol in Kigali, Rwanda [34]. A global agreement among 200 countries was reached to phase down the consumption and production of hydrofluorocarbons (HFC) by the middle of the 21st Century. This commitment requires the international community to take practical actions to achieve the goals of the 2015 Paris Agreement. That Agreement would limit global warming to 2°C, and aim for the more ambitious target of 1.5°C. Importantly, China, the biggest consumer of HFC's, agreed to an early phase down; India (another big consumer and producer) decided to come down from its earlier position of 2031 to 2024-2026.

After seven years of negotiations, Parties to the Montreal Protocol arrived at an agreement, under which developed countries (US, Japan, Europe) will begin phasing down HFC's by 2019, while the developing countries, including India, Pakistan, Iran and Iraq,

Table 8.7 Schedule of phase-out of HCFC's according to the US EPA [35].

		2020	**2021**	**2022**	**2023**	**2024**	**2025**	**2026**	**2027**	**2028**	**2029**	**2030**
HCFC-123	Consumption	650	650	650	570	490	410	330	250	170	90	0
	Production	0	0	0	0	0	0	0	0	0	0	0
HCFC-124	Consumption	200	200	200	175	150	125	100	75	50	25	0
	Production	200	200	200	175	150	125	100	75	50	25	0

will phase out at a baseline of 2024-2026 and a freeze date of 2028. HFC's are widely used in refrigerators, air-conditioners and aerosol sprays. The primary candidate for replacing HFC's are hydrofluoroolefins (HFOs). There will also be an opportunity for natural refrigerants (ammonia, carbon dioxide, water and hydrocarbons) to replace the HFC's.

8.4.1 Status of Phase Out of HCFCs

United States Environmental Protection Agency (US EPA) issued HCFC allowances for 2020 through 2029 for HCFC-123 and HCFC-124, leading up to a complete phaseout of all HCFCs in 2030 as seen in Tables 8.7 and 8.8.

Starting in 2020, production and import of all HCFCs in the United States will be less than 0.5 percent of initial amounts. Newly-produced or imported HCFCs may only be used to service air-conditioning (AC), refrigeration, and fire suppression equipment that was manufactured before January 1, 2020 [35].

Starting on January 1, 2020, U.S. production and import of HCFC-22 ended. The EPA does not require the premature retirement of equipment and there is no ban on the continued use of existing HCFC-22 systems.

Recovered and reclaimed (e.g., recycled to required purity standards) or stockpiled HCFC-22 may continue to be used in existing systems for as long as needed.

Table 8.8 U.S. action to meet the Montreal protocol phaseout schedule [36].

Year to be Implemented	Implementation of HCFC phaseout through clean air act regulations	Year to be Implemented	Percent reduction in HCFC consumption and production from baseline
2003	No production or import of HCFC-141b	2004	35.0%
2010	No production or import of HCFC-142b and HCFC-22, except for use in equipment manufactured before January 1, 2010	2010	75.0%
2015	No production or import of any other HCFCs, except as refrigerants in equipment manufactured before January 1, 2020	2015	90.0%
2020	No production or import of HCFC-142b and HCFC-22	2020	99.5%
2030	No production or import of any HCFs	2030	100.0%

HCFC-123:HCFC-123 may be imported for servicing AC, refrigeration, and fire suppression equipment. The amounts of HCFC-123 that may be imported will decrease over time.

HCFC-124: HCFC-124: HCFC-124 may be produced and imported for servicing AC and refrigeration equipment. The amounts of HCFC-124 that may be produced and imported will decrease over time.

8.5 Summary

There has been important and significant achievements in reducing ozone depletion in the last several decades. In spite of the development of fluorocarbons with low or zero ODP and very low GWP, hydrofluorocarbons will remain in use for the next few decades. The current estimate of the completion of phase-out of HFCs is 2050. HFCs do not deplete the ozone layer because of the presence of H in their molecules. They have, however, substantial global warming potentials leading to global warming. Release of HCFs by discharge and emissions of refrigeration systems including air cooling applications contribute to that warming.

As the world warms demand for cooling is growing. Over the next three decades, it's estimated that ten air conditioners will be sold every second. Without action, these air conditioners will dramatically heat up the planet. If not properly disposed of, the gases used in the cooling equipment can dramatically exacerbate global warming, triggering a brutal cycle that increases the demand for cooling, in turn, further warming the planet. Moreover, cooling requires immense amounts of energy, gobbling ten percent of global electricity today. This electricity is often generated by coal fired power plants, major emitters of carbon and toxic pollutants. If drastic measures are not taken, emissions from cooling could increase by 90 percent by 2050.

Proper management of the entire HFC lifecycle is critical because HFCs leak out at every stage of a cooling equipment's lifespan: in factories when it's built, every day when it's in use, when it's being repaired, and at the end of its life when it is destroyed or abandoned in landfills. At a recent (2020) United Nations Climate Change meeting in Madrid, Spain, Japan's Minister of the Environment launched the Fluorocarbons Life-Cycle Management Initiative to deal with the issue. The country hopes other countries also acknowledge that HFCs are not only a great liability, they're a great opportunity for stemming catastrophic warming worthy of taking action [37].

References

1. Ashford, P. *et al.*, Chapter 7, Emissions of fluorinated substitutes for ozone depleting substance, in: *Processess and product use, vol. 3, IPCC guidelines for National Greenhouse Gas Inventories*, 2006, www.ipcc-nggip,iges.or.jp.
2. National Oceanic and Atmospheric Administration Earth System Research Laboratory, Chemical Sciences Division, www.esrl.noaa.gov/csd/assessments/ozone/1998/faq3.html, May 2016.
3. *Managing refrigerants the responsible way*, Linde Corp., August 2019, www.linde-gas.com/refrigerants.
4. National Fire Protection Association, www.nfpa.org, Aug 2019.

5. Annex A, Montreal Protocol, http://ozone.unep.org, May 2016.
6. The Linde Group, HCFC phase-out has started, document No. 71591 0713 – 1.0 LCS, www.linde-gas.com, May 2016.
7. European Space Agency, www.esa.int, May 2020.
8. Farman, J.C., Gardiner, B.G., Shanklin, J.D., Large losses of total ozone in Antarctica reveal seasonal ClO_x/NO_x interaction. *Nature*, 315, 6016, 207–210, 1985.
9. Image date, NASA, www.nasa.gov/vision/earth/lookingatearth/ozone_record.html.
10. NASA, www.nasa.gov.
11. Fahey, D.W. and Hegglin, M.I., Twenty Questions and Answers About the Ozone Layer: 2010 Update, Scientific Assessment of Ozone Depletion, World Meteorological Organ Global Ozone Res and Monitoring Project - Rep No. 52, www.esrl.noaa.gov, 2010.
12. NASA, Goddard Space Flight Center, http://ozonewatch.gsfc.nasa.gov/facts/SH.html.
13. Wuebbles, D.J., Chlorocarbon emission scenarios: potential impact on stratospheric ozone. *J. Geophys. Res.*, 88, 1433–1443, 1983.
14. Ozone Layer-3: Mechanisms of Ozone Production and Destruction, Department of Earth & Climate Sciences, SF State University, May 2016.
15. Fahey, D.W. *et al.*, *Twenty Questions about the Ozone Layer: Scientific Assessment of Ozone Depletion*, WMO, Geneva, Switzerland, www.esrl.noaa.gov, 2002.
16. Rowland, F.S., Blake, D.R., Tyler, S.C., Makide, Y., Increasing Concentrations of Perhalocarbons, Methylchloroform and Methane in the Atmosphere. *Stud. Environ. Sci.*, 26, 305–339, 1986.
17. NASA Earth Observatory, http://earthobservatory.nasa.gov, May 2016.
18. *Global Warming Solutions: Reduce Emissions*, Union of Concerned Scientists, www.ucsusa.org, May 2016.
19. Tavener, S.J. and Clark, J.H., Chapter 5: Friend or foe? A green chemist's perspective, in: *Flourine and the environment*. A, Tressaud (ed.), vol. 3, Elsevier, 2006.
20. McCulloch, A., *J. Fluorine Chem.*, 123, 21, 2003.
21. Lucas, J., *J. Fluorine Chem.*, 41, 1, 1988.
22. Neilson, A.H. (Ed.), *The Hdbk of Environmental Chemistry*, vol. 3, Part N: Organofluorines, Springer, Berlin, 2002.
23. Friend, J.P., Natural Chlorine and Fluorine in the Atmosphere, Water and Precipitation, Scientific Assessment of Stratospheric Ozone, in: *World Meteorological Organ Global Ozone Project Rep*, No. 20, Vol. 2, p. 429, 1989.
24. International Panel on Climate Change, Chap 4, in: *3rd Assessment Report – Climate Change*, Cambridge Univ Press, Cambridge, 2001.
25. Riemer, J.E., *Refrigerants Transitions in the HVAC Industry*, International District Energy Association, www.districtenergy.org, June 1, 2016.
26. *Refrigeration, Air Conditioning and Heat Pumps*, Elsevier, doi: http://dx.doi.org/10.1016/B978-0-08-100647-4.00003-6, 2016.
27. *Methods of calculating Total Equivalent Warming Impact*, Australian Inst of Refrigeration, Air Conditioning and Heating, AIRAH, www.airah.org.au, 2012.
28. Regulation (EC) No 842/2006 of The European Parliament and of the Council of 17 May 2006 on certain fluorinated greenhouse gases. *Off. J. Eur. Union*, 2006.
29. Directive 2006/40/EC of The European Parliament and of the Council of 17 May 2006 relating to emissions from air-conditioning systems in motor vehicles and amending Council Directive 70/156/EC. *Off. J. Eur. Union*, 2006.
30. Nielsen, O.J. *et al.*, Atmospheric chemistry of $CF_3CF{=}CH_2$: Kinetics and mechanisms of gas-phase reactions with Cl atoms, OH radicals, and O_3. *Chem. Phys. Lett.*, 439, 18–22, 2007.
31. Brown, J.S., HFOs, New Low Global Warming Potential Refrigerants. *ASHRAE J.*, www.ashrae.org, August 2009.

32. What is Paris Agreement? https://unfccc.int, May 2020.
33. *U.S. Withdrawal from Paris Agreement is Short-sighted, Mark Tercek*, The Nature Conservancy, www.Nature,org, May 31, 2017.
34. Ebnesajjad, S., *Global Warming, Fluorocarbons, Montreal Protocol 2016 – all in Kigali, Rwanda*, Elsevier, https://chemical-materials.elsevier.com/new-materials-applications/global-warming-fluorocarbons-montreal-protocol-2016-kigali-rwanda/, Nov 1, 2016.
35. The Ozone-Depleting Substances Phaseout: 2020-2030, The 2020 HCFC Allocation and Other Updates, www.epa.gov/ods-phaseout, December 2019.
36. *Phaseout of Class II Ozone-Depleting Substances*, United States Environmental Protection Agency, www.epa.gov/ods-phaseout, May 2020.
37. *Japan launches Fluorocarbons Initiative*, Climate Clean Air Initiative, United Nations Environment, https://ccacoalition.org, Jan 22, 2020.

9
Fluorocarbon Blends

An important question is why are fluorocarbons blended? Blends refer to mixture of two or more single component fluorocarbon gases. They are often formulated to approach or match the performance and properties of an existing refrigerant [1]. In practice the blends rarely ever match the old refrigerant. For example, it has not been possible to find an exact single fluorocarbon gas to substitute for HFC-134a in many applications. HFCs were developed to eliminate the ozone depletion harm of the previous generations of refrigerants. They have an ozone depletion potential (ODP) but fairly large global warming potential (GWP). Here the emphasis is on the word "substitute" which refers to the use of new refrigerant in the existing systems.

The fourth generation of refrigerants are hydrofluoroolefins (HFOs) have ODP values of zero and very low GWP's (0 to <10). They do not match the properties and performance of HFCs in many applications. Blending has been used to create refrigerants with acceptable properties and performance characteristics to allow substitution. The mixtures are compromises in that ODP is zero but GWP is significantly higher than HFOs' though lower than HFCs'.

The blended gases are consumed in large in refrigeration both in existing and new systems examples of which can be found in Table 9.1. The formulation of mixed gases not only enhances performance but also reduce flammability and cost of the gas over those of HFOs. The outcome is equivalent performance or at the minimum acceptable performance in the refrigeration systems. Typically, the blends consist of one or more hydrofluorocarbons (HFCs) and hydrofluoroolefins (HFOs). The resulting blend must meet several criteria:

1. Is not toxic in small dozes.
2. Lowers the flammability of HFOs.
3. Is stable under system operations conditions.
4. Is self-lubricating or compatible with lubricants.
5. Has equivalent refrigeration performance in existing systems
6. Reduces the global warming potential (GWP) of HFCs whilst retaining zero ozone depletion potential.
7. Reduces cost – HFOs are typically five to six times more expensive than HFCs.

Chapter 10 presents examples of blends that are Acceptable or Unacceptable under Section 612 of the Clean Air Act, EPA's Significant New Alternatives Policy (SNAP) within a comparative risk framework. The Acceptable blends may be substituted for old fluorocarbons in industrial sectors, updated as of July 31, 2020.

Sina Ebnesajjad, PhD. Concise Handbook of Fluorocarbon Gases: Applications in Refrigeration and Other Industries, (241–266) © 2021 Scrivener Publishing LLC

Table 9.1 Composition (%) of substitute refrigerant blends [2].

ASHRAE Number	HCFCs			HFCs							HCs						HFOs		PFCs	
	22	124	142b	23	32	125	134a	143a	152a	227ea	Ethane (R-170)	Propane (R-290)	n-Butane (R-600)	Isobutane (R-600a)	Propylene (R-1270)	Isopentane	HFO-1234yf	HFO-1234ze(E)	PFC-116 (perfluoro-ethane)	PFC-218 (perfluoro-Propane)
R-401A	53	34							13											
R-401B	61	28							11											
R-401C	33	52							15											
R-402A	38					60						2								
R-402B	60					38						2								
R-403B	56											5								39
R-404A						44	4	52												
R-406A	55		41											4						
R-407A					20	40	40													
R-407B					10	70	20													
R-407C					23	25	52													
R-407D					15	15	70													
R-407F					30	30	40													
R-408A	47					7		46												
R-409A	60	25	15																	
R-410A					50	50														
R-410B					45	55														

(Continued)

Table 9.1 Composition (%) of substitute refrigerant blends [2]. (*Continued*)

ASHRAE Number	HCFCs			HFCs							HCs						HFOs		PFCs	
	22	124	142b	23	32	125	134a	143a	152a	227ea	Ethane (R-170)	Propane (R-290)	n-Butane (R-600)	Isobutane (R-600a)	Propylene (R-1270)	Isopentane	HFO-1234yf	HFO-1234ze(E)	PFC-116 (perfluoro-ethane)	PFC-218 (perfluoro-Propane)
R-411A	87.5								11						1.5					
R-411B	94								3						3					
R-414A	51	28.5	16.5											4						
R-414B	50	39	9.5											1.5						
R-417A						46.6	50						3.4							
R-420A			12				88													
R-421A						58	42													
R-421B						85	15													
R-422A						85.1	11.5							3.4						
R-422B						55	42							3						
R-422C						82	15							3						
R-422D						65.1	31.5							3.4						
R-424A						50.5	47						1	0.9		0.6				
R-426A						5.1	93						1.3	0.6						
R-427A					15	25	50	10												
R-428A						77.5		20				0.6		1.9						
R-434A						63.2	16	18						2.8						

(*Continued*)

Table 9.1 Composition (%) of substitute refrigerant blends [2]. (*Continued*)

ASHRAE Number	HCFCs			HFCs							HCs						HFOs		PFCs	
	22	124	142b	23	32	125	134a	143a	152a	227ea	Ethane (R-170)	Propane (R-290)	n-Butane (R-600)	Isobutane (R-600a)	Propylene (R-1270)	Isopentane	HFO-1234yf	HFO-1234ze(E)	PFC-116 (perfluoro-ethane)	PFC-218 (perfluoro-Propane)
R-438A					8.5	45	44.2						1.7			0.6				
R-441A											3.1	54.8	35.1	5						
R-442A					31	31	30		3	5										
R-448A					26	26	21										20	7		
R-449A					24.3	24.7	25.7										25.3			
R-450A							42											58		
R-507A						50		50												
R-508A				39															61	
R-508B				46															54	
R-513A							44										56			
			19				79													
			20				80													
		39					59						2							

9.1 General Blend Characteristics

Blends are composed of two or more individual refrigerants. In a blend two scenarios are possible depending on the strength of the molecules' attractive forces to on another. The first case is called *Azeotropic* in which a blend behaves as if the mixture were a single substance refrigerant. An azeotrope blend has unusual and surprising properties. Azeotropes exhibit a single boiling point, strictly speaking at one particular pressure, but nevertheless they may be treated as a single substance [3].

Many of the HFC and HFO refrigerants are mixtures or blends of two or more individual chemicals. Increasingly blends are being developed to give the best compromise in performance, GWP, and flammability for various applications. Mixtures can be azeotropes, near azeotropes or zeotropes. Azeotropic mixtures are designated a code in the 500 series. The first azeotropic refrigerant was a chlorofluorocarbon (CFC) that is R502, indicating the use of refrigerant blends is not new [4].

The third type of blends are called *Zeotropic* in that they act just like a mixture of the individual compounds. The zeotropes' properties follow the rules of the combinations of pure components' properties. A zeotropic refrigerant is one whose boiling point varies throughout the constant pressure boiling process and varying evaporating and condensing temperatures exist in the latent heat of vaporization phase (Figure 9.1).

Examples of Azeotropic and Zeotropic refrigerant blends and their properties have been provided in the present Chapter. Air Conditioning, Heating, and Refrigeration Institute, AHRI, provides complete lists of single components and blends in its website [6].

9.1.1 Azeotropic

Even though azeotropic mixtures contain two or more refrigerants, at a certain pressure an azeotropic mixture evaporates and condenses at a constant temperature. Azeotropic mixtures behave like pure refrigerants for all practical purposes. Figure 9.1 shows that the temperature is constant in the liquid-vapor mixture region for a given pressure. Tables 9.2 and 9.3 show the composition and properties of azeotropic refrigerant blends specified by Air Conditioning, Heating, and Refrigeration Institute [7].

9.1.2 Zeotropic Blends

A zeotropic blend is a mixture of two or more components (Figure 9.2). The components of this refrigerant possess different boiling points. These components evaporate and condense at different temperatures. Zeotropic blends are characterized by *fractionation* and *glide*.

Fractionation is the change in composition of a refrigerant blend during change of phase from liquid to vapor (evaporation) or from vapor to liquid (condensation). This behavior in blends explains the permanent changes to refrigerant composition causing the blend to deviate outside the tolerances of the designed composition, due to vapor charging or leaks in a refrigerant system.

Fractionation phenomenon forms the principle behind *distillation* of liquid mixtures for separation purposes. Distillation depends on the differences among the volatility of different liquids in a mixture. That is how various fractions of crude oil are separated for various

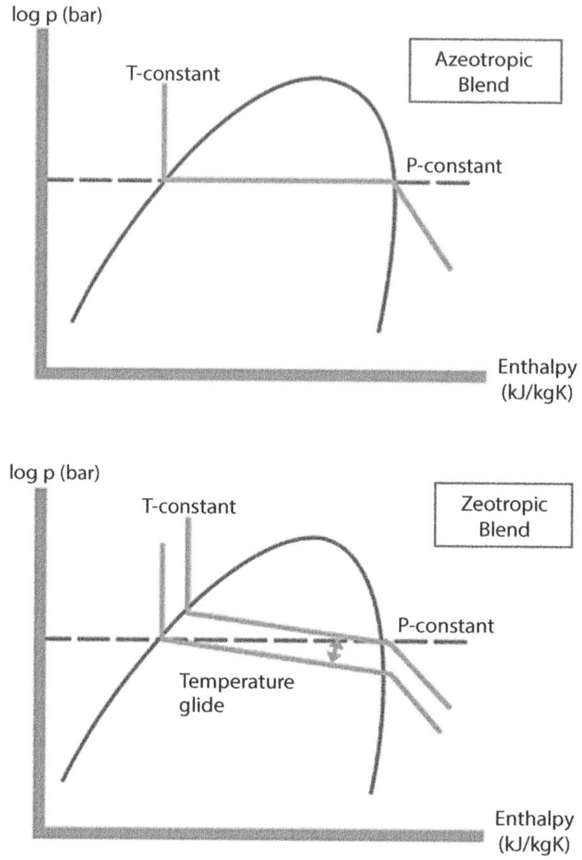

Figure 9.1 Comparison of Evaporating and condensing behavior of Azeotropic and Zeotropic Blends [5].

applications. Gases and most volatile liquids exit from the top of the distillation column while tars emerge from the bottom of the column [9].

A pure refrigerant exerts pressure in a system due to the motion of its molecules. At elevated temperatures the kinetic energy of the molecules increases resulting in an increased pressure. The reverse phenomenon occurs when temperature is reduced. The pressure exerted by a refrigerant depends on its molecular structures. The pressure generated in the system at a given temperature varies depending on the properties of a refrigerant.

Boiling liquid transfers more molecules to vapor phase; and condensing vapor transfers more from vapor to liquid. Different refrigerants transfer back and forth to the vapor at different rates. When vapor is removed from a cylinder or system containing a zeotropic blend, two events are going to take place:

1. The vapor removed is at the wrong composition. It will contain more of the more volatile component that generates higher-pressure (higher-capacity).
2. The liquid that is left behind loses more of the more volatile component to replace the vapor. In time the liquid composition changes because more of the more volatile component leaves compared to the bulk liquid composition. To avoid charging the wrong composition and fractionating the remaining

Table 9.2 Azeotropic blends (500 series refrigerants) characteristics and allowable levels of contaminants [7].

	Reporting Units	Reference Section	R-500	R-502	R-503	R-507A	R-508A	R-508B	R-509A	R-510A	R-511A	R-512A
CHARACTERISTICS:												
Refrigerant Components	N/A	N/A	R-12/ 152a	R-22/115	R-23/23	R-125/143a	R-23/116	R-23/116	R-22/218	R-E170/600a	R-290/E170	R-134a/152a
Nominal Composition	% by Weight	N/A	73.8/26.2	48.8/51.2	40.1/59.9	50.0/50.0	39.0/61.0	46.0/54.0	44.0/56.0	88.0/12.0	95.0/5.0	5.0/95.0
Allowable Composition	% by Weight	N/A	72.8-74.8/25.2-27.2	44.8-52.8/47.2-55.2	39.0-41.0/59.0-61.0	49.5-51.5/48.5-50.5	37.0-41.0/59.0-63.0	44.0-48.0/52.0-56.0	42.0-46.0/56.0-60.0	87.5-88.5/11.5-12.5	94.0-96.0/4.0-6.0	4.0-6.0/94.0-96.0
Bubble Point[1]	°C @ 101.3 kPa	N/A	-33.6	-45.2	-87.8	-46.7	-87.4	-87	-49.8	-24.9	-42	-24
Dew Point[1]	°C @ 101.3 kPa	N/A	-33.6	-45	-87.8	-46.7	-87.4	-87	-48.1	-24.9	-42	-24
Critical Temperature[1]	°C	N/A	102.1	80.2	18.4	70.6	10.8	11.8	68.6	125.7	97	112.9
VAPOR PHASE CONTAMINANTS:												
Air and Other Non-condensables, Maximum	% by volume @ 25°C	5.10	1.5	1.5	1.5	1.5	1.5	1.5	1.5	1.5	1.5	1.5

(Continued)

Table 9.2 Azeotropic blends (500 series refrigerants) characteristics and allowable levels of contaminants [7]. (*Continued*)

	Reporting Units	Reference Section	R-500	R-502	R-503	R-507A	R-508A	R-508B	R-509A	R-510A	R-511A	R-512A
LIQUID PHASE CONTAMINANTS:												
Water, Maximum	ppm by weight	5.4	10	10	10	10	10	10	10	20	20	10
All Other Volatile Impurities, Maximum	% by weight	5.11	0.5	0.5	0.5	0.5	0.5	0.5	0.5	0.5	0.5	0.5
High Boiling Residue, Maximum	% by volume or % by weight	5.8	0.01	0.01	0.01	0.01	0.01	0.01	0.01	0.01	0.01	0.01
Particulates/Solids	Pass/Fail	5.9	Visually Clean	Visually Clean	Visually Clean	Visually Clean	Visually Clean	Visually Clean	Visually Clean	Visually Clean	Visually Clean	Visually Clean
Acidity, Maximum	ppm by weight	5.7	1	1	1	1	1	1	1	1	1	1
Chloride[2]	Pass/Fail	5.6	No Visible Turbidity	No Visible Turbidity	No Visible Turbidity	No Visible Turbidity	No Visible Turbidity	No Visible Turbidity	No Visible Turbidity	No Visible Turbidity	No Visible Turbidity	No Visible Turbidity

Notes:
1. Bubble points, dew points and critical temperatures, although not required, are provided for informational purposes. Refrigerant data compiled from Refprop 9.1.
2. Recognized chloride level for pass/fail is about 3 ppm.
N/A = Not Applicable.

Table 9.3 Azeotropic Blends (500 series refrigerants) characteristics and allowable levels of contaminants [7].

	Reporting Units	Reference Section	R-513A	R-513B	R-514A	R-515A	R-516A
CHARACTERISTICS:							
Refrigerant Components	N/A	N/A	R-1234yf/134a	R-1234yf/134a	R-1336mzz(Z)/1130(E)	R-1234ze(E)/227ea	R-1234yf/134a/152a
Nominal Composition	% by weight	N/A	56.0/44.0	58.5/41.5	74.7/25.3	88.0/12.0	77.5/8.5/14.0
Allowable Composition	% by weight	N/A	55.0-57.0/43.0-45.0	58.0-59.0/41.0-42.0	74.2-76.2/23.8-25.8	86.0-89.0/11.0-14.0	76.1-78.9/7.0-9.0/12.1-14.1
Bubble Point[1]	°C @ 101.3 kPa	N/A	-29.2	-29.2	29.0	-18.9	-29.4
Dew Point[1]	°C @ 101.3 kPa	N/A	-29.1	-29.1	29.0	-18.9	-29.4
Critical Temperature[1]	°C	N/A	96.5	95.5	178.1	108.4	99.3
VAPOR PHASE CONTAMINANTS:							
Air and Other Non-condensables, Maximum	% by volume @ 25°C	5.10	1.5	1.5	N/A	1.5	1.5
LIQUID PHASE CONTAMINANTS:							
Water, Maximum	ppm by weight	5.4	10	10	20	10	10
All Other Volatile Impurities, Maximum	% by weight	5.11	0.5	0.5	0.5	0.5	0.5
High Boiling Residue, Maximum	% by volume or % by weight	5.8	0.01	0.01	0.01	0.01	0.01
Particulate/Solids	Pass or Fail	5.9	Visually clean	Visually clean	Visually clean	Visually clean	visually clean
Acidity, Maximum	ppm by weight	5.7	1	1	1	1	1
Chloride[2]	Pass or Fail	5.6	No Visible turbidity	No visible turbidity	No visible turbidity	No visible turbidity	No visible turbidity

Notes:

1. Bubble points, dew points and critical temperatures, although not required, are provided for informational purposes. Refrigerant data compiled from Refprop 9.1.

2. Recognized chloride level for pass/fail is about 3 ppm.

N/A = Not Applicable.

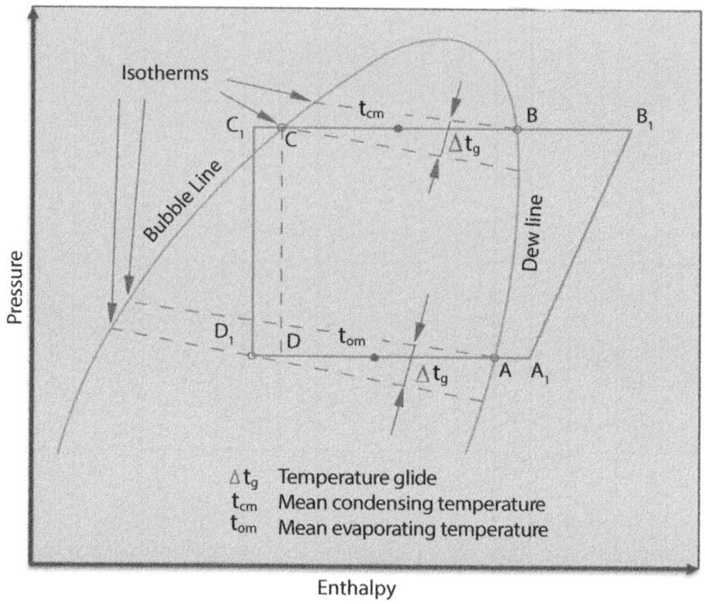

Figure 9.2 Evaporating and condensing behavior of zeotropic blends [8].

blend, zeotropic blends must be removed from the cylinder as a liquid. This can be achieved by inverting the cylinder over so the valve is at located the bottom. Alternatively, the liquid may be forced through a dip tube to the valve.

Liquid charging does not mean that liquid refrigerant should be pushed into the suction line of the system and that it be allowed to slug the compressor. After the initial charge into the high side of a system, the compressor can be started and charging can be completed by flashing the refrigerant from liquid to vapor in the charging hose or across specially designed valves. Any method which allows the refrigerant to go to vapor before it hits the compressor should work. Generally, the refrigerant needs to be added slowly at this point.

A system at rest will allow the refrigerant to pool and the vapor to come to an equilibrium concentration. Leaks that occur in vapor areas of the equipment will allow fractionation of the blend. The worst case will occur when about half of the refrigerant charge has leaked.

Recharging the system after repair will result in a blend with slightly reduced capacity and operating pressures. In smaller systems, where charge size is critical, it will be best to pull any remaining refrigerant and charge with fresh blend. In larger systems, you will need to make a decision whether the remaining charge should be pulled or not.

For low-fractionation-potential blends, one does not see much shift in composition anyway, and therefore, the charge can be topped off after repair without loss of properties.

In running systems, it has been found that the circulation composition is the bulk blend composition. In liquid and suction lines there is no second phase, and in the heat exchangers there is a lot of turbulence, so leaks will lose both vapor and liquid. Testing has shown that leaks from a running system do not cause fractionation, and a normally cycling system will not fractionate much during the off cycle [66].

Glide is the difference between the boiling point of the most volatile component compared to the least volatile component in the refrigerant blend. It is the difference in temperature between the evaporator outlet and inlet due to fractionation of the blend. Glide can be calculated by finding the difference between the dew and bubble temperatures at constant pressure. Actual measurements may differ somewhat depending on the state of the liquid refrigerant at either end of the evaporator (or condenser). Pressure losses through the evaporator may also affect glide.

Contrary to azeotropic blends which behave as single substance refrigerants with regard to evaporation and condensing processes, the phase change with zeotropic fluids occurs in a "gliding" form over a certain range of temperature. This "temperature glide" can be more or less pronounced, it depends mainly on the boiling points and the percentage proportions of the individual components. Near-azeotrope or "semiazeotrope" blends have glide value <1°C.

The temperature glide can be useful for improvement refrigeration performance, by correct design of the heat exchangers. One problem with blends is that refrigerant leakage results in a change in the component concentration of the refrigerant. However, tests have demonstrated that small changes in concentration (< 10%) have a negligible effect on refrigeration system performance.

Zeotropic mixtures are assigned numbers in the 400 series. This number designates which components are in the mixture and the following upper-case letter denotes the proportions. The sequence of the numbers are in chronological order of the refrigerant's approval by American Society of Heating, Refrigerating and Air-Conditioning Engineers (ASHRAE, www.ASHRAE.com). Tables 9.4 through 9.13 show the composition and properties of azeotropic refrigerant blends specified by Air Conditioning, Heating, and Refrigeration Institute [7].

The following considerations apply to the use of blends:

- The plant should always be charged with liquid refrigerant, or the component concentrations will shift.
- Since most blends contain at least one flammable component, the entry of air into the system should be avoided.
- Blends which have a large temperature glide, greater than 5K, should not be used for flooded-type evaporators.

9.2 Low GWP HFO and HFO/HFC Blends

This section is based on the Bitzer Refrigerant Report 2020 [11].

Aspects on the development of HFO and HFO/HFC refrigerants are discussed in this section.

The decision to use the "low GWP" refrigerant R1234yf in mobile air conditioning systems for passenger cars ("Low GWP" HFO refrigerant R1234yf) also led to the development of alternatives for further mobile applications as well as stationary refrigeration, air conditioning and heat pump systems.

Primary objectives are the use of single-component refrigerants and of mixtures with significantly reduced GWP and similar thermodynamic properties as the HFCs currently used predominantly (Table 9.14).

Table 9.4 Zeotropic blends (400 series refrigerants) characteristics and allowable levels of contaminants [7].

	Reporting Units	Reference Section	R-401A	R-401B	R-402A	R-402B	R-403A	R-403B	R-404A	R-405A	R-406A	R-407A
CHARACTERISTICS:												
Refrigerant Components	N/A	N/A	R-22/152a/124	R-22/152a/124	R-125/290/22	R-125/290/22	R-290/22/218	R-290/22/218	R-125/143a/134a	R-22/152a/142bC318	R-22/600a/142b	R-32/125/134a
Nominal Composition	% by weight	N/A	53.0/13.0/34.0	61.0/11.0/28.0	60.0/2.0/38.0	38.0/2.0/60.0	5.0/75.0/20.0	5.0/56.0/39.0	44.0/52.0/4.0	45.0/7.0/5.5/42.5	55.0/4.0/41.0	20.0/40.0/40.0
Allowable Composition	% by weight	N/A	51.0-55.0/11.5-13.5/33.0-35.0	59.0-63.0/9.5-11.5/27.0-29.0	58.0-62.0/1.0-2.1/58.0-62.0	36.0-40.0/1.0-2.1/58.0-62.0	3.0-5.2/73.0-77.0/18.0-22.0	3.0-5.2/54.0-58.0/37.0-41.0	42.0-46.0/51.0-53.0/2.0-6.0	43.0-47.0/6.0-8.0/4.5-6.5/40.5-44.5	53.0-57.0/3.0-5.0/40.0-42.0	18.0-22.0/38.0-42.0/38.0-42.0
Bubble Point[1]	°C @ 101.3 kPa	N/A	-33.3	-34.9	-49	-47	-47.8	-49.2	-46.2	-32.9	-32.7	-45.3
Dew Point[1]	°C @ 101.3 kPa	N/A	-26.4	-28.8	-46.9	-44.7	-44.3	-46.8	-45.5	-24.5	-23.5	-38.9
Critical Temperature[1]	°C	N/A	105.3	103.5	76	83	87	79.7	72.1	106	116.5	82.3
VAPOR PHASE CONTAMINANTS:												
Air and Other Non-condensables, Maximum	% by volume @ 25.0°C	5.10	1.5	1.5	1.5	1.5	1.5	1.5	1.5	1.5	1.5	1.5
LIQUID PHASE CONTAMINANTS:												
Water, Maximum	ppm by weight	5.4	10	10	10	10	10	10	10	10	10	10
All Other Volatile Impurities, Maximum	% by weight	5.11	0.5	0.5	0.5	0.5	0.5	0.5	0.5	0.5	0.5	0.5
High Boiling Residue, Maximum	% by volume or % by weight	5.8	0.01	0.01	0.01	0.01	0.01	0.01	0.01	0.01	0.01	0.01
Particulates/Solids	Pass or Fail	5.9	Visually clean	Visually clean	Visually clean	Visually clean	Visually clean	Visually clean	Visually clean	Visually clean	Visually clean	Visually clean
Acidity, Maximum	ppm by weight (as HCl)	5.7	1	1	1	1	1	1	1	1	1	1
Chloride[2]	Pass or Fail	5.6	No visible turbidity	No visible turbidity	No visible turbidity	No visible turbidity	No visible turbidity	No visible turbidity	No visible turbidity	No visible turbidity	No visible turbidity	No visible turbidity

Notes:
1. Bubble points, dew points and critical temperatures, although not required, are provided for informational purposes. Refrigerant data compiled from Refprop 9.1.
2. Recognized chloride level for pass/fail is about 3 ppm.
N/A Not Applicable.

Table 9.5 Zeotropic blends (400 series refrigerants) characteristics and allowable levels of contaminants [7].

	Reporting Units	Reference Section	R-407B	R-407C	R-407D	R-407E	R-407F	R-407G	R-407H	R-407I	R-408A	R-409A	R-409B
CHARACTERISTICS:													
Refrigerant Components	N/A	N/A	R-32/125/134a	R-32/125/134a	R-32/125/134a	R-32/125/134a	R-32/125/134a	R-32/125/134a	R-32/125/134a	R-32/125/134a	R-125/143a/22	R-22/124/142b	R-22/124/142b
Nominal Composition	% by weight	N/A	10.0/70.0/20	23.0/25.0/52	15.0/15.0/70	25.0/15.0/60	30.0/30.0/40	2.5/2.5/95.0	32.5/15.0/52.5	19.5/8.5/72.0	7.0/46.0/47	60.0/25.0/15	65.0/25.0/10
Allowable Composition	% by weight	N/A	8.0-12.0/68.0-72.0/18.0-22.0	21.0-25.0/23.0-27.0/50.0-54.0	13.0-17.0/13.0-17.0/68.0-72.0	23.0-27.0/13.0-17.0/58.0-62.0	28.0-32.0/28.0-32.0/38.0-42.0	2.0-3.0/2.0-3.0/94.0-96.0	31.5-33.5/14.0-16.0/50.5-54.5	17.5-20.5/7.5-10.5/70.0-74.0	5.0-9.0/45.0-47.0/45.0-49.0	58.0-62.0/23.0-27.0/14.0-16.0	63.0-67.0/23.0-27.0/9.0-11.0
Bubble Point[1]	°C @ 101.3 kPa	N/A	-46.8	-43.6	-39.5	-42.9	-46.1	-29.2	-44.6	-39.8	-44.6	-34.7	-35.6
Dew Point[1]	°C @ 101.3 kPa	N/A	-42.5	-36.6	-32.9	-35.8	-39.7	-27.2	-37.6	-33.0	-44.1	-26.4	-27.9
Critical Temperature[1]	°C	N/A	75	86	91.4	88.5	83	99.5	86.5	92.0	83.1	106.9	106.9
VAPOR PHASE CONTAMINANTS:													
Air and Other Non-condensables, Maximum	% by volume @ 25.0°C	5.1	1.5	1.5	1.5	1.5	1.5	1.5	1.5	1.5	1.5	1.5	1.5
LIQUID PHASE CONTAMINANTS:													
Water, Maximum	ppm by weight	5.4	10	10	10	10	10	10	10	10	10	10	10
All Other Volatile Impurities, Maximum	% by weight	5.11	0.5	0.5	0.5	0.5	0.5	0.5	0.5	0.5	0.5	0.5	0.5
High Boiling Residue, Maximum	% by volume or % by weight	5.8	0.01	0.01	0.01	0.01	0.01	0.01	0.01	0.01	0.01	0.01	0.01
Particulates/Solids	Pass or Fail	5.9	Visually clean	Visually clean	Visually clean	Visually clean	Visually clean	Visually clean	Visually clean	Visually clean	Visually clean	Visually clean	Visually clean
Acidity, Maximum	ppm by weight	5.7	1	1	1	1	1	1	1	1	1	1	1
Chloride[2]	Pass or Fail	5.6	No visible turbidity	No visible turbidity	No visible turbidity	No visible turbidity	No visible turbidity	No visible turbidity	No visible turbidity	No visible turbidity	No visible turbidity	No visible turbidity	No visible turbidity

Notes:
1. Bubble points, dew points and critical temperatures, although not required, are provided for informational purposes. Refrigerant data compiled from Refprop 9.1.
2. Recognized chloride level for pass/fail is about 3 ppm.
N/A = Not Applicable.

Table 9.6 Zeotropic blends (400 series refrigerants) characteristics and allowable levels of contaminants [7].

	Reporting Units	Reference Section	R-410A	R-410B	R-411A	R-411B	R-412A	R-413A	R-414A	R-414B	R-415A	R-415B
CHARACTERISTICS:												
Refrigerant Components	N/A	N/A	R-32/125	R-32/125	R-1270/22/152a	R-1270/22/152a	R-22/218/142b	R-22/218/134a/600a	R-22/124/600a/142b	R-22/124/600a/142b	R-22/152a	R-22/152a
Nominal Composition	% by weight	N/A	50.0/50.0	45.0/55.0	1.5/87.5/11.0	3.0/94.0/3.0	70.0/5.0/25.0	9.0/88.0/3.0	51.0/28.5/4.0/16.5	50.0/39.0/1.5/9.5	82.0/18.0	25.0/75.0
Allowable Composition	% by weight	N/A	48.5-50.5/ 49.5-51.5	44.0-46.0/ 54.0-56.0	0.5-1.5/87.5-89.5/10.0-11.0	2.0-3.0/94.0-96.0/2.0-3.0	68.0-72.0/3.0-7.0/24.0-26.0	8.0-10.0/86.0-90.0/2.0-3.0	49.0-53.0/26.5-30.5/3.5-4.5/15.5-17.0	48.0-52.0/37.0-41.0/1.0-2.0/8.5-10.0	81.0-83.0/ 17.0-19.0	24.0-26.0/ 74.0-76.0
Bubble Point[1]	°C @ 101.3 kPa	N/A	-51.4	-51.3	-39.5	-41.6	-38	-30.6	-34	-32.9	-37.5	-27.7
Dew Point[1]	°C @ 101.3 kPa	N/A	-51.4	-51.6	-36.6	-40	-28.7	-27.9	-25.8	-24.3	-34.7	-26.2
Critical Temperature[1]	°C	N/A	71.4	70.8	99.1	96	107.2	98.5	110.7	111	100	111.3
VAPOR PHASE CONTAMINANTS:												
Air and Other Non-condensables, Maximum	% by volume @ 25.0°C	5.10	1.5	1.5	1.5	1.5	1.5	1.5	1.5	1.5	1.5	1.5
LIQUID PHASE CONTAMINANTS:												
Water, Maximum	ppm by weight	5.4	10	10	10	10	10	10	10	10	10	10
All Other Volatile Impurities, Maximum	% by weight	5.11	0.5	0.5	0.5	0.5	0.5	0.5	0.5	0.5	0.5	0.5
High Boiling Residue, Maximum	% by volume or % by weight	5.8	0.01	0.01	0.01	0.01	0.01	0.01	0.01	0.01	0.01	0.01
Particulates/Solids	Pass or Fail	5.9	Visually clean	Visually clean	Visually clean	Visually clean	Visually clean	Visually clean	Visually clean	Visually clean	Visually clean	Visually clean
Acidity, Maximum	ppm by weight (as HC1)	5.7	1	1	1	1	1	1	1	1	1	1
Chloride[2]	Pass or Fail	5.6	No visible turbidity	No visible turbidity	No visible turbidity	No visible turbidity	No visible turbidity	No visible turbidity	No visible turbidity	No visible turbidity	No visible turbidity	No visible turbidity

Notes:
1. Bubble points, dew points and critical temperatures, although not required, are provided for informational purposes. Refrigerant data compiled from Refprop 9.1.
2. Recognized chloride level for pass/fail is about 3 ppm.
N/A = Not Applicable.

Table 9.7 Zeotropic blends (400 series refrigerants) characteristics and allowable levels of contaminants [7].

	Reporting Units	Reference Section	R-416A	R-417A	R-417B	R-417C	R-418A	R-419A	R-419B	R-420A	R-421A	R-421B
CHARACTERISTICS:												
Refrigerant Components	N/A	N/A	R-134a/124/600	R-125/134a/600	R-125/134a/600	R-125/134a/600	R-290/22/152a	R-125/134a/E170	R-125/134a/E170	R-134a/142b	R-125/134a	R-125/134a
Nominal Composition	% by weight	N/A	59.0/39.5/1.5	46.6/50.0/3.4	79.0/18.3/2.7	19.5/78.8/1.7	1.5/96.0/2.5	77.0/19.0/4.0	48.5/48.0/3.5	88.0/12.0	58.0/42.0	85.0/15.0
Allowable Composition	% by weight	N/A	58.0-59.5/39.0-40.5/1.3-1.6	45.5-47.7/49.0-51.0/3.0-3.5	78.0-80.0/17.3-19.3/2.2-2.8	18.5-20.5/77.8-79.8/1.2-1.8	1.0-2.0/95.0-97.0/2.0-3.0	76.0-78.0/18.0-20.0/3.0-5.0	47.5-49.5/47.0-49.0/3.0-4.0	88.0-89.0/11.0-12.0	57.0-59.0/41.0-43.0	84.0-86.0/14.0-16.0
Bubble Point[1]	°C @ 101.3 kPa	N/A	-23.4	-38	-44	-32.7	-41.2	-42.6	-37.4	-25	-40.8	-45.7
Dew Point[1]	°C @ 101.3 kPa	N/A	-21.8	-32.9	-41.5	-29.2	-40.1	-36	-31.5	-24.2	-35.5	-42.6
Critical Temperature	°C	N/A	108.2	89.9	75.2	95.4	96.7	79.1	90.4	105.4	78.5	69
VAPOR PHASE CONTAMINANTS:												
Air and Other Non-condensables, Maximum	% by volume @ 25.0°C	5.10	1.50	1.5	1.5	1.5	1.5	1.5	1.5	1.5	1.5	1.5
LIQUID PHASE CONTAMINANTS:												
Water, Maximum	ppm by weight	5.4	10	10	10	10	10	20	10	10	10	10
All Other Volatile Impurities, Maximum	% by weight	5.11	0.5	0.5	0.5	0.5	0.5	0.5	0.5	0.5	0.5	0.5
High Boiling Residue, Maximum	% by volume or % by weight	5.8	0.01	0.01	0.01	0.01	0.01	0.01	0.01	0.01	0.01	0.01
Particulates/Solids	Pass or Fail	5.9	Visually clean	Visually clean	Visually clean	Visually clean	Visually clean	Visually clean	Visually clean	Visually clean	Visually clean	Visually clean
Acidity, Maximum	ppm by weight (as HCl)	5.7	1	1	1	1	1	1	1	1	1	1
Chloride[2]	Pass or Fail	5.6	No visible turbidity	No visible turbidity	No visible turbidity	No visible turbidity	No visible turbidity	No visible turbidity	No visible turbidity	No visible turbidity	No visible turbidity	No visible turbidity

Notes:
1. Bubble points, dew points and critical temperatures, although not required, are provided for informational purposes. Refrigerant data compiled from Refprop 9.1.
2. Recognized chloride level for pass/fail is about 3 ppm.
N/A = Not Applicable.

Table 9.8 Zeotropic blends (400 series refrigerants) characteristics and allowable levels of contaminants [7].

	Reporting Units	Reference Section	R-422A	R-422B	R-422C	R-422D	R-422E	R-423A	R-424A	R-425A	R-426A	R-427A
CHARACTERISTICS:												
Refrigerant Components	N/A	N/A	R-125/134a/600a	R-125/134a/600a	R-125/134a/600a	R-125/134a/600a	R-125/134a/600a	R-134a/227ea	R-125/134a/600a/600/601a	R-32/134a/227ea	R-125/134a/600/601a	R-32/125/143a/134a
Nominal Composition	% by weight	N/A	8.5/11.5/3.4	55.0/42.0/3.0	82.0/15.0/3.0	65.1/30.5/3.4	58.0/39.3/2.7	52.7/47.5	50.5/47.0/0.9/1.0/0.6	18.5/69.5/12.0	5.1/93.0/1.3/0.6	15.0/25.0/10.0/50.0
Allowable Composition	% by weight	N/A	84.1-86.1/10.5-12.5/3.0-3.5	54.0-56.0/41.0-43.0/2.5-3.1	81.0-83.0/14.0-16.0/2.5-3.1	64.0-66.0/30.5-32.5/3.0-3.5	57.0-59.0/38.0-41.0/2.5-3.0	51.5-53.5/46.5-48.5	49.5-51.5/46.0-48.0/0.7-1.0/0.8-1.1/0.4-0.7	18.0-19.0/69.0-70.0/11.5-12.5	4.1-6.1/92.0-94.0/1.1-1.4/0.4-0.7	13.0-17.0/23.0-27.0/8.0-12.0/48.0-52.0
Bubble Point[1]	°C @ 101.3 kPa	N/A	-46.5	-40.5	-45.3	-43.2	-41.8	-24.2	-39.1	-38.1	-28.5	-43
Dew Point[1]	°C @ 101.3 kPa	N/A	-44.1	-35.6	-42.3	-38.4	-36.4	-23.5	-33.3	-31.3	-26.7	-36.3
Critical Temperature	°C	N/A	71.7	85.7	76.1	79.6	82.2	99	87.5	93.9	100.2	85.3
VAPOR PHASE CONTAMINANTS:												
Air and Other Non-condensables, Maximum	% by volume @ 25.0°C	5.10	1.5	1.5	1.5	1.5	1.5	1.5	1.5	1.5	1.5	1.5
LIQUID PHASE CONTAMINANTS:												
Water, Maximum	ppm by weight	5.4	10	10	20	10	10	10	10	10	10	10
All Other Volatile Impurities, Maximum	% by weight	5.11	0.5	0.5	0.5	0.5	0.5	0.5	0.5	0.5	0.5	0.5
High Boiling Residue, Maximum	% by volume or % by weight	5.8	0.01	0.01	0.01	0.01	0.01	0.01	0.01	0.01	0.01	0.01
Particulates/Solids	Pass or Fail	5.9	Visually clean	Visually clean	Visually clean	Visually clean	Visually clean	Visually clean	Visually clean	Visually clean	Visually clean	Visually clean
Acidity, Maximum	ppm by weight (as HCl)	5.7	1	1	1	1	1	1	1	1	1	1
Chloride[2]	Pass or Fail	5.6	No visible turbidity	No visible turbidity	No visible turbidity	No visible turbidity	No visible turbidity	No visible turbidity	No visible turbidity	No visible turbidity	No visible turbidity	No visible turbidity

Notes:
1. Bubble points, dew points and critical temperatures, although not required, are provided for informational purposes. Refrigerant data compiled from Refprop 9.1.
2. Recognized chloride level for pass/fail is about 3 ppm.
N/A = Not Applicable.

Table 9.9 Zeotropic blends (400 series refrigerants) characteristics and allowable levels of contaminants [7].

	Reporting Units	Reference Section	R-428A	R-429A	R-430A	R-431A	R-434A	R-435A	R-437A	R-438A	R-439A	R-440A
CHARACTERISTICS:												
Refrigerant Components	N/A	N/A	R-125/143a/290/600a	R-E170/152a/600a	R-152a/600a	R290/152a	R-125/143a/134a/600a	R-E170/152a	R-125/134a/600/601	R-32/125/134a/600/601a	R-32/125/600a	R-290/134a/152a
Nominal Composition	% by weight	N/A	77.5/20.0/0.6/1.9	60.0/10.0/30.0	76.0/24.0	71.0/29.0	63.2/18.0/16.0/2.8	80.0/20.0	19.5/78.5/1.4/0.6	8.5/45.0/44.2/1.7/0.6	50.0/47.0/3.0	0.6/1.6/97.8
Allowable Composition	% by weight	N/A	76.5-78.5/19.0-21.0/0.4-0.7/1.7-2.0	59.0-61.0/9.0-11.0/29.0-31.0	75.0-77.0/23.0-25.0	70.0-72.0/28.0-30.0	62.2-64.2/17.0-19.0/15.0-17.0/2.6-2.9	79.0-81.0/19.0-21.0	17.7-20.0/77.8-80.0/1.2-1.5/0.4-0.7	7.0-9.0/43.5-46.5/42.7-45.7/1.5-1.8/0.4-0.7	49.0-51.0/46.0-48.0/2.5-3.5	0.5-0.7/1.0-2.2/97.3-98.3
Bubble Point[1]	°C @ 101.3 kPa	N/A	-48.3	-25.5	-27.6	-43.2	-45.1	-26	-32.9	-43	-52	-25.5
Dew Point[1]	°C @ 101.3 kPa	N/A	-47.5	-24.9	-27.4	-43.2	-42.4	-25.8	-29.2	-36.4	-51.7	-24.3
Critical Temperature	°C	N/A	69	123.5	107	100.3	75.6	125.2	95.3	84.2	72	112.9
VAPOR PHASE CONTAMINANTS:												
Air and Other Non-condensables, Maximum	% by volume @ 25.0°C	5.10	1.5	1.5	1.5	1.5	1.5	1.5	1.5	1.5	1.5	1.5
LIQUID PHASE CONTAMINANTS:												
Water, Maximum	ppm by weight	5.4	10	20	20	10	10	20	10	10	10	10
All Other Volatile Impurities, Maximum	% by weight	5.11	0.5	0.5	0.5	0.5	0.5	0.5	0.5	0.5	0.5	0.5
High Boiling Residue, Maximum	% by volume or % by weight	5.8	0.01	0.01	0.01	0.01	0.01	0.01	0.01	0.01	0.01	0.01
Particulates/Solids	Pass or Fail	5.9	Visually clean	Visually clean	Visually clean	Visually clean	Visually clean	Visually clean	Visually clean	Visually clean	Visually clean	Visually clean
Acidity, Maximum	ppm by weight (as HC1)	5.7	1	1	1	1	1	1	1	1	1	1
Chloride[2]	Pass or Fail	5.6	No visible turbidity	No visible turbidity	No visible turbidity	No visible turbidity	No visible turbidity	No visible turbidity	No visible turbidity	No visible turbidity	No visible turbidity	No visible turbidity

Notes:
1. Bubble points, dew points and critical temperatures, although not required, are provided for informational purposes. Refrigerant data compiled from Refprop 9.1.
2. Recognized chloride level for pass/fail is about 3 ppm.
N/A = Not Applicable.

Table 9.10 Zeotropic blends (400 series refrigerants) characteristics and allowable levels of contaminants [7].

	Reporting Units	Reference Section	R-442A	R-444A	R-444B	R-445A	R-446A	R-447A	R-447B	R-448A	R-449A	R-449B
CHARACTERISTICS:												
Refrigerant Components	N/A	N/A	R-32 /125/134a/152a/227ea	R-32 /152a/1234ze(E)	R-32 /152a/1234ze(E)	R-744/134a/1234ze(E)	R-32 /1234ze(E) /600	R-32 /125 /1234ze(E)	R-32 /125 /1234ze(E)	R-32 /125 /1234yf /134a /1234ze(E)	R-32 /125 /1234yf /134a	R-32 /125 /1234yf /134a
Nominal Composition	% by weight	N/A	31.0/31.0/30.0/3.0/5.0	12.0/5.0/83.0	41.5/10.0 /48.5	6.0/9.0/85.0	68.0/29.0 /3.0	68.0/3.5 /28.5	68.0 /8.0 /24.0	26.0/26.0 /20.0/21.0 /7.0	24.3/24.7 /25.3/25.7	25.2 /24.3 /23.2 /27.3
Allowable Composition	% by weight	N/A	30.0-32.0/30.0-32.0/29.0-31.0/2.5-3.5/4.0-6.0	11.0-13.0/4.0-6.0/81.0-85.0	40.5-42.5 /9.0-11.0 /47.5-49.5	5.0-7.0 /8.0-10.0 /83.0-87.0	67.0-68.5 /28.4-31.0 /2.0-3.1	67.5-69.5 /3.0-5.0 /27.5-29.5	66.0-69.0 /7.0-10.0 /22.0-25.0	24.0-26.5 /25.5-28.0 /18.0-20.5 /20.0-23.0 /5.0-7.5	23.3-24.5 /24.5-25.7 /24.3-25.5 /25.5-26.7	23.7-25.5 /24.0-25.8 /21.7-23.5 /27.0-28.8
Bubble Point[1]	°C @ 101.3 kPa	N/A	-46.5	-34.3	-44.6	-50.3	-49.4	-49.3	-50.0	-45.9	-46	-46.1
Dew Point[1]	°C @ 101.3 kPa	N/A	-39.9	-24.3	-34.9	-23.5	-42.1	-44.2	-46.0	-39.8	-39.9	-40.2
Critical Temperature	°C	N/A	82.4	103.2	91.5	98	84.2	82.6	83.6	81.6	81.5	84.2
VAPOR PHASE CONTAMINANTS:												
Air and Other Non-condensables, Maximum	% by volume @ 25.0°C	5.10	1.5	1.5	1.5	1.5	1.5	1.5	1.5	1.5	1.5	1.5
LIQUID PHASE CONTAMINANTS:												
Water, Maximum	ppm by weight	5.4	10	10	10	10	10	10	10	10	10	10
All Other Volatile Impurities, Maximum	% by weight	5.11	0.5	0.5	0.5	0.5	0.5	0.5	0.5	0.5	0.5	0.5
High Boiling Residue, Maximum	% by volume or % by weight	5.8	0.01	0.01	0.01	0.01	0.01	0.01	0.01	0.01	0.01	0.01
Particulates/Solids	Pass or Fail	5.9	Visually clean	Visually clean	Visually clean	Visually clean	Visually clean	Visually clean	Visually clean	Visually clean	Visually clean	Visually clean
Acidity, Maximum	ppm by weight (as HC1)	5.7	1	1	1	1	1	1	1	1	1	1
Chloride[2]	Pass or Fail	5.6	No visible turbidity	No visible turbidity	No visible turbidity	No visible turbidity	No visible turbidity	No visible turbidity	No visible turbidity	No visible turbidity	No visible turbidity	No visible turbidity

Notes:

1. Bubble points, dew points and critical temperatures, although not required, are provided for informational purposes. Refrigerant data compiled from Refprop 9.1.
2. Recognized chloride level for pass/fail is about 3 ppm.

N/A = Not Applicable.

Table 9.11 Zeotropic blends (400 series refrigerants) characteristics and allowable levels of contaminants [7].

	Reporting Units	Reference Section	R-449C	R-450A	R-451A	R-451B	R-452A	R-452B	R-452C
CHARACTERISTICS:									
Refrigerant Components	N/A	N/A	R-32 /125/ 1234yf/134a	R-134a/1234ze(E)	R-1234yf/134a	R-1234yf/134a	R-32/125/1234yf	R-32/125/1234yf	R-32/125/1234yf
Nominal Composition	% by weight	N/A	20.0/20.0/31.0/29.0	42.0/58.0	89.8/10.2	88.8 /11.2	11.0 /59.0 /30.0	67.0 /7.0 /26.0	12.5/61.0/26.5
Allowable Composition	% by weight	N/A	18.5-20.5/19.5-21.5/29.5-31.5/28.5-30.5	40.0-44.0/56.0-60.0	89.6-90.0/10.0-10.4	88.6-89.0/11.0-11.4	9.3-12.7 /57.2-60.8 /29.0-30.1	65.0-69.0/5.5-8.5/24.0-28.0	11.0-13.0/60.0-62.0/25.0-27.0
Bubble Point[1]	°C @ 101.3 kPa	N/A	-44.6	-23.4	-30.8	-31	-47	-51.0	-47.5
Dew Point[1]	°C @ 101.3 kPa	N/A	-38.1	-22.8	-30.5	-30.6	-43.2	-50.3	-44.2
Critical Temperature	°C	N/A	83.5	104.4	95.4	95.5	74.9	75.7	75.8
VAPOR PHASE CONTAMINANTS:									
Air and Other Non-condensables, Maximum	% by volume @ 25.0°C	5.10	1.5	1.5	1.5	1.5	1.5	1.5	1.5
LIQUID PHASE CONTAMINANTS:									
Water, Maximum	ppm by weight	5.4	10	10	10	10	10	10	10
All Other Volatile Impurities, Maximum	% by weight	5.11	0.5	0.5	0.5	0.5	0.5	0.5	0.5
High Boiling Residue, Maximum	% by volume or % by weight	5.8	0.01	0.01	0.01	0.01	0.01	0.01	0.01
Particulates/Solids	Pass or Fail	5.9	Visually clean	Visually clean	Visually clean	Visually clean	Visually clean	Visually clean	Visually clean
Acidity, Maximum	ppm by weight (as HCl)	5.7	1	1	1	1	1	1	1
Chloride[2]	Pass or Fail	5.6	No visible turbidity	No visible turbidity	No visible turbidity	No visible turbidity	No visible turbidity	No visible turbidity	No visible turbidity

Notes:
1. Bubble points, dew points and critical temperatures, although not required, are provided for informational purposes. Refrigerant data compiled from Refprop 9.1.
2. Recognized chloride level for pass/fail is about 3 ppm.
N/A = Not Applicable.

Table 9.12 Zeotropic blends (400 series refrigerants) characteristics and allowable levels of contaminants [7].

	Reporting Units	Reference Section	R-453A	R-454A	R-454B	R-454C	R-455A	R-456A	R-457A
CHARACTERISTICS:									
Refrigerant Components	N/A	N/A	R-32/125/134a/ 227ea/600/ 601a	R-32/1234yf	R-32/1234yf	R-32/1234yf	R-744/32/ 1234yf	R-32/134a/ 1234ze(E)	R-32/ 1234yf/ 152a
Nominal Composition	% by weight	N/A	20.0/20.0 /53.8/ 5.0/0.6/0.6	35.0/65.0	68.9/31.1	21.5/78.5	3.0/21.5/75.5	6.0/45.0/49.0	18.0/70.0/ 12.0
Allowable Composition	% by weight	N/A	19.0-21.0 /19.0-21.0 /52.8-54.8 /4.5-5.5 /0.4-0.7 /0.4-0.7	33.0-37/ 63.0-67.0	67.9-69.9/ 30.1-32.1	19.5-23.5/ 76.5-80.5	2.0-5.0/19.5- 22.5/73.5- 77.5	5.0-7.0/ 44.0-46.0/ 48.0-50.0	16.5-18.5/ 68.5-70.5/ 10.1-12.1
Bubble Point[1]	°C @ 101.3 kPa	N/A	-42.2	-48.4	-50.9	-46.0	-51.6	-30.4	-42.7
Dew Point[1]	°C @ 101.3 kPa	N/A	-35	-41.6	-50.0	-37.8	-39.1	-25.6	-35.5
Critical Temperature	°C	N/A	88	86.2	76.5	82.4	82.8	102.4	92.6
VAPOR PHASE CONTAMINANTS:									
Air and Other Non-condensables, Maximum	% by volume @ 25.0°C	5.10	1.5	1.5	1.5	1.5	1.5	1.5	1.5

(Continued)

Table 9.12 Zeotropic blends (400 series refrigerants) characteristics and allowable levels of contaminants [7].

	Reporting Units	Reference Section	R-453A	R-454A	R-454B	R-454C	R-455A	R-456A	R-457A
LIQUID PHASE CONTAMINANTS:									
Water, Maximum	ppm by weight	5.4	10	10	10	10	10	10	10
All Other Volatile Impurities, Maximum	% by weight	5.11	0.5	0.5	0.5	0.5	0.5	0.5	0.5
High Boiling Residue, Maximum	% by volume or % by weight	5.8	0.01	0.01	0.01	0.01	0.01	0.01	0.01
Particulates/Solids	Pass or Fail	5.9	Visually clean	Visually clean	Visually clean	Visually clean	Visually clean	Visually clean	Visually clean
Acidity, Maximum	ppm by weight (as HCl)	5.7	1	1	1	1	N/A	1	1
Chloride[2]	Pass or Fail	5.6	No visible turbidity	No visible turbidity	No visible turbidity	No visible turbidity	No visible turbidity	No visible turbidity	No visible turbidity

Notes:
1. Bubble points, dew points and critical temperatures, although not required, are provided for informational purposes. Refrigerant data compiled from Refprop 9.1.
2. Recognized chloride level for pass/fail is about 3 ppm.
N/A = Not Applicable

Table 9.13 Zeotropic blends (400 series refrigerants) characteristics and allowable levels of contaminants [7].

	Reference Section	R-459B	R-460A	R-460B	R-460C	R-461A	R-462A	R-463A	R-464A	R-465A
CHARACTERISTICS:										
Refrigerant Components	N/A	R-32/1234yf/ 1234ze(E)	R-32/125/134a/ 1234ze(E)	R-32/125/134a/ 1234ze(E)	R-32/125/134a/ 1234ze(E)	R-125/143a/134a/ 227ea/ 600a	R-32/125/143a/134a/600	R-744/32/125/ 1234yf /134a	R-32/125/ 1234ze(E)/ 227ea	R-32/290/1234yf
Nominal Composition	N/A	21.0/69.0/ 10.0	12.0/52.0/14.0/22.0	28.0/25.0/20.0/27.0	2.5/2.5/46.0/49.0	55.0/5.0/32.0/ 5.0/ 3.0	9.0/ 42.0/2.0/44.0/3.0	6.0/36.0/30.0/ 14.0 / 14.0	27.0/27.0/40.0/6.0	21.0/7.9/71.1
Allowable Composition	N/A	20.0-21.5/ 67.0-71.0/ 9.0-11.0	11.0-13.0/51.0- 53.0/13.0-15.0/ 21.0-23.0	27.0-29.0/ 24.0- 26.0/19.0-21.0/ 26.0-28.0	2.0-3.0/2.0-3.0/ 45.0-47.0/ 48.0-50.0	54.0-56.0/4.5-5.5/ 31.0-33.0/4.5- 5.5/2.6-3.1	8.0-10.5/ 40.0-44.0/ 1.0-3.0/42.0-46.0/ 2.0-4.0	5.0-8.0 / 34.0-38.0/ 28.0-32.0/ 12.0- 16.0/ 12.0-16.0	26.0-28.0/26.0- 28.0/39.0-41.0/ 5.5-6.5	19.5-21.5/7.0-8.0/ 70.1-72.1
Bubble Point[1]	N/A	-44.0	-44.6	-45.2	-29.2	-42	-42.6	-58.4	-46.5	-51.8
Dew Point[1]	N/A	-36.1	-37.2	-37.1	-26.0	-37	-36.6	-46.9	-36.9	-40.0
Critical Temperature	N/A	89.8	81.6	85.7	99.5	81.4	83.5	75.8	82.6	89.4
VAPOR PHASE CONTAMINANTS:										
Air and Other Non-condensables, Maximum	5.10	1.5	1.5	1.5	1.5	1.5	1.5	1.5	1.5	1.5
LIQUID PHASE CONTAMINANTS:										
Water, Maximum	5.4	10	10	10	10	10	10	10	10	10
All Other Volatile Impurities, Maximum	5.11	0.5	0.5	0.5	0.5	0.5	0.5	0.5	0.5	0.5
High Boiling Residue, Maximum	5.8	0.01	0.01	0.01	0.01	0.01	0.01	0.01	0.01	0.01
Particulates/Solids	5.9	Visually clean	Visually clean	Visually clean	Visually clean	Visually clean	Visually clean	Visually clean	Visually clean	Visually clean
Acidity, Maximum	5.7	1	1	1	1	1	1	N/A	1	1
Chloride[2]	5.6	No visible turbidity	No visible turbidity	No visible turbidity	No visible turbidity	No visible turbidity	No visible turbidity	No visible turbidity	No visible turbidity	No visible turbidity

Notes:
1. Bubble points, dew points and critical temperatures, although not required, are provided for informational purposes. Refrigerant data compiled from Refprop 9.1.
2. Recognized chloride level for pass/fail is about 3 ppm.
N/A = Not Applicable.

Table 9.14 HFOs and HFO/HFC Blends [10].

Current HFC-refrigerants	Alternatives		Components/mixture components low GWP alternatives					
	Safety Group	GWP④	R1234yf A2L GWP 4	R1234ze(E) A2L 7	R32 A2L 675	R152a A2 124	R134a A1 1430	R125 A1 3500
R134a GWP 1430	A1	~600	✔	✔	✔		✔	
	A2L	<150	✔	✔	✔	✔	✔	
	A2L	<10	✔	✔				
R404A/R507A GWP 3922/3985	A1	<2500①	✔		✔			✔
	A1	~1400	✔	✔	✔		✔	✔
	A2L	<250	✔		✔	✔		
	A2L③	<150	✔		✔			
	A2	<150	✔		✔			
R22/R407C GWP 1810/1774	A1	900..1400	✔	✔	✔		✔	✔
	A2L	<250	✔		✔	✔		
	A2L	<150	✔		✔			
	A2	<150	✔		✔			
R410A GWP 2088	A1	<1500	✔		✔		✔	✔
	A1	<750			✔			✔
	A2L	<750			✔			
	A2L	~400..750	✔	✔	✔			✔

① Refrigerating capacity, mass flow, discharge gas temperature similar to R404A.
② Only low percentage – due to temperature glide (CO_2) and flammability (R290).
③ R32/HFO blends show lower refrigerating capacity than reference refrigerant, the addition of CO_2 leads to high temperature glide.
④ Approx. values according to IPCC IV (AR4).
⑤ R13I1 (CF_3I – tri-fluoroiodomethane) is an iodine-methane derivative.

An essential basic component for this is R1234yf ($CF_3CF{=}CH_2$). This refrigerant belongs to the group of hydro-fluoro-olefins (HFO), i.e. unsaturated HFCs with molecular double bonds. This group of HFOs also includes another substance called R1234ze(E), which has been mainly used as a propellant for PU foam and aerosol. R1234ze(E) differs from R1234yf in its molecular structure.

Both substances are the preferred choice in terms of their properties and are also used as basic components in HFO/HFC blends. The Global Warming Potential is very low – R1234yf with GWP 4 and R1234ze(E) with GWP 7. However, these refrigerants are flammable (safety class A2L), meaning the refrigerant quantity in the system must be considered in light of the installation location. In addition, there remain open questions concerning the long-term stability in stationary systems where long life cycles are common. Furthermore, the volumetric refrigerating capacity is relatively low; for R1234yf it is close to the level of R134a, and more than 20% lower for R1234ze(E).

There is also some uncertainty concerning flammability. In safety data sheets, R1234ze(E) is declared as non-flammable. However, this only applies to its transport and storage. When used as a refrigerant, a higher reference temperature for flammability tests of 60°C applies. At this temperature, R1234ze(E) is flammable and therefore classified in safety class A2L, like R1234yf.

R1234ze(E) is sometimes referred to as a R134a substitute, but its volumetric refrigerating capacity is more than 20% lower than that of R134a or R1234yf. The boiling point (-19°C) also greatly restricts the application at lower evaporation temperatures. Its preferred use is therefore in liquid chillers and high temperature applications. For further information Refrigerants for special applications.

The list of further potential HFO refrigerants is relatively long. However, there are only few substances that meet the requirements in terms of thermodynamic properties, flammability, toxicity, chemical stability, compatibility with materials and lubricants. These include e.g. the non-flammable (safety group A1) low-pressure refrigerants R1336mzz(Z), R1233zd(E) and R1224yd(Z). These are primarily an option for liquid chillers with large turbo-compressors, and they can be used with positive displacement compressors in high-temperature applications. R1233zd(E) and R1224yd(Z) belong to the group of HCFO (hydro-chloro-fluoro-olefins); they have a (very) low ozone depletion potential (ODP). Upon release into the atmosphere, however, the molecule rapidly disintegrates.

On the other hand, there are currently no candidates from the HFO family with similar volumetric refrigerating capacity such as R22/R407C, R404A/R507A and R410A available for commercial use. Direct alternatives for these refrigerants with significantly lower GWPs must therefore be "constructed" as a mixture of R1234yf and/or R1234ze(E) with HFC refrigerants, possibly also small proportions of hydrocarbons, CO_2 or other suitable molecules.

Though, due to the properties of the HFC refrigerants suitable as blend components, flammability and GWP are related diametrically to one another. In other words: Blends as alternatives to R22/R407C of GWP < approx. 900 are flammable. This is also true with alternatives for R404A/R507A in blends of GWP < approx. 1300 and for R410A in blends of GWP < approx. 2000. The reason for this is the high GWP of each of the required non-flammable components. There are a few exceptions, which are discussed in chapter, Further development projects with "Low GWP" refrigerants. For R134a alternatives, the situation is more favorable. Due to the already quite low GWP of R134a, a blend with R1234yf and/or R1234ze(E) allows a formulation of non-flammable refrigerants with a GWP of approx. 600.

Thus, primarily two directions for development are pursued:

Non-flammable HFC alternatives (blends) with GWP values according to the above mentioned limits – safety group A1. Regarding safety requirements, these refrigerants can then be utilized similar to currently used HFCs.

Flammable HFC alternatives (blends) with GWP values below the above mentioned possible limits – according to safety group A2L (for refrigerants of lower flammability). This group of refrigerants will be subject e.g. to charge limits according to the current standards for A2L refrigerants.

Meanwhile, there are development projects using refrigerant components with a much higher volumetric refrigerating capacity and pressure than R1234yf and R1234ze(E). These can then be used to "formulate" mixtures with R32 as an alternative to R410A, which are optimized for certain properties. See additional information, Further development projects with "Low GWP" refrigerants.

Fractionation describes the change of a blend composition when one or more of its components boil more quickly than the others. Indeed, this phenomenon forms the principle behind *distillation* of liquid mixtures for separation purposes. Distillation depends on the differences among the volatility of different liquids in a mixture. That is how various fractions of crude oil are separated for various applications. Gases and most volatile liquids exit from the top of the distillation column while tars emerge from the bottom of the column.

A pure refrigerant exerts pressure in a system due to the motion of its molecules. At elevated temperatures the kinetic energy of the molecules increases resulting in an increased pressure. The reverse phenomenon occurs when temperature is reduced. The pressure exerted by a refrigerant depends on its molecular structures. The pressure generated in the system at a given temperature varies depending on the properties of a refrigerant.

Boiling liquid transfers more from molecules to vapor phase; and condensing vapor transfers more from vapor to liquid. Different refrigerants transfer back and forth to the vapor at different rates. When vapor is removed from a cylinder or system containing a zeotropic blend, two events are going to take place:

1. The vapor removed is at the wrong composition. It will contain more of the more volatile component that generates higher-pressure (higher-capacity).
2. The liquid that is left behind loses more of the more volatile component to replace the vapor. In time the liquid composition changes because more of the more volatile component leaves compared to the bulk liquid composition. To avoid charging the wrong composition and fractionating the remaining blend, zeotropic blends must be removed from the cylinder as a liquid. This can be achieved by inverting the cylinder over so the valve is at located the bottom. Alternatively, the liquid may be forced through a dip tube to the valve.

Liquid charging does not mean that liquid refrigerant should be pushed into the suction line of the system and that it be allowed to slug the compressor. After the initial charge into the high side of a system, the compressor can be started and charging can be completed by flashing the refrigerant from liquid to vapor in the charging hose or across specially designed valves. Any method which allows the refrigerant to go to vapor before it hits the compressor should work. Generally, the refrigerant needs to be added slowly at this point.

A system at rest will allow the refrigerant to pool and the vapor to come to an equilibrium concentration. Leaks that occur in vapor areas of the equipment will allow fractionation of the blend. The worst case will occur when about half of the refrigerant charge has leaked.

Recharging the system after repair will result in a blend with slightly reduced capacity and operating pressures. In smaller systems, where charge size is critical, it will be best to pull any remaining refrigerant and charge with fresh blend. In larger systems, you will need to make a decision whether the remaining charge should be pulled or not.

Note: For low-fractionation-potential blends, you will not see much shift in composition anyway, and therefore, the charge can be topped off after repair without loss of properties.

In running systems, it has been found that the circulation composition is the bulk blend composition. In liquid and suction lines there is no second phase, and in the heat exchangers there is a lot of turbulence, so leaks will lose both vapor and liquid. Testing has shown that leaks from a running system do not cause fractionation, and a normally cycling system will not fractionate much during the off cycle.

9.3 Flammability of Blends

One of the consequences of these actions is that many of these alternatives (e.g. some hydrofluoroolefins [HFOs], hydrocarbons [HCs], etc.) are flammable. To this end, the RAC industry has spent the last several years preparing for this transition. This report will give an overview of some of the key factors that must be considered when working with flammable refrigerants.

Regulatory drivers are discussed. Safety classes and flammability parameters are reviewed, as well as their effect on refrigerant selection. Finally, codes and standards impacts are also highlighted [10].

References

1. Lavelle, J., Understanding Refrigerant Blend Performance. *Refrig. Serv. Eng. Soc. (RSES) J.*, 29–33, January 2006.
2. US EPA, Dec 2019.
3. Blends 101, An Introduction to Refrigerants, The Air Conditioning, Heating and Refrigeration NEWS, www.achrnews.com, July 13, 2000.
4. Hundy, G.F., Trott, A.R., Welch, T.C., *Refrigeration, Air Conditioning and Heat Pumps*, 5th edition, Elsevier, 2016.
5. SWEP International AB, www.swepusa.com/refrigerant-handbook/, July 2020.
6. www.ahrinet.org/search-standards, July 2020
7. 2019 Standard for Specifications for Refrigerants, AHRI Standard 700-2019, Air Conditioning, Heating, and Refrigeration Institute, www.AHRI.org, 2019.
8. *Refrigerant blends: Classification and experiences*, BITZER KÜHLMASCHINENBAU GMBH, www.bitzer-refrigerantreport.com/refrigerants/refrigerant-blends/, April 2020.
9. Lavelle, J., Understanding Refrigerant Blend Performance, Part I. *RSES J.*, 29–33, January 2006.
10. Refrigerant Report 20, BITZER Kühlmaschinenbau GmbH, www.bitzer-refrigerantreport.com, Sep 2018.
11. Bitzer Refrigerant Report 2020, BITZER Kühlmaschinenbau Gmb, www.bitzer-refrigerant report.com, www.bitzer-refrigerantreport.com, Sep 2018.

10

Substitute Fluorocarbons and Other Compounds

This chapter recites some of the Environmental Protection Agency's (EPA) Rules governing the use and replacement of fluorocarbons. Under Section 612 of the Clean Air Act, EPA's Significant New Alternatives Policy (SNAP) program reviews substitutes within a comparative risk framework in the following industrial sectors:

- Adhesives, Coatings, and Inks
- Aerosols
- Cleaning Solvents
- Fire Suppression and Explosion Protection
- Foam Blowing Agents
- Refrigeration and Air Conditioning
- Sterilants
- Tobacco Expansion

The SNAP program does not provide a static list of alternatives but instead, evolves the list as EPA makes decisions that are informed by its overall understanding of the environmental and human health impacts as well as its current knowledge about available substitutes. Section 612 also provides that EPA must prohibit the use of a substitute where EPA has determined that there are other available substitutes that pose less overall risk to human health and the environment.

In this Section some of the Substitutes for refrigerants, air conditioning and blowing agents applications of fluorocarbons are described. For a complete list of the Substitutes consult EPA Website www.epa.gov/snap/snap-substitutes-sector.

10.1 SNAP Program (EPA, www.epa.gov/snap/overview-snap)

The purpose of the SNAP program is to:

- Identify and evaluate substitutes in end-uses that have historically used ozone-depleting substances (ODS);
- Look at overall risk to human health and the environment of both existing and new substitutes;
- Publish lists of acceptable and unacceptable substitutes by end-use;
- Promote the use of acceptable substitutes; and
- Provide the public with information about the potential environmental and human health impacts of substitutes.

Sina Ebnesajjad, PhD. Concise Handbook of Fluorocarbon Gases: Applications in Refrigeration and Other Industries, (267–282)

To arrive at determinations on the acceptability of substitutes, the Agency performs a cross-media analysis of risks to human health and the environment from the use of various substitutes in different industrial and consumer uses that have historically used ODS. EPA reviews characteristics, including the following, when evaluating each proposed substitute:

- Ozone depletion potential (ODP),
- Global warming potential (GWP),
- Toxicity,
- Flammability,
- Occupational and consumer health/safety,
- Local air quality, and
- Ecosystem effects.

10.2 Guiding Principles of the SNAP Program?

The guiding principles of the SNAP program are:

- Evaluate substitutes within a comparative risk framework
- Do not require that substitutes be risk-free to be found acceptable
- Restrict only those substitutes that are significantly worse
- Evaluate risks by use
- Provide the regulated community with information as soon as possible
- Do not endorse products manufactured by specific companies
- Defer to other environmental regulations when warranted

10.3 EPA's Criteria for Evaluating Alternatives?

EPA's decision on the acceptability of new substitutes proposed by manufacturers, formulators, or users is based primarily on the potential human health and environmental risks posed by the substitutes as compared other substitutes available for a particular end-use. EPA's evaluation of each substitute in an end-use is based on the following types of information and analyses.

10.3.1 Atmospheric Effects

The SNAP program considers the ODP and 100-year integrated GWP of compounds to assess atmospheric effects.

10.3.2 Exposure Assessments

Exposure assessments are used to estimate concentration levels of substitutes to which workers, consumers, the general population, and environmental receptors may be exposed over a determined period of time. These assessments are based on personal monitoring data or area sampling data if available. Exposure assessments may be conducted for many types of releases including:

- Releases in the workplace and in homes
- Releases to ambient air and surface water
- Releases from the management of solid wastes

10.3.3 Toxicity Data

Toxicity data is used to assess the possible health and environmental effects for exposure to substitutes. The Occupational Safety and Health Administration (OSHA) or EPA approve wide health-based criteria that are available for a substitute such as:

- Permissible exposure limits (PELs for occupational exposure)
- Inhalation reference concentrations (R_fCs for non-carcinogenic effects on the general population)
- Cancer slope factors (for carcinogenic risk to members of the general population)

If OSHA has not issued a PEL for a compound, EPA also considers Workplace Environmental Exposure Limits (WEEL) set by the American Industrial Hygiene Association (AIHA), threshold limit values (TLV) set by the American Conference of Governmental Industrial Hygienists (ACGIH), or Recommended Exposure Limits (REL) set by the National Institutes of Occupational Safety and Health (NIOSH). If limits for occupational exposure or exposure to the general population are not already established, then EPA derives these values following the Agency's peer-reviewed guidelines. If a manufacturer provides a recommended exposure limit, EPA would review the information used in derivation and compare to its own exposure limit derivation. Exposure information is combined with this toxicity information to explore any basis for concern. Toxicity data is used with existing EPA guidelines to develop health-based criteria for interim use in these risk characterizations.

10.3.4 Flammability

Flammability is examined as a safety concern for workers and consumers. EPA assesses flammability risk using data on:

- Flash point and flammability limits (e.g., OSHA flammability/combustibility classifications)
- Data on testing of blends with flammable components
- Test data on flammability in consumer applications conducted by independent laboratories
- Information on flammability risk minimization techniques

10.3.5 Other Environmental Impacts

The SNAP program also examines other potential environmental impacts such as ecotoxicity and local air quality impacts. A compound that is likely to be discharged to water may be evaluated for impacts on aquatic life. Some substitutes are volatile organic compounds (VOCs), which are chemicals that increase tropospheric air pollution by contributing to ground-level ozone formation. In addition, EPA notes whenever a potential substitute is considered a hazardous air pollutant or hazardous waste.

10.4 Alternatives for Refrigeration

10.4.1 Chillers

Chillers typically cool water, which is then circulated to provide comfort cooling throughout a building or other location. Chillers can be classified by compressor type, including centrifugal and positive displacement. Chillers used to cool industrial processes are discussed under Industrial Process Refrigeration.

10.4.2 Cold Storage Warehouses

Cold storage warehouses store meat, produce, dairy products, and other perishable goods. The majority of cold storage warehouses in the United States use ammonia as the refrigerant in a vapor compression cycle, although some rely on other refrigerants.

10.4.3 Commercial Ice Machines

Commercial ice machines are used in commercial establishments (e.g., hotels, restaurants, convenience stores) to produce ice for consumer use. Ice machines produce ice in various sizes and shapes, and with different retrieval mechanisms (e.g., dispensers or self-retrieval from bins).

10.4.4 Household Refrigerators and Freezers

Household refrigerators and freezers are intended primarily for residential use, although they may be used outside the home. Household freezers only offer storage space at freezing temperatures. Products with both a refrigerator and freezer in a single unit are most common. Small refrigerated household appliances may also include chilled kitchen drawers, wine coolers, mini fridges, household beverage centers, ice makers that are part of a household refrigerator-freezer, and stand-alone ice makers for household use.

A few of the Substitutes for residential Refrigerators and Freezers use can be seen in Table 10.1. The data in the Table show whether a Substitute is New or Retrofit and also its Listing Status which indicates whether the compounds can be used. For a complete and up-to-date list of the Substitutes consult EPA Website www.epa.gov/snap/snap-substitutes-sector.

10.4.5 Ice Skating Rinks

Ice skating rinks are used by the general public for recreational purposes and also include professional rinks. These systems frequently use secondary loop refrigeration systems.

Table 10.1 Substitutes in household refrigerators and freezers.

Substitute	Trade name(s)	Retrofit/ new	ODP	GWP	ASHRAE designation (safety classification)	SNAP listing date	Listing status
R-290 (Propane)		N	0	3	A3	April 10, 2015; August 8, 2018	Acceptable with Use Conditions: See rule for detailed conditions.
R-401A	SUVA MP-39	N	0.037	1,182	A1	January 13, 1995	Acceptable
R-401B	SUVA MP-66	N	0.040	1,288	A1	January 13, 1995	Acceptable
R-402A	HP-80	R	0.019	2,746	A1	August 26, 1994	Acceptable
R-402B	HP-81	R	0.03	2,379	A1	August 26, 1994	Acceptable
R-404A	SUVA HP-62	R/N	0	3,920	A1	March 18, 1994; January 13, 1995; December 20, 2002; December 1, 2016	Unacceptable in new equipment, as of January 1, 2021.
R-407F	Genetron, Performax, LT	R/N	0	1,820	A1	October 4, 2011; December 1, 2016	Unacceptable in new equipment, as of January 1, 2021.
R-409A	HCFC Blend Gamma	N	0.047	1,558	A1	August 26, 1994	Acceptable
R-410A	AZ-20, Suva 9100, Puron	N	0	2,090	A1	February 8, 1996; December 20, 2002 December 1, 2016	Unacceptable in new equipment, as of January 1, 2021.
R-410B		N	0	2,230	A1	February 8, 1996; December 1, 2016	Unacceptable in new equipment, as of January 1, 2021.

(Continued)

Table 10.1 Substitutes in household refrigerators and freezers. (*Continued*)

Substitute	Trade name(s)	Retrofit/ new	ODP	GWP	ASHRAE designation (safety classification)	SNAP listing date	Listing status
R-414A	GHG-X4, Autofrost, Chill-it, HCFC Blend Xi	R/N	0.045	1,478	A1	September 5, 1996	Acceptable
R-414B	Hot Shot, Kar Kool, HCFC Blend Omicron	R/N	0.098	3,337	A1	September 5, 1996	Acceptable
R-416A	FRIGC FR-12, HCFC Blend Beta	R/N	0.009	1,081	A1	February 8, 1996	Acceptable

R-401A = Chlorodifluoromethane (R-22), Difluoroethane (R-152a), Chlorotetrafluoroethane (R124), in the ratio of 53/13/34.
R-401B = Chlorodifluoromethane (R-22), Difluoroethane (R-152a), Chlorotetrafluoroethane (R124), in the ratio of 61/11/28.
R-402A = Chlorodifluoromethane (R-22), Pentafluoroethane (R-125), Propane (R-290), in the ratio of 38/60/2.
R-402B = Chlorodifluoromethane (R-22), Pentafluoroethane (R-125), Propane (R-290), in the ratio of 60/38/2.
R-404A = Pentafluoroethane (R-125), Trifluoroethane (R-143A), Tetrafluoroethane (R-134A), in the ratio of 44/52/4
R-407F = Pentafluoroethane (R-125), Difluoromethane (R-32), Tetrafluoroethane (R-134a), in the ratio of 30/30/40
R-409A = Chlorodifluoromethane (R-22), Chlorodifluoroethane (R-142b), Chlorotetrafluoroethane (R-124), in the ratio of 60/15/25.
R-410A = Difluoromethane (R-32), Pentafluoroethane (R-125), in the ratio of 50/50.
R-410B = Difluoromethane (R-32), Pentafluoroethane (R-125), in the ratio of 45/55.
R-414A = Chlorodifluoromethane (R-22), Chlorotetrafluoroethane (R-124), Isobutane (R-600a), Chlorodifluoroethane (R-142b), in the ratio of 51/28.5/4/16.5.
R-414B = Blend Chlorodifluoromethane (R-22), Chlorotetrafluoroethane (R124), Chlorodifluoroethane (R142b), Isobutane (R-600a), in the ratio of 50/39/1.5/9.5.
R-416A = Tetrafluoroethane (HFC-134a), Chlorotetrafluoroethane (R124), Butane (R600), in the ratio of 59/39.5/1.5.

10.4.6 Industrial Process Refrigeration

Industrial process refrigeration systems cool process streams in industrial applications. The choice of substitute for specific applications depends on ambient and required operating temperatures and pressures.

10.4.7 Refrigerated Transport

Refrigerated transport moves products (e.g., perishable goods) from one place to another by various modes of transportation while maintaining necessary temperatures, including refrigerated ship holds, truck trailers, railway freight cars, ships, and other shipping containers (Table 10.2).

10.4.8 Retail Food Refrigeration

Retail food refrigeration, or commercial refrigeration, includes equipment designed to store, display, process, or dispense chilled or frozen goods for commercial sale. This end-use includes these categories of equipment: stand-alone equipment, remote condensing units, supermarket systems, and refrigerated food processing and dispensing equipment (Table 10.3).

10.4.9 Vending Machines

Vending machines are self-contained units that dispense goods that must be kept cold or frozen.

10.4.10 Very Low Temperature Refrigeration

Very low temperature refrigeration systems require maintaining temperatures at approximately -80 degrees Fahrenheit (-62 degrees Celsius) or lower. Examples include medical freezers and freeze-dryers, which generally require extremely reliable refrigeration cycles to maintain low temperatures and must meet stringent technical standards that do not normally apply to refrigeration systems.

10.4.11 Water Coolers

Water coolers are self-contained units providing chilled water, and possibly heated water, for drinking. They may or may not feature detachable containers of water.

10.5 Alternatives for Air Conditioning

10.5.1 Industrial Process Air Conditioning

Industrial process air-conditioning units, which are distinct from commercial and residential air conditioning, provide comfort cooling for operators and protect process equipment.

Table 10.2 Substitutes in refrigerated transport.

Substitute	Trade name(s)	Retrofit/new	ODP	GWP	ASHRAE designation (safety classification)	SNAP listing date	Listing status
Cryogenic Transport Refrigeration System		N	0	0	AI	December 6, 1999; June 16, 2010	Acceptable
Direct Nitrogen Expansion		N	0	0	AI	March 18, 1994; June 16, 2010	Acceptable
R-401A	SUVA MP-39	N	0.037	1,182	A1	January 13, 1995	Acceptable
R-401B	SUVA MO-66	N	0.040	1,288	A1	January 13, 1995	Acceptable
R-402A	HP-80	R/N	0.019	2,746	A1	March 18, 1994	Acceptable
R-402B	HP-81	R/N	0.03	2,379	A1	March 18, 1994	Acceptable
R-404A	SUVA HP-62	R/N	0	3,920	A1	March 18, 1994; December 20, 2002	Acceptable
R-406A	GHG-12, GHG-X3, McCool, Autofrost X3	R	0.057	1,900	A2	August 26, 1994	Acceptable
R-407A	Klea 60, Klea 407A	R/N	0	2,110	A1	August 26, 1994; January 2, 2009	Acceptable
R-422B	ICOR XAC1, NU-22B	R/N	0	2,530	A1	March 29, 2006	Acceptable
R-422C	ICOR XLT1	R/N	0	3,390	A1	March 29, 2006	Acceptable
R-422D	ISCEON MO29	R/N	0	2,730	A1	September 28, 2006	Acceptable
R-424A	RS-44	R/N	0	2,440	A1	September 28, 2006	Acceptable
R-426A	RS-24	R/N	0	1,510	A1	September 28, 2006; June 16, 2010	Acceptable

(Continued)

Table 10.2 Substitutes in refrigerated transport. (*Continued*)

Substitute	Trade name(s)	Retrofit/new	ODP	GWP	ASHRAE designation (safety classification)	SNAP listing date	Listing status
R-428A	RS-52	R/N	0	3,610	A1	October 4, 2007	Acceptable
R-434A	RS-45	R/N	0	3,250	A1	October 4, 2007	Acceptable
R-437A	KDD6, ISCEON MO49 Plus	R/N	0	1,810	A1	January 2, 2009	Acceptable
R-438A	KDD5, ISCEON MO99	R/N	0	2,270	A1	October 4, 2007; June 16, 2010	Acceptable

R-401A = Chlorodifluoromethane (R-22), Difluoroethane (R-152a), Chlorotetrafluoroethane (R124), in the ratio of 53/13/34.
R-401B = Chlorodifluoromethane (R-22), Difluoroethane (R-152a), Chlorotetrafluoroethane (R124), in the ratio of 61/11/28.
R-402A = Chlorodifluoromethane (R-22), Pentafluoroethane (R-125), Propane (R-290), in the ratio of 38/60/2.
R-402B = Chlorodifluoromethane (R-22), Pentafluoroethane (R-125), Propane (R-290), in the ratio of 60/38/2.
R-404A = Trifluoroethane (HFC-143a), Pentafluoroethane (HFC-125), Tetrafluoroethane (R-134a), in the ratio of 52/44/2.
R-406A = Blend Chlorodifluoromethane (R-22), Isobutane (R-600a) and Chlorodifluoroethane (R-142b) in the ratio of 55/42/3.
R-407A = Pentafluoroethane (HFC-125), Tetrafluoroethane (R-134a), Difluoromethane (HFC-32), in the ratio of 40/40/20.
R-422A = Tetrafluoroethane (R-134a), Pentafluoroethane (R-125), Isobutane (R-600a), in the ratio of 31.5/65.1/3.4.
R-422B = Tetrafluoroethane (R-134a), Pentafluoroethane (R-125), Isobutane (R-600a), in the ratio of 42/55/3.
R-422C = Tetrafluoroethane (R-134a), Pentafluoroethane (R-125), Isobutane (R-600a), in the ratio of 15/82/3.
R-422D = Tetrafluoroethane (R-134a), Pentafluoroethane (R-125), Isobutane (R-600a), in the ratio of 31.5/65.1/3.4.
R-424A = Pentafluoroethane (R125), Tetrafluoroethane (R-134a), Butane (R-600), Isopentane (R-601a), in the ratio of 50.5/47/0.9/1/0.6.
R-426A = Pentafluoroethane (R125), Tetrafluoroethane (R-134a), Isobutane (R-600a), Butane (R600), Isopentane (R601a), in the ratio of 5.1/93/1.3/0.6.
R-428A = Trifluoroethane (R-143a), Pentafluoroethane (HFC-125), Isopentane (R600a), Propane (R-290) in the ratio of 20/77.5/1.9/0.6.
R-434A = Pentafluoroethane (R125), Trifluoroethane (HFC-143a), Tetrafluoroethane (R-134a), Isobutane (R600a), in the ratio of 63/18/15.7/3.3.
R-437A = Pentafluoroethane (HFC-125), Tetrafluoroethane (R-134a), Butane (R-600), Pentane (R-601), in the ratio of 78.5/19.5/1.4/0.6.
R-438A = Pentafluoroethane (HFC-125), Tetrafluoroethane (R-134a), Difluoromethane (R-32), Butane (R-600), Isopentane (R-601), in the ratio of 45/44.2/8.5/1.7/0.6.

Table 10.3 Substitutes in retail food refrigeration.

Substitute	Trade name(s)	Retrofit/new	ODP	GWP	ASHRAE designation (safety classification)	SNAP listing date	Listing status
R-417A	ISCEON 59, NU-22	R/N	0	2,350	A1	December 6, 1999; December 20, 2002; June 16, 2010; July 20, 2015	Unacceptable in new Stand-Alone Medium-Temperature Units with a compressor capacity equal to or greater than 2,200 Btu/hour and Stand-Alone Medium-Temperature Units
R-290 (Propane)		N	0	3	A3	December 20, 2011	Acceptable with Use Conditions: See rule for detailed conditions.
R-401A	SUVA MP-39	N	0.037	1,182	A1	January 13, 1995	Acceptable
R-401B	SUVA MP-66	N	0.040	1,288	A1	January 13, 1995	Acceptable
R-402A	HP-80	R/N	0.019	2,746	A1	March 18, 1994	Acceptable
R-402B	HP-81	R/N	0.03	2,379	A1	March 18, 1994	Acceptable

(Continued)

Table 10.3 Substitutes in retail food refrigeration. (*Continued*)

Substitute	Trade name(s)	Retrofit/new	ODP	GWP	ASHRAE designation (safety classification)	SNAP listing date	Listing status
R-406A	GHG-12, GHG-X3, McCool, Autofrost X3	R	0.057	1,900	A2	August 26, 1994	Acceptable
R-408A	HCFC Blend Epsilon	R	0.026	55	A1	August 26, 1994	Acceptable
R-409A	HCFC Blend Gamma	R	0.047	1,558	A1	August 26, 1994	Acceptable
R-411A		R/N	0.048	1,600	A2	February 8, 1996	Acceptable
R-411B		R/N	0.052	1,700	A2	February 8, 1996	Acceptable
R-414A	GHG-X4, Autofrost, Chill-it, HCFC Blend Xi	R/N	0.045	1,478	A1	September 5, 1996	Acceptable
R-414B	Hot Shot, Kar Kool, HCFC Blend Omicron	R/N	0.098	3,337	A1	September 5, 1996	Acceptable

(*Continued*)

Table 10.3 Substitutes in retail food refrigeration. (*Continued*)

Substitute	Trade name(s)	Retrofit/new	ODP	GWP	ASHRAE designation (safety classification)	SNAP listing date	Listing status
R-416A	FRIGC FR-12, HCFC Blend Beta	R/N	0.009	1,081	A1	February 8, 1996; February 24, 1998	Acceptable
R-420A	Choice R-420A	R/N	0.008	1,536	A1	October 1, 2004	Acceptable
R-744 (Carbon Dioxide, CO_2)		N	0	1	A1	September 30, 2009	Acceptable

R-401A = Chlorodifluoromethane (R-22), Difluoroethane (R-152a), Chlorotetrafluoroethane (R124), in the ratio of 53/13/34.
R-401B = Chlorodifluoromethane (R-22), Difluoroethane (R-152a), Chlorotetrafluoroethane (R124), in the ratio of 61/11/28.
R-402A = Chlorodifluoromethane (R-22), Pentafluoroethane (R-125), Propane (R-290), in the ratio of 38/60/2.
R-402B = Chlorodifluoromethane (R-22), Pentafluoroethane (R-125), Propane (R-290), in the ratio of 60/38/2.
R-406A = Blend Chlorodifluoromethane (R-22), Isobutane (R-600a) and Chlorodifluoroethane (R-142b) in the ratio of 55/4/41.
R-408A = Chlorodifluoromethane (R-22), Trifluoroethane (R-143a), Pentafluoroethane (R-125), in the ratio of 47/46/7.
R-409A = Chlorodifluoromethane (R-22), Chlorodifluoroethane (R-142b), Chlorotetrafluoroethane (R-124), in the ratio of 60/15/25.
R-411A = Chlorodifluoromethane (R-22), Difluoroethane (R-152a), Propylene (R-1270), in the ratio of 87.5/11/1.5.
R-411B = Chlorodifluoromethane (R-22), Difluoroethane (R-152a), Propylene (R-1270), in the ratio of 94/3/13.
R-414A = Blend Chlorodifluoromethane (R22), Chlorotetrafluoroethane (R124), Isobutane (R-600a), 1-Chloro-1,1-difluoroethane (R142b), in the ratio of 51/28.5/4/16.5.
R-414B = Blend Chlorodifluoromethane (R-22), Chlorotetrafluoroethane (R124)and 1-Chlorodifluoroethane (R142b), Isobutane (R-600a), in the ratio of 50/39/1.5/9.5.
R-416A = Tetrafluoroethane (HFC-134a), 1-Chlorotetrafluoroethane (R124), Butane (R600), in the ratio of 59/39.5/1.5.
R-417A = Pentafluoroethane (R-125), Tetrafluoroethane (R-134a), Butane (R600), in the ratio of 46.6/3.4/50.
R-420A = Tetrafluoroethane (R-134a), Chlorodifluoroethane (142b), in the ratio of 88/12.

This end-use is often used when ambient temperatures approach 200°F (93°C) and corrosive conditions exist.

10.5.2 Motor Vehicle Air Conditioning

Motor vehicle air-conditioning systems (MVACs) provide comfort cooling for passengers in light-duty cars and trucks, buses, trains, and other forms of transportation. A few of the Substitutes for MVACs can be seen in Table 10.4. The data in the Table show whether the Substitutes is New or Retrofit and its Listing Status which simply means whether the compounds can be used. For a complete and up-to-date list of the Substitutes consult EPA Website www.epa.gov/snap/snap-substitutes-sector (Table 10.4).

10.5.3 Non-Mechanical Heat Transfer Systems

Non-mechanical heat transfer systems include cooling systems that do not rely on a vapor compression cycle, such as those using convection to remove heat from an area. Two types of such systems are recirculating coolers, i.e., systems with fluid pumps, and thermosiphons, i.e., those that rely on natural convection currents. This end-use also includes Organic Rankine Cycle (ORC) devices that typically pump refrigerant to recover and utilize energy from lower-temperature heat rejected from other processes.

10.5.4 Residential and Light Commercial Air Conditioning and Heat Pumps

This end-use includes equipment that cools enclosed spaces in households and commercial industries, but excludes chillers which include room air conditioning such as window units, packaged terminal air conditioners (PTAC) and heat pumps (PTHP), and portable air conditioners; central air conditioners (i.e., ducted); non-ducted systems (both mini and multi splits); packaged rooftop units; water-source and ground-source heat pumps; and other products. Residential and light commercial air conditioning and heat pumps are often distinguished from chillers by the fact that they condition the air directly, rather than cool (or heat) water that is then used to condition air.

10.5.5 Residential Dehumidifiers

Residential dehumidifiers are primarily used to remove water vapor from ambient air or directly from indoor air for comfort or material preservation purposes. While air-conditioning systems often combine cooling and dehumidification, this application serves only the latter purpose.

Table 10.4 Passenger air conditioning in light-duty, medium-duty, heavy-duty and off-road vehicles.

Substitute	Trade name(s)	Retrofit/ new	ODP	GWP	ASHRAE designation (safety classification)	SNAP listing date	Listing status
HFC-152a		N	0	124	A2	June 12, 2008	Acceptable with Use Conditions: See rule for detailed conditions
HFO-1234yf	Opteon™ XL10, Solstice® yf	N	0	4	A2L	March 29, 2011; December 1, 2016	Acceptable with Use Conditions: For use in newly manufactured passengers cars, light-duty trucks, medium-duty passenger vehicles, heavy-duty pickup trucks, and complete heavy-duty vans; see rules for detailed conditions.
R-401C		R/N	0	933	A1	June 13, 1995	Acceptable
R-406A	GHG-12, GHG-X3, McCool, Autofrost X3	R/N	0.057	1,900	A2	October 16, 1996; July 20, 2015	Unacceptable as of MY 2017. Detailed conditions apply-see rule.
R-414A	GHG-X4, HCFC Blend Xi, Autofrost, Chill-it	R	0.045	1,478	A1	October 16, 1996; July 20, 2015	Unacceptable as of MY 2017. Detailed conditions apply-see rule.

(Continued)

Table 10.4 Passenger air conditioning in light-duty, medium-duty, heavy-duty and off-road vehicles. (*Continued*)

Substitute	Trade name(s)	Retrofit/ new	ODP	GWP	ASHRAE designation (safety classification)	SNAP listing date	Listing status
R-414B	Hot Shot, Kar Kool, HCFC Blend Omicron	R/N	0.098	3,337	A1	October 16, 1996; July 20, 2015	Unacceptable as of MY 2017. Detailed conditions apply-see rule.
R-416A	FRIGC FR-12, HCFC Blend Beta	R/N	0.009	1,081	A1	June 13, 1995; July 20, 2015	Unacceptable as of MY 2017. Detailed conditions apply-see rule.
R-426A	RS-24	R/N	0	1,510	A1	September 28, 2006; July 20, 2015	Unacceptable as of MY 2017.[1] Detailed conditions apply-see rule.
R-744 (Carbon Dioxide, CO_2)		N	0	1	A1	June 6, 2012	Acceptable with Use Conditions: See rule for detailed conditions.

[1] As per EPA's April 27, 2018 Guidance (83 FR 18431), based on the court's partial vacatur, in the near term EPA will not apply these HFC listings in the 2015 Rule, pending a rulemaking.
HFC-152a = Difluoroethane HFO-1234yf= Tetrafluoropropene R-401C = Chlorodifluoromethane
R-406A = Blend Chlorodifluoromethane (R-22), Isobutane (R-600a) and Chlorodifluoroethane (R-142b) in the ratio of 55/4/41
R-414A = Blend Chlorodifluoromethane (R22), Chlorotetrafluoroethane (R124), Isobutane (R-600a), -Chlorodifluoroethane (R142b), in the ratio of 51/28.5/4/16.5.
R-414B = Blend Chlorodifluoromethane (R-22), Chlorotetrafluoroethane (R-124)and Chlorodifluoroethane (R-142b), Isobutane (R-600a), in the ratio of 50/39/1.5/9.5.
R-416A = Tetrafluoroethane (HFC-134a), Chlorotetrafluoroethane (R124), Butane (R600), in the ratio of 59/39.5/1.5.
R-426A = Pentafluoroethane (R-125), Tetrafluoroethane (R-134a), Butane (R-600), Isopentane (R-601a), in the ratio of 5.1/93/1.3/0.6.

11

Future Directions of Fluorocarbons

11.1 Introduction

There are a great deal of certainties and facts about fluorocarbon and non-fluorinated fluids that guide one to project into the future. Air conditioning and refrigeration were the original applications for which fluorocarbons were developed. This chapter focuses on refrigerant fluids to illustrate the evolution of fluorocarbons. Non-fluorocarbons refrigerants were in use long before the first chlorofluororcarbon (CFC) was developed.

Refrigeration is critically important to people's survival and health. Refrigerants are consumed in large volumes that continue to grow (Figure 11.1). Refrigeration use can be categorized into four applications:

- Domestic refrigeration and freezers are used for the storage of food and beverages typically in households.
- Commercial refrigeration and freezers are used for the storage and display of food and beverage products in retail businesses.
- Industrial refrigeration systems are used for the processing and storage of food and beverages, and in the manufacturing of a number of chemicals.
- Transport refrigeration and freezer systems are used for the carriage of food and beverage products.

Commercial refrigeration is overwhelmingly the largest consumer of refrigerants. Figure 11.2 shows the progression of commercial refrigerant development from the beginning to the present.

There is no single "ideal" refrigerant. Due to the phase-down under the Kigali Amendment, the target refrigerants for main applications will include low GWP refrigerants such as R-717, R-744, hydrocarbons (HCs), unsaturated halochemicals such as hydrofluoroolefins (unsaturated HFCs often referred to as HFOs) and hydrofluorochloroolefins (unsaturated HCFCs, often referred to as HCFOs), and blends of these refrigerants, some even with traditional refrigerant fluids (HFCs, some HCs). Alternatives to high GWP refrigerants exist and new more efficient refrigerants are being proposed which creates a challenge to finding the right refrigerant for each application [3].

A number of gases including carbon dioxide, hydrocarbons (HC's) and ammonia had been used for refrigeration until the advent of fluorocarbon gases in 1920's. Starting in the 1930 a series of fluorocarbon gases were developed to replace the original refrigeration gases until the present. Since the 1970's most of the developments have been aimed at

Sina Ebnesajjad, PhD. Concise Handbook of Fluorocarbon Gases: Applications in Refrigeration and Other Industries, (283–316) © 2021 Scrivener Publishing LLC

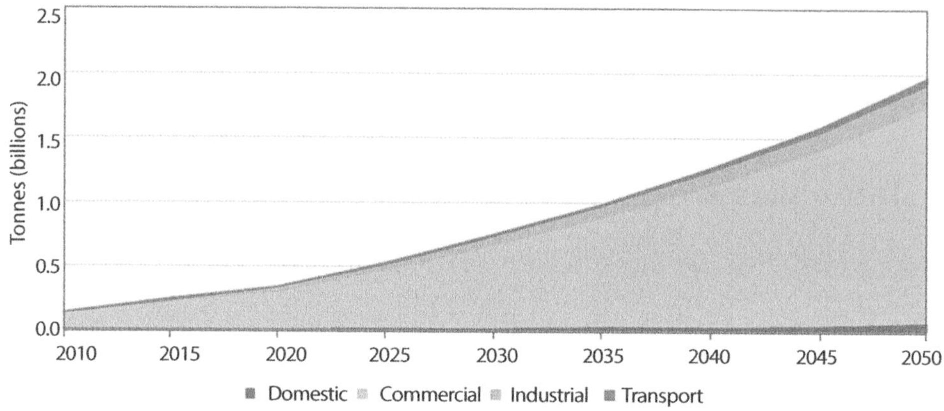

Figure 11.1 Refrigeration demand in its four applications [1].

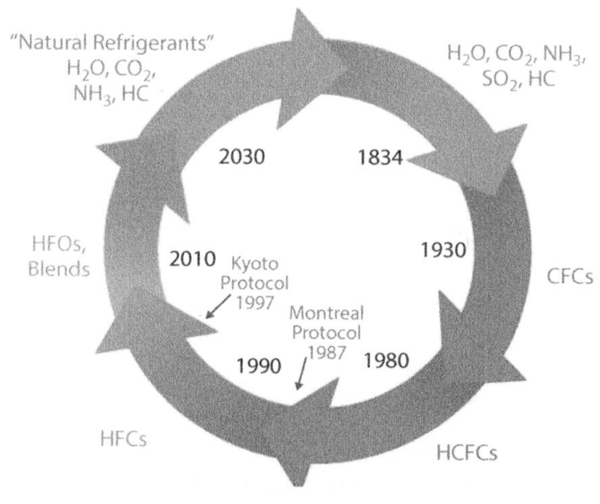

Figure 11.2 Refrigerants – historical development [2].

reducing ozone depletion and global warming through the discovery of new fluorocarbons to replace the old ones. Nearly for 90 years fluorocarbon have been developed to replace the old refrigerants (CO_2, HC's and NH_3). There are some attempts to close the circle (see Figure 11.2) by returning to the same old compounds and away from fluorocarbons.

11.2 Inception and Evolution of Fluorocarbons

Initially, fluorocarbons were developed to overcome flammability and toxicity of old refrigerant gases. That was just the start and much more would be coming. The original CFC morphed into a massive number of fluorinated gases and polymers that are at the foundation of human life in the twenty first century Table 11.1.

Some of the applications required a reliable refrigerant for air conditioning systems of automobiles that were becoming ever more popular. Another important need was for commercial and household refrigeration. Building of large ships and expansion of railroads then allowed shipping of native foods across regions, countries and oceans. No longer people were limited to the "local" food products, transport of which was limited by their perishability.

Development of massive refrigerated ships and aircrafts has contributed to the current state of availability of every imaginable food, fruit and vegetable everywhere. Mobile refrigeration has allowed Australian lamb chops, Chilean nectarines and Hawaiian pineapples to become available disregarding the season and distance. The underdeveloped corners of Africa and Asia have been significant beneficiaries of modern refrigeration.

Prior to the 1930's selection of refrigerants relied on "whatever worked" without many constraints. In early twentieth century automobile manufacturing technologies progressed resulting in more reliable and less costly vehicles. Cars were becoming popular among the masses who could afford them. An impediment to the use of cars in summer was the heat in the passenger compartment. Air conditioning was a solution hampered by lack of suitable refrigerants that were non-toxic, inflammable, efficient and economical.

One convenient way to discuss fluorocarbons and refrigerants is to divide them into classes. Traditionally, refrigerants have been divided into four generations based on the decrease in ozone depletion and global warming potentials. Different authors and organizations have proposed different variations in the classification of refrigerants. The most significant difference is the inclusion of the so called Natural Refrigerants such as a NH_3 and CO_2 in the First or Fourth Generation.

Table 11.1 Examples of proliferation of fluorocarbons in the last ~100 years.

Product	Size of molecule	Example of compound	Example of application
Fluorocarbons	Relatively Small	HFC-134a (1,1,1,2-tetrafluoroethane), HFO-1234yf (2,3,3,3-tetrafluoropropen)	Refrigerant, Propellant for aerosols, Blowing agent for foams
Thermoplastic Fluoropolymers (TPF) and Fluoroelastomers (FE)	Large macromolecules	Polyvinylidene fluoride (TPF), copolymer of vinylidene fluoride/ tetrafluoroethylene	Tubes, pump bodies (TPF), seals and gaskets (FE)
Fluorinated Coating	Dispersion (D), Powder (P)	Polytetrafluoroethylene coatings, Tetrafluoroethylene/ perfluoroalkyl vinyl ether copolymer powder	Interior surface of oil pipelines (D), Corrosion resistant (P)

11.3 Classification of Refrigerants

This section describes the classification of refrigerants in which non-fluorocarbon refrigerants are listed as the First Generation. Table 11.2 is a simple illustration of the impact of the four generations of refrigerants with an emphasis on fluorocarbons roles. The second generation consisted of refrigerants that were safe and not flammable but they had both high ozone depletion and global warming potentials.

11.3.1 First Generation (Prior to 1930)

During the first century solvents and other volatile fluids as long as it worked and was available. Nearly all of those early refrigerants were flammable, toxic and even chemically reactive (Tables 11.3 and 11.4). Accidents were common with fluids such as propane (R-290) because of flammability. In contrast ammonia was preferred because of its higher flame resistance than propane [4]. Significantly, the first generation have literally no Ozone Depletion Potential (ODP) and low Global warming Potential (GWP). But they suffer from poor properties such as insufficient flammability resistance and toxicity and in some cases poor refrigeration efficiency.

Carrier and Waterfill [5] assessed the performance of carbon dioxide (R-744) and concluded it depended on the cycle and amount of liquid subcooling. However, it yielded the lowest predicted performance of the fluids analyzed. They also stated ammonia and water (R-718) would require excessive stages for centrifugal compressors to achieve the desired conditions. Water "gives a low efficiency of performance."

Sulfur dioxide (R-764) came short for safety reasons and carbon tetrachloride (R-10) for incompatibility with metals in presence of moisture. The remaining candidate was 1,2-dichloroethene (R-1130) for the first centrifugal machine. This selection was flawed because of high flammability, harsh and sharp odor at low concentrations [6]. There began research to develop new refrigerants.

Table 11.2 The impact of four generations of refrigerants on ozone depletion and global warming.

	Time period	Chemical family	ODP	GWP
First Generation	<1930	Natural Refrigerants (NR)	Zero	Low
Second Generation	1931-1990	CFCs/HCFCs	High	High
Third Generation	1990-2010	HFCs	Zero	High
Fourth Generation	>2010	HFO and NR	Zero	Low

Table 11.3 Properties of first generation refrigerants [4].

Substance	R number	MW	Boiling point (°C)	ODP	GWP
Carbon Dioxide	R-744	44	-55.6	0	1
Ammonia	R-717	17	-33.3	0	0
Sulphur Dioxide	R-764	64	-10	0	0
Ethyl Ether	R-610	74	35	0	0
Dimethyl Ether	R-70	46	-25	0	0
Methyl Chloride	R-40	50.50	-24.2	0.02	16

Table 11.4 Application ranges of natural refrigerants [2].

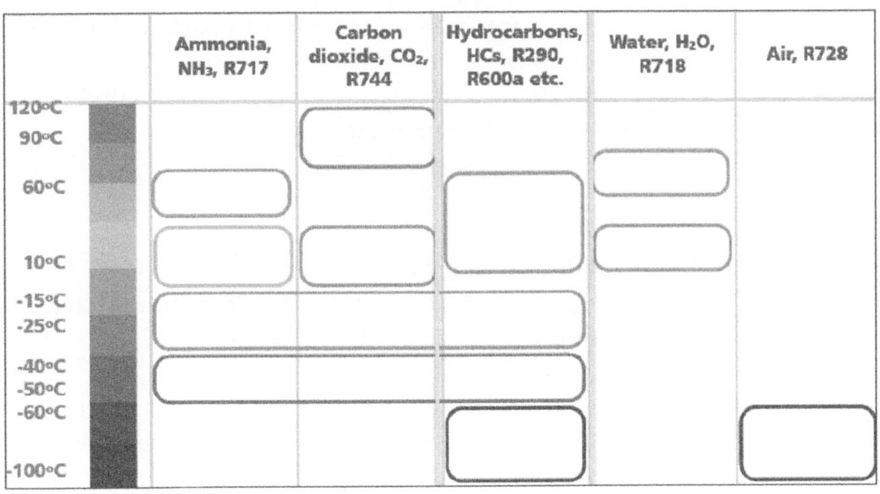

Ammonia is used extensively commercially in large freezing and refrigeration plants. It in anhydrous form because it contains almost no water by being 99.98% pure...Ammonia is a more efficient refrigerant than CFCs by 3-10%. That implies an ammonia-based system requires less electricity, resulting in lower operating costs. It is a colorless, highly irritating gas with a pungent and suffocating odor. Ammonia is not chemically inert because it is alkaline and corrosive. For example, it reacts with copper thus requiring non-copper tubing. It is toxic and combustible but not highly flammable. Ammonia containers may explode when exposed to high heat.

11.3.2 Second Generation (1931–1990)

Having not found a suitable fluid for refrigeration, research and development on new chemistries began. The second generation of refrigerants included chlorofluorocarbons (CFCs), which became widely used because they were efficient, chemically inert, non-flammable and non-toxic [7].

With the directive stating that "the refrigeration industry needs a new refrigerant if they expect to get anywhere," Thomas Midgley, Jr., and his associates Albert L. Henne and Robert R. McNary searched for chemicals with the desired boiling point and other properties. They restricted the search to those compounds known to be stable, that were neither toxic nor flammable. The second generation refrigerants utilized chlorofluorocarbon and hydrofluorocarbon gases because of their safety and resistance to degradation [8, 9].

Neither compounds were toxic or flammable gases with relatively high mass. They were good refrigerants because they could be compressed readily to liquid and carries away a significant amount of heat upon evaporation. They were quite stable and were only degraded by ultraviolet rays thus were good fits for a number of applications. They did not react with almost any substance, worked well as solvents, blowing agent, fire extinguishing agent and aerosol propellants.

Commercial production of R-12 began in 1931 followed by R-11 in 1932. Chlorofluorocarbons (CFCs) and later – especially starting in the 1950s in residential and small commercial air conditioners and heat pumps – hydrochlorofluorocarbons (HCFCs) dominated the second generation of refrigerants. Ammonia continued as, and remains today, the most popular refrigerant in large, industrial systems especially for food and beverage processing and storage [6]. Characteristics and properties of second generation of refrigerant and their properties have been summarized in Table 11.5.

11.3.3 Third Generation (1990–2010)

After the signing of the Montreal Protocol in 1987 switch from CFCs to HCFCs began, due to lower ODP of HCFCs, as a temporary measure began. Fluorochemicals continued to be the primary choice, with emphasis on HCFCs for interim (transitional) use and hydrofluorocarbons (HFCs) for the longer term (Table 11.6).

In the decade following the Montreal Protocol manufacturers developed a number of hydrofluorocarbons (HFCs) and their derivatives or blends that could meet the specifications

Table 11.5 Properties of second generation refrigerants [11].

Substance	R no.	M Kg/kmol	NBP (°C)	ODP	GWP
Trichlorofluoro methane	R-11	137.4	23.71	1	4000
Dichlorodifluoro methane	R-12	120.91	-29.75	1	8500
Chlorotrifluoro methane	R-13	104.5	-81.3	1	11700
Chlorodifluoro methane	R-22	86.47	-40.8	0.055	1700
R-22/R115	R-502	111.6	-45.3	0.33	5600

Table 11.6 Properties of the third generation refrigerants (NBP = Normal Boiling Point, M = Molecular Weight) [11].

R number	M Kg/kmol	NBP(°C)	GWP
R-32	52.02	-51.65	580
R-134A	102.03	-26.07	1300
R-404A	97.6	-46.6	3800
R-407C	86.2	-43.8	1600
R-410A	72.59	-51.6	1900
R-507	98.86	-47.1	4000
R-508A	100.1	-87.4	13000

R-404A blend = HFC-125, HFC-143A and HFC-134A (44/52/4 wt.%).
R-407C blend = HFC-32, HFC-125, and HFC-134A (23/25/52 wt.%).
R-410A blend = R-125 and R-143A (50/50 wt.%).
R-508A blend = R-23 and R-116 (39/61 wt.%).

of various refrigeration applications. Care was taken to ensure reduced service and emissions of refrigerants during service and disposal of old fluorocarbons. This class of refrigerants was thus considered as third generation. In 2018, HCFCs and HFCs contributed only about 2% of the total global warming. They are on a schedule to be phased out by the mid 21st Century [10].

Hydrofluorocarbons (HFC's) form the third generation of fluorine-based refrigerants and include: R-134a, R-404A, R-407C, R-410A, R-413A, R-417A, and more. Compared to HCFC's, HFC's have low ODP (zero or close to zero) and relatively, though significant, low GWP. Commercialization of the Third Generation of the fluorocarbons began to reverse the damage to the ozone layer as discussed in the earlier chapters of this book. Nevertheless, the estimate indicate it will take over a century to repair the damage to the ozone layer. Unfortunately, there was limited impact on the global warming and additional developments were required.

11.3.4 Fourth Generation (Beyond 2010)

The Intergovernmental Panel on Climate Change (IPCC) Fourth Assessment Report set forth the latest scientific consensus, that stated "warming of the climate system is unequivocal, as is now evident from observations of increases in global average air and ocean temperatures, widespread melting of snow and ice, and rising global average sea level" [12].

The assessment concluded that "most of the observed increase in globally averaged temperatures since the mid-20th Century is very likely due to the observed increase in anthropogenic greenhouse gas concentrations" and that "discernible human influences now extend to other aspects of climate, including ocean warming, continental-average temperatures, temperature extremes and wind patterns".

The effort to reduce the emission of Green House Gases (GHG) began by the Kyoto Protocol that was adopted in Kyoto, Japan, on 11 December 1997. It entered into force in

February 2005 to expire in 2012. There were 192 parties to the Protocol. The targets for the first commitment period of the Kyoto Protocol cover emissions of the six main greenhouse gases, namely:

- Carbon dioxide (CO_2);
- Methane (CH_4);
- Nitrous oxide (N_2O);
- Hydrofluoroolefins (HFCs);
- Perfluorocarbons (PFCs);
- Sulphur hexafluoride (SF_6); and
- Nitrogen trifluoride (NF_3)

The maximum amount of emissions (measured as the equivalent in carbon dioxide) that a Party may emit over a commitment period in order to comply with its emissions target is known as a Party's assigned amount. The individual targets for Annex I Parties are listed in the Kyoto Protocol's Annex B [13].

In Doha, Qatar, on 8 December 2012, the "Doha Amendment to the Kyoto Protocol" was adopted [14]. The amendment includes:

1. New commitments for Annex I Parties to the Kyoto Protocol who agreed to take on commitments in a second commitment period from 1 January 2013 to 31 December 2020;
2. A revised list of greenhouse gases (GHG) to be reported on by Parties in the second commitment period; and
3. Amendments to several articles of the Kyoto Protocol which specifically referenced issues pertaining to the first commitment period and which needed to be updated for the second commitment period.

A second commitment period was agreed in 2012, known as the Doha Amendment to the Kyoto Protocol, in which 37 countries have binding targets. Some countries withdrew from the Kyoto Protocol or did not put into legal force the Amendment with second round targets. Countries such as Japan, New Zealand, and Russia participated in Kyoto's first-round but have not taken on new targets in the second commitment period.

Other developed countries without second-round targets are Canada (which withdrew from the Kyoto Protocol in 2012) and the United States (which has not ratified). As of January 2020, 136 countries have accepted the Doha Amendment, while entry into force requires the acceptances of 144 states. Of the 37 countries with binding commitments, 7 have ratified [14]. The second commitment period affects only 14% of global emissions because only EU Member States, other European countries and Australia have commitments.

The Fourth Generation of refrigerants had to have both low ODP and GWP to be acceptable replacements for the HFCs [6]. The fluorocarbon industry had anticipated the global action to require refrigerants that met both requirements. Early development overcame the once insurmountable task of developing refrigerants that met both low ODP and GWP requirements.

11.3.5 Hydrofluoroolefin Fluorocarbons

Unsaturated HFCs known as HFOs contain carbon, hydrogen and fluorine which are the elements present in other fluorocarbons. HFO do not contain chlorine (Table 11.7). These molecules that are called hydrofluoroolefins (HFOs) contain one carbon-carbon double bond. They are based on the propylene molecule (CH_3-CH=CH_2) as alternatives to high-GWP HFCs.

Unsaturated HFOs are used as a single substance such as HFO-1234yf for automotive air-conditioning systems, or in mixtures with HFCs to reduce the GWP of the blend. Unsaturated HFOs have high reactivity thus degrade in the atmosphere and have shorter lifetimes. Table 11.8 shows examples of hydrofluoroolefins some of which are now in commercial applications. For example, HFO-1234yf and HFO-1234ze have proven to be the most promising fourth-generation refrigerants [16]. Table 11.9 shows the boiling points a few HFO fluorocarbons.

Table 11.10 shows a list of global warming potentials for select fluorocarbons from the CFC, HCFC, HFC and HFO fluorocarbon groups. To put it in context, CFCs contribute anywhere from four to fourteen thousand times than HFOs. In other words, one ton of CFC-13 is equivalent to emission of 14,000 tons because almost all HFO's have GWP <1. This is why the hydrofluoroolefins are being commercialized in spite of cost and somewhat lower resistance to flammability. Several properties of two HFOs are summarized in Table 11.11.

ASTM E681 standard defines flammability limits and provides methods for its measurement. Flammable refrigerants, whether having low or high flammability propagate flame thus have limits. The Lower (LFL) and Upper (UFL) and Flammability Limits refer to the minimum and maximum concentrations of a substance in air that can propagate a flame. At concentrations lower than LFL there is insufficient combustible gas to sustain flame. Above UFL the concentration of combustible gas is too high and oxygen too high for sustaining a flame. The lower the LFL, the higher the risk, because a flammable concentration can be more easily reached due to gas leakage. Another factor is the difference between LFL and UFL. As this difference increases the concentration window where ignition could take place grows.

Table 11.7 Chemical composition of fluorinated hydrocarbons [15].

Chemical prefix	Common name	Atoms in molecule			
		H	Cl	F	C
CFC	Chlorofluorocarbon		X	X	X
HCFC	Hydrochlorofluorocarbon	X	X	X	X
HFC	Hydrofluorocarbon	X		X	X
HC	Hydrocarbon	X			X
HFO	Hydrofluoroolefin	X		X	X

Table 11.8 Examples of hydrofluoroolefins.

HFO 216 HFP	Hexafluoropropene	$CF_3C{=}CF_2$
HFO-1225yeZ	1, 2, 3, 3, 3-pentafluoropropene Z isomer	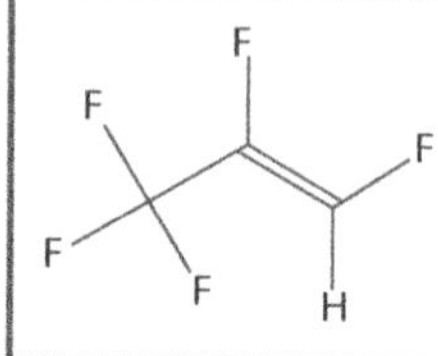
HFO-1225yeE	1, 2, 3, 3, 3-pentafluoropropene E isomer	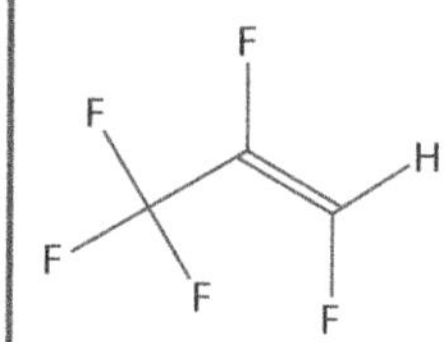
HFO-1225zc	1, 1, 3, 3, 3-pentafluoropropene	$CF_3CH{=}F_2$
HFO-1225yc	1, 1, 2, 3, 3-pentafluoropropene	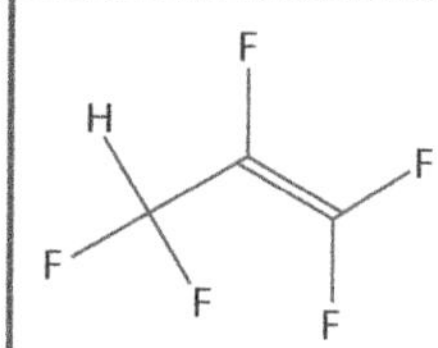
HFO-1234yf	2, 3, 3, 3-tetrafluoropropene	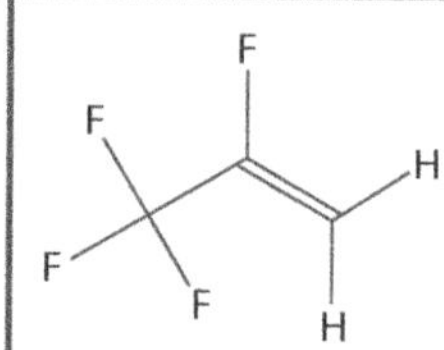
HFO-1234zeE	Trans-1, 3, 3, 3-tetrafluoropropene	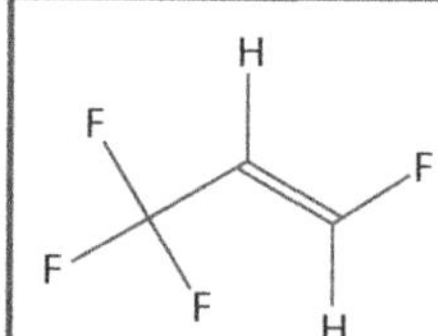
HFO-1234zeZ	Cis-1, 3, 3, 3-tetrafluoropropene	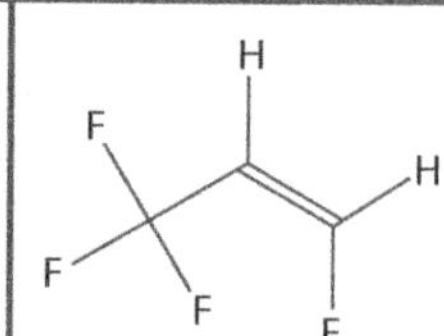
HFO-1243zf	3, 3, 3-trifluoropropene	$CF_3CH{=}CH_2$

Table 11.9 Examples of boiling points of HFO refrigerants.

Selected HFOs	Boiling point °C
HFO-216	-29.6
HFO-1225yeZ	-21
HFO-1225yeE	-16
HFO-1234yf	-29.4
HFO-1234zeE	-19
HFO-1234zeZ	-6
HFO-1243zf	-26

Table 11.10 Range of global warming potential of the second, third and fourth generation of refrigerants [17].

	Lowest GWP		Highest GWP	
CFCs	CFC-11	4,660	CFC-13	13,900
HCFCs	HCFC-122	59	HCFC-142b	1,980
HFCs	HFC-152a	138	HFC-23	12,400
HFOs	HFO-1132a	<1		
	HFO-1141	<1		
	HFO-1225ye	<1		
	HFO-1234yf	<1		
	HFO-1234ze	<1		
	HFO-1132a	<1		
	HFO-1132a	<1		
	HFO-1345zfc	<1		

R-290 (propane), as seen in Table 11.12 and Figure 11.3, has a much lower LFL than both R-32 and R-1234yf. Therefore, it is easier to reach a flammable concentration from a leak with R-290. This is typical of A3s versus A2Ls, as hydrocarbons (A3s) tend to have lower flammability limits than A2Ls (Figure 11.3). Additionally, the molecular weights of these molecules also tend to be lower than those of A2Ls, meaning less mass is required to reach a flammable concentration. This is critical when designing equipment, as it plays largely into system charge size [19].

One of the benefits of blending HFOs with previous generous agents is to increase its flammability at the expense of increased GWP. R-448A is an example of a blend of HFC and HFO fluorocarbons for use in refrigeration. It was suitable for the replacement of a number

Table 11.11 Physical properties of HFO-1123 and HFO-1234yf refrigerants [18].

	HFO-1123	HFO-1234yf
Molecular weight, g/mol	82.0	114.0
Normal boiling point, K	214.0	243.7
Critical temperature, K	331.9	367.9
Critical pressure, MPa	4.5	3.38
Critical density, kg/m^3	492	476
Vapor pressure (273K), kPa	1084	316
Liquid density (273K), kg/m^3	1126	1176
Vapor density (273K),kg/m^3	48.6	17.7
Flammable limits (296K, 101.3kPa), vol%	10.4-29.3[a,b]	6.2-12.3
LFL, kg/m^3	0.34	0.29
Burning velocity, cm/s	6.6*	1.5
Nominal heat of combustion, MJ/kg	9.9	10.7
Atmospheric lifetime, Year	1.5 Days[a]	11 Days
GWP(ITH = 100), CO_2 = 1	0.0054[a]	4
Ames test	Negative	Negative
LC_{50}, ppm	>200,000	>400,000
OEL, ppm	750	500
ASHRAE Classification	(A2L) (Expected)	A2L

[a]Measured by the National Institute of Advanced Industrial Science and Technology.
[b]35°C, dry air.

of the Second and Third generation single and blend fluorocarbons. A commercial example of R-448A is Soltice® N40 by Honeywell (Table 11.13 and 11.14). It is a zeotropic blend designed to serve as a replacement for R-22 and R-404A in vending machines, supermarket refrigeration retrofits or new systems. It has high energy efficiency.

Effect of blending HFO and HFC fluorocarbons on the physical properties of the blends is summarized in Table 11.15, based on calculations according to Lemmon *et al.* [22]. HFO-1123 density at the compressor inlet is higher than HFC's like R-410A and HFC-32 thus the capacity is larger than both. However, the coefficient of performance (COP) of HFO-1123 is inferior due to the lower critical point of HFO-1123. To improve the COP HFO-1123 and HFC-32 were then blended together and studied as refrigerants. According to Table 11.15, the evaporating and condensing pressures of HFO-1123, HFC-32, and HFO-1123/HFC-32 mixtures were higher than those of R-410A. By adding HFO-1234yf, the evaporating and

Table 11.12 Safety groups & flammability test requirements per ANSI/ASHRAE standard 34-2016 [19].

Refrigerant ASHRAE designation #	**R-1234yf**	**R-32**	**R-290**
ASHRAE Safety Group	A2L	A2L	A3
Lower Flammability Limit (LFL) (vol. % in air/ kg/m^3)	6.2/0.289	14.4/0.307	2.2/0.038
Upper Flammability Limit (UFL) (vol. % in air)	12.3	29.3	10.0
UFL – LFL (vol. % in air)	6.1	14.9	7.8
Minimum Ignition Energy (MIE) (mJ)	>5,000	30-100	0.25
Burning Velocity (S_u) (cm/s)	1.5	6.7	46
Heat of Combustion (HOC) (kJ/g)	10.7	9.4	46.3
	$C_3H_2F_4$ (HFO)	CF_2H_2 (HCF)	C_3H_8 (HC)

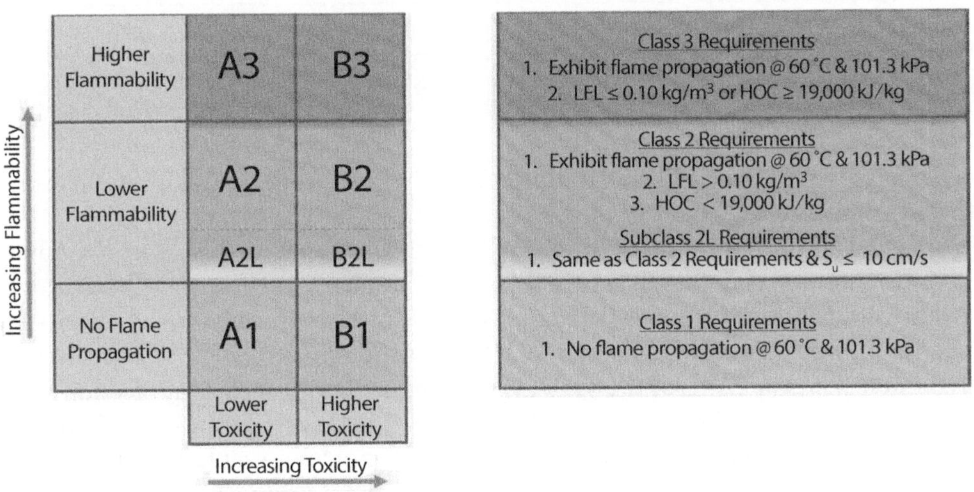

Figure 11.3 Safety groups & flammability test requirements ANSI/ASHRAE standard 34-2016 [19].

Table 11.13 Composition of R-448A is an example of HFO and HFC blend for commercial refrigeration [20].

Refrigerant	**Type**	**Replaces**	**Applications**	**Comments**
Solstice N40 (R-448A)	Blend R-32 R-125 R-1234yf R-134a R-1234ze	 R-22 R-404A R-507 R-402A R-408A	Supermarket freezers, coolers, display cases. Liquid (DX) chillers. Warehouse refrigeration	Among the most efficient refrigerant replacements. Lowest GWP value compared to available nonflammable alternatives. Make appropriate control adjustments.

Table 11.14 Properties of R-448A (HFO and HFC blend) [21].

Class/Type	**Zeotropic blend**
Formula	26%/26%/21%/7%/20%
	R-32/R-125/R-134a/R-1234ze/R-1234yf
Kind	HFC/HFO
Appearance	Colourless
ODP (ODP-R11=1)	0
GWP rev 3rd/4th/5th IPCC	1300/1387/1273
Flammability Limits – ASTM E681-04 @ 21°C	Non Flammable
ASHRAE Std. 34 Safety Class	A1
ATEL/ODL (kg/m^3)	0.390
Practical limit kg/m^3	0.390
LFL (% vol)	Non flammable
REACH	Registered

condensing pressures were adjusted to the level of R-410A, while maintaining their relative COPs and relative capacities.

Figures 11.4 and 11.5 are plots of the effect of HFC-32 content on the Lower Flammability Limit (LFL) and GWP of blends with HFOs. The numbers in these have been extracted from Table 11.15. Both LFL and GWP increase as HFC-32 content increases, as expected.

There is a trade-off in the inclusion of HFOs in the refrigerant blends because of the differences in the properties of the gases. For example, the drop-in test results for HFO-1123/HFC-32 and HFO-1123/HFC-32/HFO-1234yf mixed refrigerants (See Table 11.15) are shown in Figure 11.6. APF (Annual performance factor) represents heating and cooling capacity per kWh over one year of use of an air conditioner under specific conditions. The higher the APF, the greater an air conditioner's energy efficiency [23]. Here APF was calculated according to ISO 16358–3 (2013): *Air-cooled Air Conditioners and Air-to-air Heat Pumps – Testing and Calculating Methods for Seasonal Performance Factors – Part 3: Annual Performance Factor.*

In summary, it is clear that blends of HFO-1123/HFC-32 refrigerants and HFO-1123/HFC-32/HFO-1234yf refrigerants are somewhat flammable. They have low-toxicity, similar to the properties of R-410A, which render the HFO-1123 mixtures appropriate as substitutes for R-410A.

11.4 Natural Refrigerants

The word "natural" in natural refrigerants implies to the origin of the fluids is nature. That is, they occur as a result of geological or biological reactions. The most common compounds

Table 11.15 Physical properties of HFO-1123 and HFO-1234yf refrigerant mixtures (A1, A2, A3, B1, B2) [24].

	HFO-1123	HFC-32	HFO-1234yf	A1	A2	A3	B1	B2	
	100	0	0	32	40	45	19	40	
	0	100	0	68	60	55	55	44	
	0	0	100	0	0	0	26	16	R-410A
Boiling point, K	214.0	221.5	243.7	217.6	216.8	216.4	220.1	217.6	221.8
Vapor pressure (298K), kPa	2123	1690	683	1915	1967	1997	1710	1881	1657
Liquid density (298K), kg/m^3	1006	961	1092	980	985	988	992	999	1059
Vapor density (298K), kg/m^3	103	47.3	37.9	61.2	65.6	68.6	57.3	66.5	66.0
Temperature glide, K	0.0	0.0	0.0	1.7	1.7	1.6	2.6	2.7	0.1
Flammability limits, vol%	10.4–29.3a	13.5 – 27.5	6.7 – 11.7	11.7 – 25.2	11.5 – 25.9	11.2 – 25.9	ND	ND	–
Burning velocity, cm/s	6.6a	6.7	1.5	<4.0	<4.0a	<4.0a	(<4.0)	(<4.0)	–
GWP(ITH = 100), CO_2 = 1	<1a	675	<1	<461	<405	<373	<373	<298	2090

HFC-32 = CH_2F_2.
HFO-1123 = $CHCl{=}CH(CF_3)$.
HFC-410A = mixture 50% CH_2F_2 and 50% CHF_2CF_3.
HFC-125 = CHF_2CF_3.

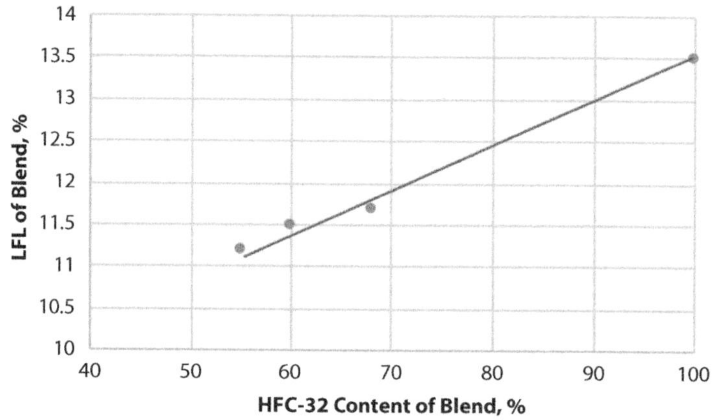

Figure 11.4 Effect of HFC-32 on the LFL of the mixture with HFOs [18].

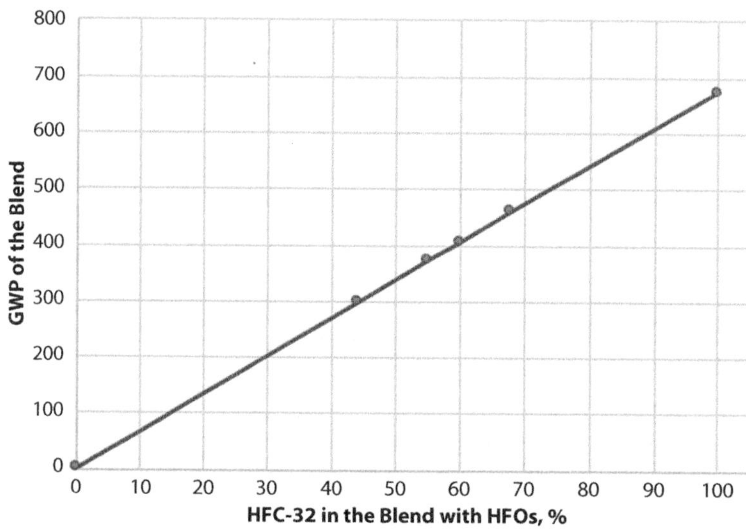

Figure 11.5 Effect of HFC-32 on the GWP of the blends with HFOs [18].

include carbon dioxide, ammonia and hydrocarbons which in reality are industrial gases produced in large chemical processing plants. These facilities consume energy to produce and purify those gases. They also consume feedstocks leading to generate of wastes similar to other chemical manufacturing processes. All three chemicals have appreciable carbon foot prints. The word natural must thus be used with that understanding.

There are a number of other viable natural refrigerants, such as water, air, helium, nitrogen, et ct, and a variety of types of systems that they can be used in (sorption, gas cycles, Stirling machines, and so on) Properties of some of the candidate gases can be found in Table 11.16. Only three main refrigerant types are of most interest in compression systems. These three fluids, ammonia, carbon dioxide and hydrocarbons, are more environmentally friendly than many other synthetic refrigerants [25].

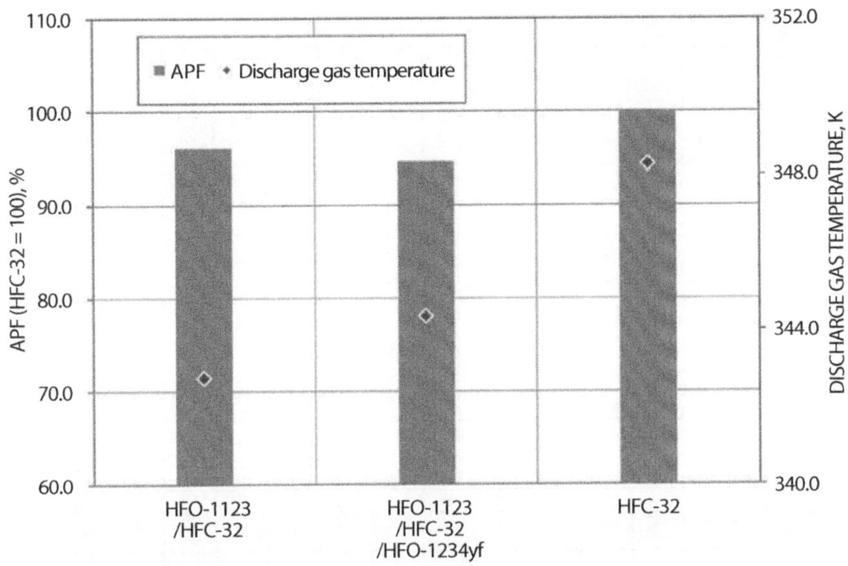

Figure 11.6 The drop-in test results of HFO-1123 mixed refrigerants [24].

Ammonia (NH_3), carbon dioxide (CO_2) and hydrocarbons (HC), water and air have been used as refrigerants for vapor-compression refrigeration during the second half of the 19th century (Table 11.17). Consequently, a great deal of experience and knowledge has been accumulated. However, the use of natural refrigerants requires careful and special handling. In this section three main types of natural refrigerants are described: 1. CO_2 (R744); 2. HCs, e.g. propane (R290); and 3. NH_3 (R717) [26]. Important applications of those refrigerants are summarized in Table 11.18.

11.4.1 Carbon Dioxide

Carbon dioxide known as R744 is not a new comer to refrigeration. It has a longstanding history as a refrigerant going back to mid 1800s. Carbon dioxide is economical due to its broad availability. It has low GWP and is safe because of inflammability. Further it is inexpensive, non-corrosive, low acute toxicity thus meets many requirements of desirable refrigerants. Consumption of R744 declined after the introduction of fluorocarbon refrigerants. Some of the other important characteristics of R744 include excellent heat transfer coefficients, even higher than many fluorocarbons. It can also be used as foam blowing agent and as a fire extinguishing gas [28].

The most noticeable characteristics of CO_2 that differ from other commonly used refrigerants are critical temperature and working pressure. Figure 11.7 shows that CO_2 reaches its critical temperature at 31°C (87.7°F) with a saturated pressure of 7.27 MPa (1,055 psig) [29].

CO_2 has the highest saturation pressure among all natural refrigerant candidates (see Table 11.19) and even fluorocarbons. Refrigeration using carbon dioxide requires substantially higher pressures than other natural refrigerants. The refrigeration equipment must use stronger construction materials along with increased wall thickness of process surfaces. On the mitigating side, carbon dioxide has a significantly larger refrigeration capacity than

Table 11.16 Standard parameters of some ASHREA natural refrigerant candidates – R-170= ethane, R-290=propane, R-600=n-butane, R-600a=iso-butane, R-1270= propylene, R-717=ammonia and R-744=CO_2 [26].

Type	Groups parameters	Natural refrigerants						
		(R-170)	(R-290)	(R-600)	(R-600a)	(R-1270)	(R-717)	(R-744)
Environment	ODP[1]	0	0	0	0	0	0	0
	$GWP_{(100\text{-}Years)2}$	20	20	20	20	20	0	1
	Life (Yrs)	12	12	12	12	12	0.019	1
	MROA (°C)	401.8	376.8	301.8	301.85	301.85	426.85	1726.9
Physical	T_{FP}(°C)	−183	−188	−140	−159.6	−59	−77.73	−78
	T_{NBP}(°C)	−88.58	−42.11	−0.49	−11.75	−47.62	−33.33	-78.46
	T_{CP}(°C)	32.17	96.74	151.98	134.66	91.06	132.25	31.1
	T_{IG} (°C)	505	470	420	460	460	850	∞
	$P_{Evap}{}^{3}$(bar)	16.29	2.916	0.563	0.890	3.6302	2.3617	22.908
	$P_{Conds}{}^{4}$(bar)	46.55	10.790	2.834	4.0472	13.050	11.672	72.137
	P_{crit}(bar)	48.72	42.512	37.96	36.29	45.55	113.33	73.77
	$h_{fg\,evap}$(kJ/kg)	350.6	394.65	398.23	367.97	399.09	1313.2	270.93
	$V_{SPevap}{}^{5}$(m^3/kg)	0.0334	0.1538	0.6375	0.39904	0.12815	0.50868	0.01646

(Continued)

Table 11.16 Standard parameters of some ASHREA natural refrigerant candidates – R-170= ethane, R-290=propane, R-600=n-butane, R-600a=iso-butane, R-1270= propylene, R-717=ammonia and R-744=CO_2 [26]. (*Continued*)

	Groups parameters	**Natural refrigerants**						
		(R-170)	**(R-290)**	**(R-600)**	**(R-600a)**	**(R-1270)**	**(R-717)**	**(R-744)**
Transport Properties at 30°C	$\mu_{1(\mu Pa\text{-}s)}$	292.42	92.188	151.07	143.43	91.479	125.45	43.768
	$\mu v_{(\mu Pa\text{-}s)}$	8.15	8.4628	7.5182	7.6308	9.1707	9.9953	25.170
	kl (m-w/m-K)	153.12	91.409	102.66	87.489	110.14	471.35	95.356
	kv (m-w/m-K)	14.561	19.724	17.114	17.371	19.722	26.846	98.023
	hfg (kJ/kg)	529.10	326.70	356.30	323.33	324.98	1144.4	60.575
	ρ_1 (kg/m^3)	779.02	484.39	566.98	544.31	497.45	595.17	593.31
	ρ_v (kg/m^3)	0.900	23.451	7.1366	10.480	27.718	9.0533	345.10
	P_{reduc}	0.955	0.253	0.074	0.111	0.286	0.103	0.978

(*Continued*)

Table 11.16 Standard parameters of some ASHREA natural refrigerant candidates – R-170= ethane, R-290=propane, R-600=n-butane, R-600a=iso-butane, R-1270= propylene, R-717=ammonia and R-744=CO_2 [26]. (*Continued*)

	Groups parameters	**Natural refrigerants**						
		(R-170)	**(R-290)**	**(R-600)**	**(R-600a)**	**(R-1270)**	**(R-717)**	**(R-744)**
Performance properties	C°p (298K)	52.456	73.307	98.442	96.598	64.376	35.637	37.134
	(J- mol-^{1} k-^{1})	30.069	44.096	58.122	58.122	42.08	17.03	44.01
	M (kg/kmol)	2.85	3.70	5.033	4.54	3.59	4.94	3.14
	CRω	0.0995	0.1521	0.201	0.184	0.146	0.2560	0.22394
Chemical	ASHREA-34	A3	A3	A3	A3	A3	B2L	A1

Nomenclature

P_E/pc Pressure ratio
pr Reduced Pressure=p_{Conds}/P_{crit}
$\mu_{(\mu Pa\text{-}s)}$ Dynamic Viscosity
$h_{fg\,(KJ/kg)}$ heat of vaporation
K(m-w/m-K Thermal Conductivity
T (°C) Temperature
P (bar) Pressure
C°p (J-mol^{-1} k^{-1}) Specific heat at constant pressure
ρ (kg/m^3) Dynamic Viscosity
P_{reduc} p/pcrit Reduce Pressure
P_r Prandtl no.
P_{Cond}/P_{Evap} pressure ratio
MROA (°C) Maximum Range of Applicability(Tmax)
M (kg/kmol) Mass of refrigerants
COP Coefficient of Performance
V_{sp}(m^3/kg) Specific Volume

ω Acentric factor
LFL Lower Flammability Limit
P_{Cond}/P_{Evap} pressure ratio
CR Compression ratio pConds/pEvap

Subscripts

FP Freezing point
NBP Normal boiling point (1.01 bar)
Evap Evaporator
crit Critical
v vapour
n normalized
red Reduced pressure
IG Auto ignition temperature
Conds Condensor
l liquid
a actual

Table 11.17 Natural and other alternative refrigerant characteristics [2].

Refrigerant	Refrigerant number	Chemical formula	GWP (100 years)	ODP	Normal boiling point (°C)	Critical temperature (°C)	Critical pressure (bar)	Safety group	Molecular weight (g/mol)
Ammonia	R17	NH_3	0	0	-33.3	132.4	114.2	B2	17.03
Carbon dioxide	R744	CO_2	1	0	-78	31.4	73.8	A1	44.0
Propane	R290	C_3H_8	3.3	0	-42.1	96.7	42.5	A3	44.10
Isobutene	R600a	C_4H_{10}	4	0	-11.8	134.7	36.48	A3	58.12
Propylene	R1270	C_3H_6	1.8	0	-48	91	46.1	A3	42.08
Water	R718	H_2O	0	0	100	373.9	217.7	A1	18.0
Air	R729	–	0	0	-192.97	–	–	–	28.97

Table 11.18 Applications of natural refrigerants [27].

Refrigerant	Sectors of application
CO_2 (R744)	Commercial and industrial refrigeration, chillers, heat pumps, transport refrigeration, marine refrigeration, vehicle air conditioning, integrals
Propane (R290)	Commercial and industrial refrigeration, chillers, heat pumps, split air conditioners (non-ducted), integrals
Iso-butane (R600a)	Domestic refrigerators, small commercial refrigeration systems, heat pumps
Propylene (R1270)	Chillers, heat pumps, integrals
NH_3(R717)	Industrial and large commercial refrigeration, chillers, industrial heat pumps

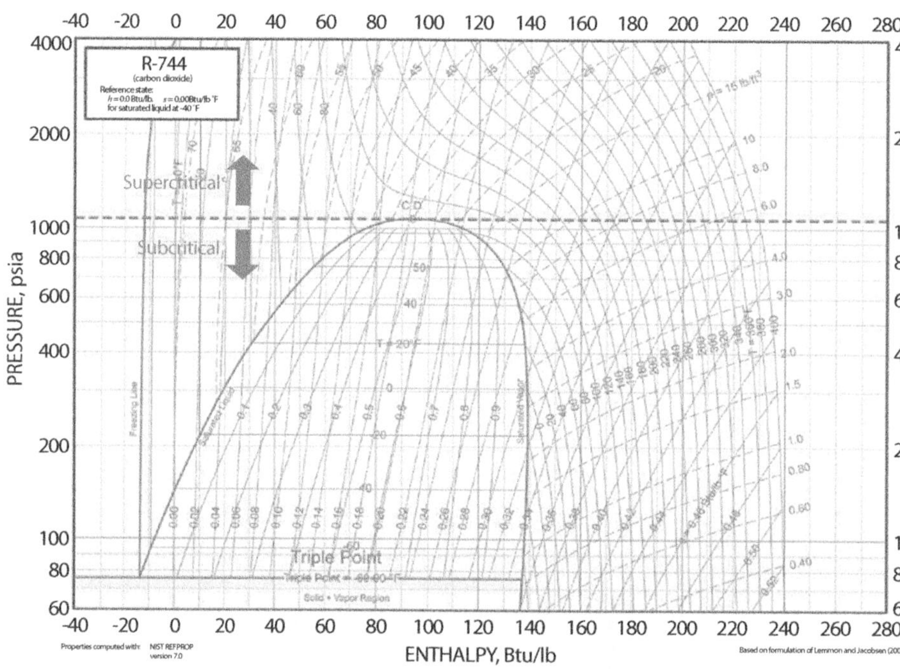

Figure 11.7 Refrigerant classification of CO_2 pressure-enthalpy diagram [29].

most other refrigerants. That allows the refrigeration systems and their ancillary components to have smaller volumes [28].

Carbon dioxide is compatible with conventional piping materials and lubricants. Its high volumetric cooling capacity – 5 to 8 times higher than HFCs allows the system charge to be reduced by 50%. It, therefore, allows significantly less compression work in addition to equipment size-reduction.

The high working pressure of CO_2 during condensation and evaporation results in a weaker performance under high ambient temperature conditions. The low critical temperature of (31°C) has resulted in CO_2 refrigeration cycles to often operate in the trans-critical

Table 11.19 Properties of natural refrigerants indicating high saturation pressure of CO_2 [28].

	Boiling temperature at atmospheric pressure in °C	**Critical temperature in °C**	**Saturation pressure in bar at boiling temperature of**			**Flammable**	**Toxic**	**GWP* (IPCC AR4, except water and ammonia)**
			-30°C	**0°C**	**40°C**			
Isobutane	-11.7	134.7	0.5	1.6	5.3	yes	no	3.3
Propane	-42.1	96.7	1.7	4.7	13.7	yes	no	3.3
Propylene	-47.6	91.1	2.1	5.9	16.5	yes	no	1.8
Pentane	36.1	196.6	0.05	0.25	1.16	yes	no	3.3
Ammonia	-33.3	132.3	1.2	4.3	15.5	(yes)	yes	0
Water	100	373.9	Ice	6mbar	0.0738	no	no	0
CO_2	(-78.4)	31.0	14.3	34.8	not applicable	no	<10%no	1

state. In typical refrigeration systems the trans-critical state is less energy-efficient and is encountered during the summer months. Investigations have been undertaken to enhance the energy efficiency in warmer climates.

A refrigerant undergoes four cyclic states in a refrigeration cycle in sub-critical state. Evaporation is followed by compression, condensation and expansion. Under trans-critical conditions, the CO_2 does not condense because the gas is at above its critical temperature. Rather the gas is cooled in the absence of phase change. The pressure in systems operating with CO_2 in trans-critical state require significantly higher pressures than fluorocarbon alternatives [27].

11.4.2 Hydrocarbons

Hydrocarbon refrigerants include propane (R290), iso-butane (R600a) and propylene (R1270). They have been among the first refrigerants to be used. In spite of being non-toxic they are highly flammable. Flammability has limited their use was decreased after the advent of CFCs and HCFCs. The risk of higher flammability has been mitigated by limiting the refrigerant charge. Hydrocarbon refrigerant charges are severely limited in comparison with fluorocarbons. They offer high efficiency even at high ambient temperatures. Hydrocarbons are compatible with most compressor lubricants and materials used in conventional refrigerating systems [27].

11.4.3 Ammonia

Ammonia (R717) was commercially used for cooling in the second half of the 19th century and has been in use ever since. Ammonia has excellent thermodynamic properties and high energy efficiencies. It is thus a viable alternative to fluorocarbon refrigerants. Application of ammonia is concentrated on large systems with high charge sizes. NH_3 has a low flammability but is highly toxic. NH_3 systems are usually located away from populated places due to its high toxicity. There are few incidents because of the leakage of R-717. The pungent odor of ammonia renders the detection leakages fairly easy. Care has to be taken to prevent blending of ammonia and water because of corrosive copper and other non-ferrous metals by ammonium solution. Use of stainless pipe in ammonia system is a convenient solution

Table 11.20 provides a comparison of the properties of natural and a few fluorocarbon refrigerants.

11.5 Phase Out of Fluorocarbon Gases

In 2016 a breakthrough took place among the signatories of the Montreal Protocol in Kigali, Rwanda [**31**]. A global agreement was reached to phase down the consumption and production of hydrofluorocarbons. This commitment requires the international community to take practical actions to achieve the goals of the 2015 Paris Agreement. That Agreement would limit global warming to 2°C, and aim for the more ambitious target of 1.5°C. Importantly, China, the biggest consumer of HFC's, agreed to an early phase down;

Table 11.20 General safety properties and characteristics of some synthetic and natural refrigerants [27].

	HCFC	HFC					Natural refrigerant			
Refrigerant[a]	R22	R134A	R407C	R404A	R410A	R774 (CO_2)	R290 (propane)	R600a (iso-butane)	R1270 (propylene)	R717 (NH_3)
Chemical formula/ composition	$CHClF_2$	CH_2FCF_3	R-32/125/134a (23/25/52)	R-125/143a/134a (44/52/4)	R-32/125 (50/50)	CO_2	$CH_3CH_2CH_3$	$CH(CH_3)_3$	$CH_3CH{=}CH_2$	NH_3
Molecular weight (g/mol)	124.0	102.0	86.2	97.6	72.6	44.0	44.1	58.1	42.1	17.0
Heavier than air?	Heavier	Heavier	Heavier	Heavier	Heavier	Heavier	Heavier	Heavier	Heavier	Lighter
Safety classification[b]	A1	A1	A1	A1	A1	A3	A3	A3	A3	B2(L[c])
Flammability	–	–	–	–	–	–	High	High	High	Low
Temperature class (ATEX)[d]	–	–	–	–	–	–	T1	T1	T1	T1
Explosion group[e]	–	–	–	–	–	–	IIA	IIA	IIA	IIA
PED fluid group[f]	2	2	2	2	2	2	1	1	1	1
PL (kg/m^3)[g]	0.30	0.25	0.31	0.48	0.44	0.1	0.008	0.0086	0.008	0.00035
ATEL/ODL (kg/m^3)[h]	0.30	0.25	0.31	0.48	0.44	0.036	0.09	0.06	0.0010	0.00035
LFL(kg/m^3)[i]	–	–	–	–	–	–	0.038	0.043	0.040	0.104

(*Continued*)

Table 11.20 General safety properties and characteristics of some synthetic and natural refrigerants [27]. (*Continued*)

	HCFC	**HFC**					**Natural refrigerant**			
LFL/UFL[h](Vol.-%)	–	–	–	–	–	–	1.7/10.8	1.5/9.4	1.8/11.2	15.4/33.6
Auto-ignition temperature (°C)	635	743	704	728	–	–	450	460	455	630
Normal boiling point (°C)	-40.8	-26.2	-43.8 to -36.7	-46.5 to -45.7	-51.6 to -51.5	-78.5°C	-42.1°C	-11.7	-47.7°C	-33.4°C
Critical temperature (°C)	96.1°C	101.1°C	86.0°C	72.0°C	71.4°C	31.0°C	96.8°C	135.0°C	92.4°C	132.4°C
Critical pressure (bar)	49.8	41.7	46.2	35.7	49.5	73.8	42.6	36.5	46.6	112.8

Sources: http://gestis.itrust.de/, EN 378-1.

[a] R-22 is the only refrigerant with an ODP.

[b] According to ISO 817 and EN 378-1: Refrigerant safety classification.

[c] proposed in prEN 378-1:2013.

[d] ATEX temperature class T1 – Ignition temperature of the flammable substance >450°C.

[e] ATEX explosion group-classification for use of electrical equipment.

[f] Group 1: flammable to explosive, oxidizing to very toxic substance; Group 2: other.

[g] Practical Concentration Limit-highest to concentration limit in an occupied space which will not exhibit any negative effect.

[h] Acute Toxicity Exposure Limit/Oxygen Deprivation Limit.

[i] Lower Flammability Limit – The lower concentration limit of a flammable substance for which air/vapour mixtures can ignite.

[j] Upper Flammability Limit – Concentration limit of a flammable substance above which there is insufficient oxygen for combustion.

India (another big consumer and producer) decided to come down from its earlier position of 2031 to 2024-2026.

After seven years of negotiations, Parties to the Montreal Protocol arrived at an agreement, under which developed countries will begin phasing down HFC's by 2019, while the developing countries, including India, Pakistan, Iran and Iraq, will phase out at a baseline of 2024-2026 and a freeze date of 2028. HFC's are widely used in refrigerators, air-conditioners and aerosol sprays. According to the US EPA as a result of a July 2015 rulemaking, by model year 2021, the HVAC systems in newly manufactured light-duty vehicles in the United States will no longer use HFC-134a. They will be converted to HFO's like R-1234yf, R-1234ze, and their blends, which have seen adoption in various refrigeration equipment, room air-conditioner, package air-conditioner, blowing agent, solvent, aerosol propellants and others.

There appears to be opportunity for the old refrigerants (ammonia, carbon dioxide and hydrocarbons) as replacement to fluorocarbons. In the 1930's fluorinated hydrocarbon refrigerants were developed to replace the old gases. They were considered superior than the incumbent gases including ammonia, sulfur dioxide and carbon dioxide by being safer, more efficient, nonflammable, nontoxic, colorless, and odorless. Little did the World know that the safer fluorocarbon gases would end up having a big hand in destroying the Ozone Layer. Nearly a century later, ironically the world may have to return, at least in part, to the *old* refrigerant gases.

11.6 Future Directions of Refrigerants

11.6.1 Introduction

China and the United States are the two largest consumers of fluorocarbon gases and refrigerants. Any change of policy or delay in implementation of global agreements for conversion to lower ODP and GWP by those governments has implications for progress towards those goals. The most recent fluctuation has taken place in the US environmental policies post 2016. Weakening and revocation of environmental and loosening air and water standards have regressed the progress achieved in the previous decades.

In spite of the publicity to the contrary, conversion to zero ODP has been completed and the transition toward refrigerants with a low GWP continues. Manufacturers, researchers, and facility owners worldwide are making significant progress, because of consistent public policies but also from recent development of proven technologies. In the US, some states, large cities and multinational corporations drive the conversion to lower GWP alternatives.

11.6.2 Towards the Future

The rest of Section 12 is based on an excellent article in August 6 issue of 2019 article published in the Air Conditioning, Heating & Refrigeration News Magazine [30]. It offers a comprehensive discussion of the anticipated forward path of the refrigeration applications including air conditioning. Those applications are both varied and consume a majority of fluorocarbons quantitatively. A large number of single component and blended fluorocarbons serve this segment. Consequently, the focus on the future of refrigeration also paints a picture of the expected future directions of the entire fluorocarbons gases and their substitute.

11.6.2.1 Innovation

It is the key that is unlocking the potential of the new refrigerants to create a low-GWP future. Important developments include: Utilizing the thermodynamics of CO_2 refrigerant—specifically, its expansion energy and heating properties—to make CO_2 even more feasible for commercial refrigeration; the ongoing decarbonizing of the electric power supply; and boosting flexibility in electric consumption with demand response and thermal storage technologies. These developments are driving forward the adoption of low-GWP refrigerants in food and commercial refrigeration today.

11.6.2.2 Innovation Accelerating Transition

In the earlier transition from high to low ODP refrigerants, manufacturers could generally use a "drop-in" approach—that is, simply replacing one refrigerant for another after modifying seals and other minor components. That is not always possible with the switch to low-GWP refrigerants. Natural refrigerants, such as ammonia and CO_2, operate with different parameters compared to HFCs and require specially designed system components. Other natural refrigerants, such as propane, use components similar to those employed with HFCs, but special safety precautions need to be taken.

Depending on corporate goals and local regulations, manufacturers and end users worldwide are adopting low-GWP-refrigerant systems employing various technological innovations.

As a practical example of how innovation accelerates the transition, it's worth examining the case of a military supermarket that successfully implemented a CO_2 refrigeration system in a hot-climate region. The region experiences 32°C and higher ambient temperatures in the summer. Compared to using HFCs at those temperatures, employing CO_2 refrigerant within its normally high operating pressures was not energy efficient. Nevertheless, several advanced technologies were employed to go beyond CO_2's traditional ambient temperature limitations. The system used the Danfoss Multi Ejector Solution™, [www.Danfoss.com] consisting of a high-pressure valve, integrated pressure transmitter, and an intelligent system controller in conjunction with the compressor rack.

The system operates reliably in the transcritical zone of the compression cycle, where pressures can exceed 1,067 psia. In the transcritical zone, heat rejection can efficiently occur even at high ambient temperatures—even at 36°C. The bottom line: The CO_2 system supplying display cabinets has an extremely low GWP of 1.0 and provides up to 30 percent energy savings compared to the incumbent solutions based on R-22 and R-404A refrigerants.

11.6.2.3 Speed Bumps

Several HFO refrigerant formulations have been readily adopted for a number of applications. R-1234yf is used by automotive manufacturers for car air conditioners, and

R-1234ze(E) is used in innovative centrifugal compressors employing magnetic bearings, such as Danfoss Turbocor® compressors.

Some low-GWP refrigerants are flammable, some are toxic, and some are both. ASHRAE 34 and International Standards Organization (ISO) 817 standards have been developed to classify flammability and toxicity. Lower GWP properties tend to correlate with higher flammability/toxicity characteristics. For example, HFC refrigerant R-410A (with a 2,088 GWP) is non-toxic, non-flammable, and classified A1. The HFO refrigerant R-1234yf (with a 4 GWP) is non-toxic, mildly flammable, and classified A2L. Hydrocarbon refrigerants (R-600a isobutane and R-290 propane) are highly flammable (A3). Ammonia (R-717) has higher toxicity and flammability (B2), while CO_2 (R-744) has no flammability or toxicity (A1).

The correlation between GWPs, refrigerant density (weight per volume), and safety classifications of various refrigerants is of great concern to HVACR system manufacturers and specifiers. Blends of HFOs and HFCs are being developed to optimize performance and safety. Whether using HFOs, natural refrigerants, or blends, HVACR manufacturers must innovate technologies that maximize efficiency and minimize risk for the selected refrigerant.

Development of new technologies makes it easier for manufacturers to balance safety and environmental responsibility. Some leading retailers have been using R-290 (propane) in equipment for over a decade. Taking advantage of R-290's excellent thermodynamic properties, systems are operating successfully with a charge limited to 150 grams (5 ounces) for safety. Further research and product development led the International Electrotechnical Commission (IEC) to propose a 500-gram (1.1-pound) limit for R-290 in single commercial refrigeration appliances. The new IEC 60335-2-89 standard was passed in May 2019.

As time passes, it's logical to think that Europe and the U.S. will continue to explore the balance between safety, efficiency and the environment—but not at the expense of endangering users and technicians. The service industry, in particular, will need to become more familiar with flammable refrigerants.

"Installation and service are the areas where proper guidelines, training, certification, and standards are the most important," explains Bill Goetzler of Navigant Consulting [https://guidehouse.com]. "The lines carrying flammable refrigerants may be opened during various servicing operations, and service personnel also may be working with a high-temperature ignition source. Those are really [the] places to be very careful and to be sure we've got standards and guidelines in place to minimize the risk."

Safety is a vital factor relative in a refrigerant sustainability triangle that includes environmental and economic factors. Other factors come into play depending on the properties of the refrigerant. With some HFO refrigerants, for example, flammability can present a safety issue, as previously discussed. Cost can also be a factor, as HFO formulations are more expensive to produce, and supplies are constrained. Finally, environmental concerns can still be an issue in several countries, because some HFOs break down in the lower atmosphere, forming fluorinated products. Trifluoroacetic acid (TFA) is an HFO breakdown substance but also occurs naturally in seawater. Hydrogen fluoride (HF) is another HFO breakdown substance that is very toxic. The small amounts of TFA and HF are not expected to pose global or regional problems.

11.6.2.4 New Developments

The search for refrigerants that perfectly balance environmental, economic, and safety concerns is continuing. Unfortunately, this quest throws customers and end-users off balance – planning is stressful when the future is subject to change. Fortunately, in recent years, technology has developed to the point that it is possible to build a long-lasting platform for low-GWP commercial refrigeration. Proven technologies are now available in four areas:

1. Developments in CO_2 refrigeration

Until recently, it's been impossible to apply the energy efficiency, safety, and environmental benefits of trans-critical CO_2 systems in all climates. However, as previously explained, Danfoss Multi Ejector technology enables a CO_2 refrigeration system to operate in a transcritical zone, even in warm climates. The result is up to 10 percent annualized savings compared to prior transcritical parallel compression systems and up to 30 percent energy savings during the hottest hours of the year compared to booster systems. Moreover, compressor discharge temperatures in transcritical operation can exceed 250 °F. This heat, which is usually rejected by a gas cooler, can be reclaimed by heat exchangers mounted before the gas cooler. This combination of technologies brings all the safety, economic, and environmental benefits of CO_2 to many large and mid-size U.S. commercial refrigeration applications, regardless of location.

2. Developments in demand response

The high usage of electricity by supermarkets can make it very advantageous to participate in demand response programs. In this type of program, the utility sends a signal to a central control unit such as the Danfoss ADAP-KOOL® System Manager AK-SM 800 Series, which can then reduce power to motors, compressors, and other electrical equipment. The signal triggers a reduction in electric consumption for brief periods ranging from less than one minute to long periods within a 24-hour interval, depending on the application.

Not every utility offers demand response programs, but those that do pass on the prospective savings to participants. Economic benefits are based on the customer's commitment to respond. A carefully designed demand response program using appropriate refrigeration control and management systems can achieve substantial energy savings while protecting food integrity.

3. Developments in thermal storage and related thermal-shifting technologies

Thermal storage makes it possible to shift electricity consumption from expensive peak rates during the day to lower rate periods during the night. Thermal storage tanks hold a liquid medium chilled at night for cooling display cases, reducing equipment run time.

In a supermarket application, the display cases can themselves function as a form of thermal storage. Lowering display-case temperatures during off-peak hours enables compressors to run less during peak hours. Thermal storage capacity can be used to produce a revenue stream when it's connected to other stores through a district heating and cooling (DHC) system. Heat recovery technology can also be used to capture heat otherwise rejected by the refrigeration system's condensers. For stores using CO_2 with a heat-reclaim system, the size of a separate heat source—such as a boiler—can be greatly reduced or even eliminated.

4. Developments in the decarbonized grid

Called "the largest machine ever built," the more than century-old electric grid is the largest producer of greenhouse gasses. Fortunately, technology exists today to re-engineer that machine to cut CO_2 emissions dramatically. Increased use of renewable power (wind and solar), continued use of stable nuclear power, and use of responsive hydropower are low-carbon resources that are gradually creating a decarbonized electricity supply. Models of a moderately decarbonized electricity system for the year 2050 estimate CO_2 emissions intensity could be 60 to 80 percent lower than today's U.S. grid.

11.7 Conclusions

Recent technological developments are supporting the transition to low-GWP refrigerants. While policies may fluctuate, research and innovation are bringing proven products and platforms to market that can be implemented today in more food and commercial refrigeration applications than ever before.

HFOs in conjunction with oil-free system developments are very viable options in air conditioning and, eventually, heat pump applications. Advances in transcritical CO_2 systems are also providing an attractive value proposition for many supermarkets. Other developments—demand response, thermal storage, and heat reclaim—complement the trend toward a decarbonized electric grid and significantly reduce commercial refrigeration's carbon footprint.

Utilities and regulators play a vital role in creating an HFC-free future. Incentives and policies can reassure supermarket owners operating on tight margins that their investment will pay off by boosting profits in the short term, as well as the environment in the long run.

References

1. Seidel, S., Ye, J., Andersen, S.O., Hillbrand, A., *Not-in-kind Alternatives to high Global Warming HFCs*, Center for Climate and Energy Solutions, www.c2es.org, October 2016.
2. Huehren, R., *Hydrocarbon Refrigerants: Features, advantages, security aspects, applications*, Ozone Cap Depleting Substances, Ministry of Environment, Germany, 2018.
3. *2018 Report of the Refrigeration,Air Conditioning and Heat Pumps*, Technical Options Committee, 2018 Assessment Report, United Nations Environment Program, February 2019.
4. Emani, M.S., Roy, R., Mandal, B.K., Development of Refrigerant: A Brief Review. *Indian J. Sci. Res.*, 14, 2, 175–181, 2017.
5. Carrier, W.H. and Waterfill, R.W., Comparison of thermodynamic characteristics of various refrigerating fluids. *Refrig. Eng.*, 1924.
6. Calm, J.M., The next generation of refrigerants – Historical review, considerations, and outlook. *Int. J. Refrig.*, 31, 1123–1133, 2008.
7. www.energy.gov/sites/prod/files/2014/12/f19/Refrigerants%20Roadmap%20Final%20Report%202014.pdf.

8. Midgley, T., Jr., From the periodic table to production. *Ind. Eng. Chem.*, 29, 2, 239–244, 1937.
9. Midgley, T., Jr. and Henne, A.L., Organic fluorides as refrigerants. *Ind. Eng. Chem.*, 22, 542–545, 1930.
10. Arora, P., Seshadri, G., Tyagi, A.K., *Curr. Sci.*, 115, 8, 25 Oct 2018.
11. Emani, M.S., Roy, R., Mandal, B.K., Development of Refrigerant: A Brief Review. *Indian J. Sci. Res.*, 14, 2, 175–181, 2017.
12. Intergovernmental Panel on Climate Change (IPCC), *Climate Change 2007: The Physical Science Basis — Summary for Policymakers, Contribution of Working Group I to the Fourth Assessment Report*, World Meteorological Organization (WMO) and United Nations Environment Program (UNEP), IPCC Secretariat, Geneva, Switzerland, 2007.
13. Kyoto Protocol the United Nations Framework Convention on Climate Change, https://unfccc.int/resource/docs/convkp/kpeng.html, 2020.
14. Doha Amendment to the Kyoto Protocol, https://unfccc.int/kyoto_protocol, January 2020.
15. Carvalho, S., Andersen, S.O., Brack, D., Sherman, N.J., *Alternatives to High GWP Hydrofluorocarbons*, Institute for Governance and Sustainable Development, http://www.igsd.org/documents/HFCSharpeningReport.pdf, 2014.
16. Nielsen, O.J., Javadi, M.S., Andersen, M.P.S., Hurley, M.D., Wallington, T.J., Singh, R., Atmospheric chemistry of CF3CFCH2: kinetics and mechanisms of gas-phase reactions with Cl atoms, OH radicals, and O3. *Chem. Phys. Lett.*, 439, 18–22, 2007.
17. Hydrocarbons GWP, US EPA, www.EPA.gov/ozone/downloads/EPA_HFC_ComRef.pdf, January 2020.
18. Hashimoto, M. and Otsuka *et al*, T., Development of New Low-GWP Refrigerants–Refrigerant Mixtures Including HFO-1123. *Sci. Technol. Built Environ.*, 25, 776–783, 2019.
19. Spletzer, S., Coughlan, P., Gerstel, J., An overview of how flammable refrigerants are changing the RAC landscape. *1st IIR international conference on the application of HFO refrigerants*, Birmingham, September 2-5 20, 18.
20. *HCFC to HFO Blend Retrofit Guidelines Comm Refrig. Syst*, Honeywell Refrigerants, www.honeywell-refrigerants.com, August 2018.
21. Solstice® N40 (R-448A), Technical Data Sheet, www.honeywell-refrigerants.com/europe, January 2020.
22. Lemmon, E.W., Huber, M.L., McLinden, M.O., REFPROP. 2013. Version 9.1: NIST Reference Fluid Thermodynamic and Transport Properties, 2013.
23. Daikin Corp, Increasing Air Conditioner Efficiency, www.daikin.com/csr/environment/climatechange/air_conditioner.html, April 2020.
24. Hashimoto, M., Otsuka, T. *et al.*, Development of New Low-GWP Refrigerants–Refrigerant Mixtures Including HFO-1123. *Sci. Technol. Built Environ.*, 25, 776–783, 2019.
25. Hasse, V., Ederberg, L., Colbourne, D., *Natural Refrigerants*, PROKLIMA, Deutsche Gesellschaft für Technische Zusam-menarbeit GmbH, German Federal Ministry for Economic Cooperation and Development (BMZ) Environment and Sustainable Useof Natural Resources Division, Bonn, Germany, 2008.
26. Abas, N., Kalair, A.R., Khan, N., Haider, A., Saleem, M.S., Saleemd, Z., Natural and synthetic refrigerants, global warming: A review. *Renewable Sustainable Energy Rev.*, 90, 557–569, 2018.
27. Radermacher, P., Both, J., Papst, I., (HEAT GmbH), Gschrey, B., Zeiger, B., (Öko-Recherche GmbH), *Recommendations to safety guidelines and standards for the use of natural refrigerants*, Advisory Assistance Program, German Environment Agency, www.umweltbundesamt.de/en/publications, November 2016.
28. Kauffeld, M., *Availability of low GWP alternatives to HFCs, Feasibility of an early phase-out of HFCs by 2020*, Environmental Investigation Agency, www.eia-international.org, May 2012.

29. *Standard 15-2013 (packaged w/Standard 34-2013) – Safety Standard for Refrigeration Systems and Designation and Classification of Refrigerants (ANSI Approved)*, ASHRAE, www.ashrae.org/resources–publications/bookstore/standards-, 15–34, 2013.
30. Tryson, L. and Funder-Kristensen, T., *New Developments Drive Adoption of Low-GWPf Refrigerants In Commercial Refrigeration*, www.achrnews.com/articles/141693-new-developments-drive-adoption-of-low-gwp-refrigerants-in-commercial-refrigeration, August 6, 2019.
31. Ebnesajjad, S., Global Warming, Fluorocarbons, *Montreal Protocol 2016 - all in Kigali, Rwanda*, Elsevier, https://chemical-materials.elsevier.com/new-materials-applications/global-warming-fluorocarbons-montreal-protocol-2016-kigali-rwanda/, Nov 1, 2016.

Appendix I

Section 1.1
The Montreal Protocol on Substances that Deplete the Ozone Layer

as adjusted and amended by the
Second Meeting of the Parties
(London, 27-29 June 1990)

and by the
Fourth Meeting of the Parties
(Copenhagen, 23-25 November 1992)

and further adjusted by the
Seventh Meeting of the Parties
(Vienna, 5-7 December 1995)

and further adjusted and amended by the
Ninth Meeting of the Parties
(Montreal, 15-17 September 1997)

and by the
Eleventh Meeting of the Parties
(Beijing, 29 November - 3 December 1999)

and further adjusted by the
Nineteenth Meeting of the Parties
(Montreal, 17-21 September 2007)

and further amended by the
Twenty-Eight Meeting of the Parties
(Kigali, 10-15 October 2016)

Preamble

The Parties to this Protocol,

Being Parties to the Vienna Convention for the Protection of the Ozone Layer,

Mindful of their obligation under that Convention to take appropriate measures to protect human health and the environment against adverse effects resulting or likely to result from human activities which modify or are likely to modify the ozone layer,

Recognizing that world-wide emissions of certain substances can significantly deplete and otherwise modify the ozone layer in a manner that is likely to result in adverse effects on human health and the environment,

Conscious of the potential climatic effects of emissions of these substances,

Aware that measures taken to protect the ozone layer from depletion should be based on relevant scientific knowledge, taking into account technical and economic considerations,

Determined to protect the ozone layer by taking precautionary measures to control equitably total global emissions of substances that deplete it, with the ultimate objective of their elimination on the basis of developments in scientific knowledge, taking into account technical and economic considerations and bearing in mind the developmental needs of developing countries,

Acknowledging that special provision is required to meet the needs of developing countries, including the provision of additional financial resources and access to relevant technologies, bearing in mind that the magnitude of funds necessary is predictable, and the funds can be expected to make a substantial difference in the world's ability to address the scientifically established problem of ozone depletion and its harmful effects,

Noting the precautionary measures for controlling emissions of certain chlorofluorocarbons that have already been taken at national and regional levels,

Considering the importance of promoting international co-operation in the research, development and transfer of alternative technologies relating to the control and reduction of emissions of substances that deplete the ozone layer, bearing in mind in particular the needs of developing countries,

HAVE AGREED AS FOLLOWS:

Article 1: Definitions

For the purposes of this Protocol:

1. "Convention" means the Vienna Convention for the Protection of the Ozone Layer, adopted on 22 March 1985.

2. "Parties" means, unless the text otherwise indicates, Parties to this Protocol.

3. "Secretariat" means the Secretariat of the Convention.

4. "Controlled substance" means a substance in Annex A, Annex B, Annex C, Annex E or Annex F to this Protocol, whether existing alone or in a mixture. It includes the isomers of any such substance, except as specified in the relevant Annex, but excludes any controlled substance or mixture which is in a manufactured product other than a container used for the transportation or storage of that substance.

5. "Production" means the amount of controlled substances produced, minus the amount destroyed by technologies to be approved by the Parties and minus the amount entirely used as feedstock in the manufacture of other chemicals. The amount recycled and reused is not to be considered as "production"

6. "Consumption" means production plus imports minus exports of controlled substances.

7. "Calculated levels" of production, imports, exports and consumption means levels determined in accordance with Article 3.

8. "Industrial rationalization" means the transfer of all or a portion of the calculated level of production of one Party to another, for the purpose of achieving economic efficiencies or responding to anticipated shortfalls in supply as a result of plant closures.

Article 2: Controlled Measures

1. *Incorporated in Article 2A.*

2. *Replaced by Article 2B.*

3. *Replaced by Article 2A.*

4. *Replaced by Article 2A.*

5. Any Party may, for one or more control periods, transfer to another Party any portion of its calculated level of production set out in Articles 2A to 2F, Articles 2H and 2J, provided that the total combined calculated levels of production of the Parties concerned for any group of controlled substances do not exceed the production limits set out in those Articles for that group. Such transfer of production shall be notified to the Secretariat by each of the Parties concerned, stating the terms of such transfer and the period for which it is to apply.

5 *bis.* Any Party not operating under paragraph 1 of Article 5 may, for one or more control periods, transfer to another such Party any portion of its calculated level of consumption set out in Article 2F, provided that the calculated level of consumption of controlled substances in Group I of Annex A of the Party transferring the portion of its calculated level of consumption did not exceed 0.25 kilograms per capita in 1989 and that the total combined calculated levels of consumption of the Parties concerned do not exceed the consumption limits set out in Article 2F. Such transfer of consumption shall be notified to the Secretariat by each of the Parties concerned, stating the terms of such transfer and the period for which it is to apply.

6. Any Party not operating under Article 5, that has facilities for the production of Annex A or Annex B controlled substances under construction, or contracted for, prior to 16 September 1987, and provided for in national legislation prior to 1 January 1987, may add the production from such facilities to its 1986 production of such substances for the purposes of determining its calculated level of production for 1986, provided that such facilities are completed by 31 December 1990 and that such production does not raise that Party's annual calculated level of consumption of the controlled substances above 0.5 kilograms per capita.

7. Any transfer of production pursuant to paragraph 5 or any addition of production pursuant to paragraph 6 shall be notified to the Secretariat, no later than the time of the transfer or addition.

8. (a) Any Parties which are Member States of a regional economic integration organization as defined in Article 1 (6) of the Convention may agree that they shall jointly fulfil their obligations respecting consumption under this Article and Articles 2A to 2J provided that their total combined calculated level of consumption does not exceed the levels required by this Article and Articles 2A to 2J. Any such agreement may be extended to include obligations respecting consumption or production under Article 2J provided that the total combined calculated level of consumption or production of the Parties concerned does not exceed the levels required by Article 2J.

   (b) The Parties to any such agreement shall inform the Secretariat of the terms of the agreement before the date of the reduction in consumption with which the agreement is concerned.

   (c) Such agreement will become operative only if all Member States of the regional economic integration organization and the organization concerned are Parties to the Protocol and have notified the Secretariat of their manner of implementation.

9. (a) Based on the assessments made pursuant to Article 6, the Parties may decide whether:
   (i) Adjustments to the ozone depleting potentials specified in Annex A, Annex B, Annex C and/or Annex E should be made and, if so, what the adjustments should be;
   (ii) Adjustments to the global warming potentials specified in Group I of Annex A, Annex C and Annex F should be made and, if so, what the adjustments should be: and
   (iii) Further adjustments and reductions of production or consumption of the controlled substances should be undertaken and, if so, what the scope, amount and timing of any such adjustments and reduction should be;

   (b) Proposals for such adjustments shall be communicated to the Parties by the Secretariat at least six months before the meeting of the Parties at which they are proposed for adoption;

   (c) In taking such decisions, the Parties shall make every effort to reach agreement by consensus. If all efforts at consensus have been exhausted, and no agreement reached,such decisions shall, as a last resort, be adopted by a two-thirds majority vote of the Parties present and voting representing a majority of the Parties operating under Paragraph 1 of Article 5 present and voting and a majority of the Parties not so operating present and voting;

   (d) The decisions, which shall be binding on all Parties, shall forthwith be communicated to the Parties by the Depositary. Unless otherwise provided in the decisions, they shall enter into force on the expiry of six months from the date of the circulation of the communication by the Depositary.

10. Based on the assessments made pursuant to Article 6 of this Protocol and in accordance with the procedure set out in Article 9 of the Convention, the Parties may decide:

    (a) whether any substances, and if so which, should be added to or removed from any annex to this Protocol, and

    (b) the mechanism, scope and timing of the control measures that should apply to those substances;

11. Notwithstanding the provisions contained in this Article and Articles 2A to 2J Parties may take more stringent measures than those required by this Article and Articles 2A to 2J.

Article 2A: CFCs

1. Each Party shall ensure that for the twelve-month period commencing on the first day of the seventh month following the date of entry into force of this Protocol, and in each twelve-month period thereafter, its calculated level of consumption of the controlled substances in Group I of Annex A does not exceed its calculated level of consumption in 1986. By the end of the same period, each Party producing one or more of these substances shall ensure that its calculated level of production of the substances does not exceed its calculated level of production in 1986, except that such level may have increased by no more than ten per cent based on the 1986 level. Such increase shall be permitted only so as to satisfy the basic domestic needs of the Parties operating under Article 5 and for the purposes of industrial rationalization between Parties.

2. Each Party shall ensure that for the period from 1 July 1991 to 31 December 1992 its calculated levels of consumption and production of the controlled substances in Group I of Annex A do not exceed 150 per cent of its calculated levels of production and consumption of those substances in 1986; with effect from 1 January 1993, the twelve-month control period for these controlled substances shall run from 1 January to 31 December each year.

3. Each Party shall ensure that for the twelve-month period commencing on 1 January 1994, and in each twelve-month period thereafter, its calculated level of consumption of the controlled substances in Group I of Annex A does not exceed, annually, twenty-five per cent of its calculated level of consumption in 1986. Each Party producing one or more of these substances shall, for the same periods, ensure that its calculated level of production of the substances does not exceed, annually, twenty-five per cent of its calculated level of production in 1986. However, in order to satisfy the basic domestic needs of the Parties operating under paragraph 1 of Article 5, its calculated level of production may exceed that limit by up to ten per cent of its calculated level of production in 1986.

4. Each Party shall ensure that for the twelve-month period commencing on 1 January 1996, and in each twelve-month period thereafter, its calculated level of consumption of the controlled substances in Group I of Annex A does not exceed zero. Each Party producing one or more of these substances shall, for the same periods, ensure that its calculated level of production of the substances does not exceed zero. However, in order to satisfy the basic domestic needs of the Parties operating under paragraph 1 of Article 5, its calculated level of production may exceed that limit by a quantity equal to the annual average of its production of the controlled substances in Group I of Annex A for basic domestic needs for the period 1995 to 1997 inclusive. This paragraph will apply save to the extent that the Parties decide to permit the

level of production or consumption that is necessary to satisfy uses agreed by them to be essential.

5. Each Party shall ensure that for the twelve-month period commencing on 1 January 2003 and in each twelve-month period thereafter, its calculated level of production of the controlled substances in Group I of Annex A for the basic domestic needs of the Parties operating under paragraph 1 of Article 5 does not exceed eighty per cent of the annual average of its production of those substances for basic domestic needs for the period 1995 to 1997 inclusive.

6. Each Party shall ensure that for the twelve-month period commencing on 1 January 2005 and in each twelve-month period thereafter, its calculated level of production of the controlled substances in Group I of Annex A for the basic domestic needs of the Parties operating under paragraph 1 of the Article 5 does not exceed fifty per cent of the annual average of its production of those substances for basic domestic needs for the period 1995 to 1997 inclusive.

7. Each Party shall ensure that for the twelve-month period commencing on 1 January 2007 and in each twelve-month period thereafter, its calculated level of production of the controlled substances in Group I of Annex A for the basic domestic needs of the Parties operating under paragraph 1 of Article 5 does not exceed fifteen per cent of the annual average of its production of those substances for basic domestic needs for the period 1995 to 1997 inclusive.

8. Each Party shall ensure that for the twelve-month period commencing on 1 January 2010 and in each twelve-month period thereafter, its calculated level of production of the controlled substances in Group I of Annex A for for the basic domestic needs of the Parties operating under paragraph 1 of Article 5 does not exceed zero.

9. For the purposes of calculating basic domestic needs under paragraph 4 to 8 of this Article, the calculation of the annual average of production by a Party includes any production entitlements that it has transferred in accordance with paragraph 5 of Article 2, and excludes any production entitlements that it has acquired in accordance with paragraph 5 of Article 2.

Article 2B: Halons

1. Each Party shall ensure that the twelve-month period commencing on 1 January 1992, and in each twelve-month period thereafter, its calculated level of consumption of the controlled substances in Group II of Annex A does not exceed, annually, its calculated level of consumption in 1986. Each

Party producing one or more of these substances shall, for the same periods, ensure that its calculated level of production of the substances does not exceed, annually, its calculated level of production in 1986. However, in order to satisfy the basic domestic needs of the Parties operating under paragraph 1 of Article 5, its calculated level of production may exceed that limit by up to ten per cent of its calculated level of production in 1986.

2. Each Party shall ensure that for the twelve-month period commencing on 1 January 1994, and in each twelve-month period thereafter, its calculated level of consumption of the controlled substances in Group II of Annex A does not exceed zero. Each Party producing one or more of these substances shall, for the same periods, ensure that its calculated level of production of the substances does not exceed zero. However, in order to satisfy the basic domestic needs of the Parties operating under paragraph 1 of Article 5, its calculated level of production may, until 1 January 2002 exceed that limit by up to fifteen per cent of its calculated level of production in 1986; thereafter, it may exceed that limit by a quantity equal to the annual average of its production of the controlled substances in Group II of Annex A for basic domestic needs for the period 1995 to 1997 inclusive. This paragraph will apply save to the extent that the Parties decide to permit the level of production or consumption that is necessary to satisfy uses agreed by them to be essential.

3. Each Party shall ensure that for the twelve-month period commencing on 1 January 2005 and in each twelve-month period thereafter, its calculated level of production of the controlled substances in Group II of Annex A for the basic domestic needs of the Parties operating under paragraph 1 of Article 5 does not exceed fifty per cent of the annual average of its production of those substances for basic domestic needs for the period 1995 to197 inclusive.

4. Each Party shall that for the twelve-month period commencing on 1 January 2010 and in each twelve-month period thereafter, its calculated level of production of the controlled substances in Group II of Annex A for the basic domestic needs of the Parties operating under paragraph 1 of Article 5 does not exceed zero.

Article 2C: Other fully halogenated CFCs

1. Each Party shall ensure that for the twelve-month period commencing on 1 January 1993, its calculated level of consumption of the controlled substances in Group I of Annex B does not exceed, annually, eighty per cent of its calculated level of consumption in 1989. Each Party producing one or more of these substances shall, for the same period, ensure that its calculated level of production of the substances does not exceed, annually, eighty per cent of its calculated level of production in 1989. However, in order to satisfy the

basic domestic needs of the Parties operating under paragraph 1 of Article 5, its calculated level of production may exceed that limit by up to ten per cent of its calculated level of production in 1989.

2. Each Party shall ensure that for the twelve-month period commencing on 1 January 1994, and in each twelve-month period thereafter, its calculated level of consumption of the controlled substances in Group I of Annex does not exceed, annually, twenty-five per cent of its calculated level of consumption in 1989. Each Party producing one or more of these substances shall, for the same periods, ensure that its calculated level of production of the substances does not exceed, annually, twenty-five per cent of its calculated level of production in 1989. However, in order to satisfy the basic domestic needs of the Parties operating under paragraph 1 of Article 5, its calculated level of production may exceed that limit by up to ten per cent of its calculated level of production in 1989.

3. Each Party shall ensure that for the twelve-month period commencing on 1 January 1996, and in each twelve-month period thereafter, its calculated level of consumption of the controlled substances in Group I of Annex B does not exceed zero. Each Party producing one or more of these substances shall, for the same periods, ensure that its calculated level of production of the substances does not exceed zero. However, in order to satisfy the basic domestic needs of the Parties operating under paragraph 1 of Article 5, its calculated level of production may, until 1 January 2003 exceed that limit by up to fifteen per cent of its calculated level of production in 1989; thereafter, it may exceed that limit by a quantity equal to eighty per cent of the annual average of its production of the controlled substances in Group I of Annex B for basic domestic needs for the period 1998 to 2000 inclusive. This paragraph will apply save to the extent that the Parties decide to permit the level of production or consumption that is necessary to satisfy uses agreed by them to be essential.

4. Each Party shall ensure that for the twelve-month period commencing on 1 January 2007 and in each twelve-month period thereafter, its calculated level of production of the controlled substances in Group I of Annex B for the basic domestic needs of the Parties operating under paragraph 1 of Article 5 does not exceed fifteen per cent of the annual average of its production of those substances for basic domestic needs for the period 1998 to 2000 inclusive.

5. Each Party shall ensure that for the twelve-month period commencing on 1 January 2010 and in each twelve-month period thereafter, its calculated level of production of the controlled substances in Group I of Annex B for the basic domestic needs of the Parties operating under under paragraph 1 of Article 5 does not exceed zero.

Article 2D: Carbon tetrachloride

1. Each Party shall ensure that for the twelve-month period commencing on 1 January 1995, its calculated level of consumption of the controlled substance in Group II of Annex B does not exceed, annually, fifteen per cent of its calculated level of consumption in 1989. Each Party producing the substance shall, for the same period, ensure that its calculated level of production of the substance does not exceed, annually, fifteen per cent of its calculated level of production in 1989. However, in order to satisfy the basic domestic needs of the Parties operating under paragraph 1 of Article 5, its calculated level of production may exceed that limit by up ten per cent of its calculated level of production in 1989.

2. Each Party shall ensure that for the twelve-month period commencing on 1 January 1996, and in each twelve-month period thereafter, its calculated level of consumption of the controlled substance in Group II of Annex B does not exceed zero. Each Party producing the substance shall, for the same periods, ensure that its calculated level of production of the substance does not exceed zero. However, in order to satisfy the basic domestic needs of the Parties operating under paragraph 1 of Article 5, its calculated level of production may exceed that limit by up to fifteen per cent of its calculated level of production in 1989. This paragraph will apply save to the extent that the Parties decide to permit the level of production or consumption that is necessary to satisfy uses agreed by them to be essential.

Article 2E: 1,1,1-Trichloroethane (methyl chloroform)

1. Each Party shall ensure that for the twelve-month period commencing on 1 January 1993, its calculated level of consumption of the controlled substance in Group III of Annex B does not exceed, annually, its calculated level of consumption in 1989. Each Party producing the substance shall, for the same period, ensure that its calculated level of production of the substance does not exceed, annually, is calculated level of production in 1989. However, in order to satisfy the basic domestic needs of the Parties operating under paragraph 1 of Article 5, its calculated level of production may exceed that limit by up to ten per cent of its calculated level of production in 1989.

2. Each Party shall ensure that for the twelve-month period commencing on 1 January 1994, and in each twelve-month period thereafter, its calculated level of consumption of the controlled substance in Group III of Annex B does not exceed, annually, fifty per cent of its calculated level of consumption in 1989. Each Party producing the substance shall, for the same periods, ensure that its calculated level of production of the substance does not exceed, annually, fifty per cent of its calculated level of production in 1989. However, in order

to satisfy the basic domestic needs of the Parties operating under paragraph 1 of Articles 5, its calculated level of production may exceed that limit by up to ten per cent of its calculated level of production in 1989.

3. Each Party shall ensure that for the twelve-month period commencing on 1 January 1996, and in each twelve-month period thereafter, its calculated level of consumption of the controlled substance in Group III of Annex B does not exceed zero, Each Party producing the substance shall, for the same periods, ensure that its calculated level of production of the substance does not exceed zero. However, in order to satisfy the basic domestic needs of the Parties operating under paragraph 1 of Article 5, its calculated level of production may exceed that limit by up to fifteen per cent of its calculated level of production for 1989. This paragraph will apply save to the extend that the Parties decide to permit the level of production or consumption that is necessary to satisfy uses agreed by them to be essential.

Article 2F: Hydrochlorofluorocarbons

1. Each Party shall ensure that for the twelve-month period commencing on 1 January 1996, and in each twelve-month period thereafter, its calculated level of consumption of the controlled substances in Group I of Annex C does not exceed, annually, the sum of:

   (a) Two point eight per cent of its calculated level of consumption in 1989 of the controlled substances in Group I of Annex A; and

   (b) Its calculated level of consumption in 1989 of the controlled substances in Group I of Annex C.

2. Each Party producing one or more of these substances shall ensure that for the twelve-month period commencing on 1 January 2004, and in each twelve-month period thereafter, its calculated level of production of the controlled substances in Group I of Annex C does not exceed, annually, the average of:

   (a) The sum of its calculated level of consumption in 1989 of the controlled substances in Group I of Annex C and two point eight per cent of its calculated level of consumption in 1989 of the controlled substances in Group I of Annex A; and

   (b) The sum of its calculated level of production in 1989 of the controlled substances in Group I of Annex and two point eight per cent of its calculated level of production in 1989 of the controlled substances in Group I of Annex A.

However, in order to satisfy the basic domestic needs of the Parties operating under paragraph 1 of Article 5, its calculated level of production may exceed that limit by up to fifteen per cent of its calculated level of production of the controlled substances in Group I of Annex C as defined above.

3. Each Party shall ensure that for the twelve month period commencing on 1 January 2004, and in each twelve-month period thereafter, its calculated level of consumption of the controlled substances in Group I of Annex C does not exceed, annually, sixty-five per cent of the sum referred to in paragraph 1 of this Article.

4. Each Party shall ensure that for the twelve-month period commencing on 1 January 2010, an in each twelve-month period thereafter, its calculated level of consumption of the controlled substances in Group I of Annex C does not exceed, annually, twenty-five per cent of the sum referred to in paragraph 1 of this Article. Each Party producing one or more of these substances shall, for the same periods, ensure that its calculated level of production of the controlled substances in Group I of Annex C does not exceed, annually, twenty-five per cent of the calculated level referred to in paragraph 2 of this Article. However, in order to satisfy the basic domestic needs of the Parties operating under paragraph 1 of Article 5, its calculated level of production may exceed that limit by up to ten per cent of its calculated level of production of the controlled substances in Group I of Annex C as referred to in paragraph 2.

5. Each Party shall ensure that for the twelve-month period commencing on 1 January 2015, and in each twelve-month period thereafter, its calculated level of consumption of the controlled substances in Group I of Annex C does not exceed, annually, ten per cent of the sum referred to in paragraph 1 of this Article. Each party producing one or more of these substances shall, for the same periods, ensure that its calculated level of production of the controlled substances in Group I of Annex C does not exceed, annually, ten per cent of the calculated level referred in paragraph 2 of this Article. However, in order to satisfy the basic domestic needs of the Parties operating under paragraph 1 of Article 5, its calculated level of production may exceed that limit by up to ten per cent of its calculated level of production of the controlled substances in Group I of Annex C as referred to in paragraph 2.

6. Each Party ensure that for the twelve-month period commencing on 1 January 2020, and in each twelve-month period thereafter, its calculated level of consumption of the controlled substances in Group I of Annex C does not exceed zero. Each Party producing one or more of theses substances shall, for the same periods, ensure that its calculated level of production of the controlled substances in Group I of Annex C does not exceed zero. However:

(a) Each Party may exceed that limit on consumption by up to zero point five per cent of the sum referred to in paragraph 1 of this Article in any such twelve-month period ending before 1 January 2030, provided that such consumption shall be restricted to the servicing of refrigeration and air-conditioning equipment existing on 1 January 2020;

(b) Each Party may exceed that limit on production by up to zero point five per cent of the average referred to in paragraph 2 of this Article in any such twelve-month period ending before 1 January 2030, provided that such production shall be restricted to the servicing of refrigeration and air-conditioning equipment existing on 1 January 2020;

7. As of 1 January 1996, each Party shall endeavour to ensure that:

(a) The use of controlled substances in Group I of Annex C is limited to those applications where other more environmentally suitable alternative substances or technologies are not available;

(b) The use of controlled substances in Group I of Annex C is not outside the areas of application currently met by controlled substances in Annexes A, B and C, except in rare cases for the protection of human life or human health; and

(c) Controlled substances in Group I of Annex C are selected for use in a manner that minimizes ozone depletion, in addition to meeting other environmental, safety and economic considerations.

Article 2G: Hydrobromofluorocarbons

Each Party shall ensure that for the twelve-month period commencing on 1 January 1996, and in each twelve-month period thereafter, its calculated level of consumption of the controlled substances in Group II of Annex C does not exceed zero. Each Party producing the substances shall, for the same periods, ensure that its calculated level of production of the substances does not exceed zero. This paragraph will apply save to the extent that the Parties decide to permit the level of production or consumption that is necessary to satisfy uses agreed by them to be essential.

Article 2H: Methyl bromide

1. Each Party shall ensure that for the twelve-month period commencing on 1 January 1995, and in each twelve-month period thereafter, its calculated level of consumption of the controlled substance in Annex E does not exceed,

annually its calculated level of consumption in 1991. Each Party producing the substance shall, for the same period, ensure that its calculated level of production of the substance does not exceed annually, its calculated level of production in 1991. However, in order to satisfy the basic domestic needs of the Parties operating under paragraph 1 of Article 5, its calculated level of production may exceed that limit by up to ten per cent of its calculated level of production in 1991.

2. Each Party shall ensure that for the twelve-month period commencing on 1 January 1999, and in the twelve-month period thereafter, its calculated level of consumption of the controlled substance in Annex E does not exceed, annually, seventy-five per cent of its calculated level of consumption in 1991. Each Party producing the substance shall, for the same periods, ensure that its calculated level of production of the substance does not exceed, annually, seventy-five per cent of its calculated level of production in 1991. However, in order to satisfy the basic domestic needs of the Parties operating under paragraph 1 of Article 5, its calculated level of production may exceed that limit by up to ten per cent of its calculated level of production in 1991.

3. Each Party shall ensure that for the twelve-month period commencing on 1 January 2001, and in the twelve-month period thereafter, its calculated level of consumption of the controlled substance in Annex E does not exceed, annually, fifty per cent of its calculated level of consumption in 1991. Each Party producing the substance shall, for the same periods, ensure that its calculated level of production of the substance does not exceed, annually, fifty per cent of its calculated level of production in 1991. However, in order to satisfy the basic domestic needs of the Parties operating under paragraph 1 of Article 5, its calculated level of production may exceed that limit by up to ten per cent of its calculated level of production in 1991.

4. Each Party shall ensure that for the twelve-month period commencing on 1 January 2003, and in the twelve-month period thereafter, its calculated level of consumption of the controlled substance in Annex E does not exceed, annually, thirty per cent of its calculated level of consumption in 1991. Each Party producing the substance shall, for the same periods, ensure that its calculated level of production of the substance does not exceed, annually, thirty per cent of its calculated level of production in 1991. However, in order to satisfy the basic domestic needs of the Parties operating under paragraph 1 of Article 5, its calculated level of production may exceed that limit by up to ten per cent of its calculated level of production in 1991.

5. Each Party shall ensure that for the twelve-month period commencing on 1 January 2005, and in each twelve-month period thereafter, its calculated level of consumption of the controlled substance in Annex E does not exceed zero. Each Party producing the substance shall, for the same periods, ensure that its calculated level of production of the substance does not exceed zero. However,

in order to satisfy the basic domestic needs of the Parties operating under paragraph 1 of Article 5, its calculated level of production may, until 1 January 2002 exceed that limit by up to fifteen per cent of its calculated level of production in 1991; thereafter, it may exceed that limit by a quantity equal to the annual average of its production of the controlled substance in Annex E for basic domestic needs for the period 1995 to 1998 inclusive. This paragraph will apply save to the extent that the Parties decide to permit the level of production or consumption that is necessary to satisfy uses agreed by them to be critical uses.

5 *bis.* Each Party shall ensure that for the twelve-month period commencing on 1 January 2005 and in each twelve-month period thereafter, its calculated level of production of the controlled substance in Annex E for the basic domestic needs of the Parties operating under paragraph 1 of Article 5 does not exceed eighty per cent of the annual average of its production of the substance for basic domestic needs for the period 1995 to 1998 inclusive.

5 *ter.* Each Party shall ensure that for the twelve-month period commencing on 1 January 2015 and in each twelve-month period thereafter, its calculated level of production of the controlled substance in Annex E for the basic domestic needs of the Parties operating under paragraph 1 of Article 5 does not exceed zero.

6. The calculated levels of consumption and production under this Article shall not include the amounts used by the Party for quarantine and pre-shipment applications.

Article 2I: Bromochloromethane

Each Party shall ensure that for the twelve-month period commencing on 1 January 2002, and in each twelve-month period thereafter, its calculated level of consumption and production of the controlled substance in Group III of Annex C does not exceed zero. This paragraph will apply save to the extent that the Parties decide to permit the level of production or consumption that is necessary to satisfy uses agreed by them to be essential.

Article 2J: Hydrofluorocarbons

1. Each Party shall ensure that for the twelve-month period commencing on 1 January 2019, and in each twelve-month period thereafter, its calculated level of consumption of the controlled substances in Annex F, expressed in CO_2 equivalents, does not exceed the per centage, set out for the respective range of years specified in subparagraphs (a) to (e) below, of the annual average of its calculated levels of consumption of Annex F controlled substances for the years 2011, 2012 and 2013, plus fifteen per cent of its calculated level of consumption of Annex C, Group I, controlled substances as set out in paragraph 1 of Article 2F, expresses in CO_2 equivalents:

(a) 2019 to 2023: 90 per cent

(b) 2024 to 2028: 60 per cent

(c) 2029 to 2033: 30 per cent

(d) 2034 to 2035: 20 per cent

(e) 2036 and thereafter: 15 per cent

2. Notwithstanding paragraph 1 of this Article, the Parties may decide that a Party shall ensure that, for the twelve-month period commencing on 1 January 2020, and in each twelve-month period thereafter, its calculated level of consumption of the controlled substances in Annex F, expressed in CO_2 equivalents, does not exceed the per centage, set out for the respective range of years specified in subparagraphs (a) to (e) below, of the annual average of its calculated levels of consumption of Annex F controlled substances for the years 2011, 2012 and 2013, plus twenty-five per cent of its calculated level of consumption of Annex C, Group I, controlled substances as set out in paragraph 1 of Article 2F, expressed in CO_2 equivalents:

(a) 2020 to 2024: 95 per cent

(b) 2025 to 2028: 65 per cent

(c) 2029 to 2033: 30 per cent

(d) 2034 to 2035: 20 per cent

(e) 2036 and thereafter: 15 per cent

3. Each Party producing the controlled substances in Annex F shall ensure that for the twelve-month period commencing on 1 January 2019, and in each twelve-month period thereafter, its calculated level of production of the controlled substances in Annex F, expressed in CO_2 equivalents, does not exceed the per centage, set out for the respective range of years specified in subparagraphs (a) to (e) below, of the annual average of its calculated levels of production of Annex F controlled substances for the years 2011, 2012 and 2013, plus fifteen per cent of its calculated level of production of Annex C, Group I, controlled substances as set out in paragraph 2 of Article 2F, expressed in CO_2 equivalents:

(a) 2019 to 2023: 90 per cent

(b) 2024 to 2028: 60 per cent

(c) 2029 to 2033: 30 per cent

(d) 2034 to 2035: 20 per cent

(e) 2036 and thereafter: 15 per cent

4. Notwithstanding paragraph 3 of this Article, the Parties may decide that a Party producing the controlled substances in Annex F shall ensure that for the twelve-month period commencing on 1 January 2020, and in each twelve-month period thereafter, its calculated level of production of the controlled substances in Annex F, expressed in CO_2 equivalents, does not exceed the per centage, set out for the respective range of years specified in subparagraphs (a) to (e) below, of the annual average of its calculated levels of production of Annex F controlled substances for the years 2011, 2012 and 2013, plus twenty-five per cent of its calculated level of production of Annex C, Group I, controlled substances as set out in paragraph 2 of Article 2F, expressed in CO_2 equivalents

(a) 2020 to 2024: 95 per cent

(b) 2025 to 2028: 65 per cent

(c) 2029 to 2033: 30 per cent

(d) 2034 to 2035: 20 per cent

(e) 2036 and thereafter: 15 per cent

5. Paragraphs 1 to 4 of this Article will apply save to the extent that the Parties decide to permit the level of production or consumption that is necessary to satisfy uses agreed by the Parties to be exempted uses.

6. Each Party manufacturing Annex C, Group I, or Annex F substances shall ensure that for the twelve-month period commencing on 1 January 2020, and in each twelve-month period thereafter, its emissions of Annex F, Group II, substances generated in each production facility that manufactures Annex C, Group I, or Annex F substances are destroyed to the extent practicable using technology approved by the Parties in the same twelve-month period.

7. Each Party shall ensure that any destruction of Annex F, Group II, substances generated by facilities that produce Annex C, Group I, or Annex F substances shall occur only by technologies approved by the Parties.

Article 3: Calculation of control levels

1. For the purposes of Articles 2, 2A to 2J and 5, each Party shall, for each group of substances in Annex A, Annex B, Annex C, Annex E or Annex F, determine its calculated levels of:

(a) Production by:
  (i) multiplying its annual production of each controlled substance by the ozone depleting potential specified in respect of it in Annex A, Annex B, Annex C or Annex E, except as otherwise specified in paragraph 2;
  (ii) adding together, for each such Group, the resulting figures;

(b) Imports and exports, respectively, by following, *mutatis mutandis*, the procedure set out in subparagraph (a); and

(c) Consumption by adding together its calculated levels of production and imports and subtracting its calculated level of exports as determined in accordance with subparagraphs (a) and (b). However, beginning on 1 January 1993, any export of controlled substances to non-Parties shall not be subtracted in calculating the consumption level of the exporting Party; and

(d) Emissions of Annex F, Group II, substances generated in each facility that generates Annex C, Group I, or Annex F substances by including, among other things, amounts emitted from equipment leaks, process vents and destruction devices, but excluding amounts captured for use, destruction or storage.

2. When calculating levels, expressed in CO_2 equivalents, of production, consumption imports, exports and emissions of Annex F and Annex C, Group I, substances for the purposes of Article 2J, paragraph 5 of Article 2 and paragraph 1 (d) of Article 3, each Party shall use the global warming potentials of those substances specified in Group I of Annex A, Annex C and Annex F.

Article 4: Control of trade with non-Parties

1. As of 1 January 1990, each Party shall ban the import of the controlled substances in Annex A from any State not party to this Protocol.

1 *bis*. Within one year of the date of the entry into force of this paragraph, each Party shall ban the import of the controlled substances in Annex B from any State not party to this Protocol.

1 *ter*. Within one year of the date of entry into force of this paragraph, each Party shall ban the import of any controlled substances in Group II of Annex C from any State not party to this Protocol.

1 *qua*. Within one year of the date of entry into force of this paragraph, each Party shall ban the import of the controlled substance in Annex E from any State not party to this Protocol.

1 *quin.* As of 1 January 2004, each Party shall ban the import of the controlled substances in Group I of Annex C from any State not party to this Protocol.

1 *sex.* Within one year of the date of entry into force of this paragraph, each Party shall ban the import of the controlled substance in Group III of Annex C from any State not party to this Protocol.

1 *sept.* Upon entry into force of this paragraph, each Party shall ban the import of the controlled substances in Annex F from any State not Party to this Protocol.

2. As of 1 January 1993, each Party shall ban the export of any controlled substances in Annex A to any State not party to this Protocol.

2 *bis.* Commencing one year after the date of entry into force of this paragraph, each Party shall ban the export of any controlled substances in Annex B to any State not party to this Protocol.

2 *ter.* Commencing one year after the date of entry into force of this paragraph, each Party shall ban the export of any controlled substance in Group II of Annex C to any State not party to this Protocol.

2 *qua.* Commencing one year of the date of entry into force of this paragraph, each Party shall ban the export of the controlled substance in Annex E to any State not party to this Protocol.

2 *quin.* As of 1 January 2004, each Party shall ban the export of the controlled substances in Group I of Annex C to any State not party to this Protocol.

2 *sex.* Within one year of the date of entry into force of this paragraph, each Party shall ban the export of the controlled substance in Group III of Annex C to any State not party to this Protocol.

2 *sept.* Upon entry into force of this paragraph, each Party shall ban the export of the controlled substances in Annex F to any State not Party to this Protocol.

3. By 1 January 1992, the Parties shall, following the procedures in Article 10 of the Convention, elaborate in an annex a list of products containing controlled substances in Annex A. Parties that have not objected to the annex in accordance with those procedures shall ban, within one year of the annex having become effective, the import of those products from any State not party to this Protocol.

3 *bis.* Within three years of the date of the entry into force of this paragraph, the Parties shall, following the procedures in Article 10 of the Convention, elaborate in an annex a list of products containing controlled substances in Annex B. Parties that have not objected to the annex in accordance with

those procedures shall ban, within one year of the annex having become effective, the import of those products from any State not party to this Protocol.

3 *ter.* Within three years of the date of entry into force of this paragraph, the Parties shall, following the procedures in Article 10 of the Convention, elaborate in an annex a list of products containing controlled substances in Group II of Annex C. Parties that have not objected to the annex in accordance with those procedures shall ban, within one year of the annex having become effective, the import of those products from any State not party to this Protocol.

4. By 1 January 1994, the Parties shall determine the feasibility of banning or restricting, from States not party to this Protocol, the import of products produced with, but not containing, controlled substances in Annex A. If determined feasible, the Parties shall, following the procedures in Article 10 of the Convention, elaborate in an annex a list of such products. Parties that have not objected to the annex in accordance with those procedures shall ban or restrict, within one year of the annex having become effective, the import of those products from any State not party to this Protocol.

4 *bis.* Within five years of the date of the entry into force of this paragraph, the Parties shall determine the feasibility of banning of banning or restricting, from States not party to this Protocol, the import of products produced with, but not containing, controlled substances in Annex B. If determined feasible, the Parties shall, following the procedures in Article 10 of the Convention, elaborate in an annex a list of such products. Parties that have not objected to the annex in accordance with those procedures shall ban or restrict, within one year of the annex having become effective, the import of those products from any State not party to this Protocol.

4 *ter.* Within five years of the date of entry into force of this paragraph, the Parties shall determine the feasibility of banning or restricting, from States not party to this Protocol, the import of products produced with, but not containing, controlled substances in Group II of Annex C. If determined feasible, the Parties shall, following the procedures in Article 10 of the Convention, elaborate in an annex a list of such products. Parties that have not objected to the annex in accordance with those procedures shall ban or restrict, within one year of the annex having become effective, the import of those products from any State not party to this Protocol.

5. Each Party undertakes to the fullest practicable extent to discourage the export to any State not party to this Protocol of technology for producing and for utilizing controlled substances in Annexes A, B, C, E and F.

6. Each Party shall refrain from providing new subsidies, aid, credits, guarantees or insurance programmes for the export to State not party to this Protocol of products, equipment, plants or technology that would facilitate the production of controlled substances in Annexes A, B, C, E and F.

7. Paragraph 5 and 6 shall not apply to products, equipment, plants or technology that improve the containment, recovery, recycling or destruction of controlled substances, promote the development of alternative substances, or otherwise contribute to the reduction of emissions of controlled substances in Annexes A, B, C, E and F.

8. Notwithstanding the provisions of this Article, imports and exports referred to in paragraph 1 to 4 *ter* of this Article may be permitted from, or to, any State not party to this Protocol, if that State is determined, by a meeting of the Parties, to be in full compliance with Article 2, Article 2A to 2J and this Article, and have submitted data to that effect as specified in Article 7.

9. For the purposes of this Article, the term "State not party to this Protocol" shall include, with respect to a particular controlled substance, a State or regional economic integration organization that has not agreed to be bound by the control measures in effect for that substance.

10. By 1 January 1996, the Parties shall consider whether to amend this Protocol in order to extend the measures in this Article to trade in controlled substances in Group I of Annex C and in Annex E with States not party to the Protocol.

Article 4A: Control of trade with Parties

1. Where, after the phase-out date applicable to it for a controlled substance, a Party is unable, despite having taken all practicable steps to comply with its obligation under the Protocol, to cease production of that substance for domestic consumption, other than for uses agreed by the Parties to be essential, it shall ban the export of used, recycled and reclaimed quantities of that substance, other than for the purpose of destruction.

2. Paragraph 1 of this Article shall apply without prejudice to the operation of Article 11 of the Convention and the non-compliance procedure developed under Article 8 of the Protocol.

Article 4B: Licensing

1. Each Party shall, by 1 January 2000 or within three months of the date of entry into force of this Article for it, whichever is the later, establish and

implement a system for licensing the import and export of new, used, recycled and reclaimed controlled substances in Annexes A, B, C and E.

2. Notwithstanding paragraph 1 of this Article, any Party operating under paragraph 1 of Article 5 which decides it is not in a position to establish and implement a system for licensing the import and export of controlled substances in Annexes C and E, may delay taking those actions until 1 January 2005 and 1 January 2002, respectively.

2 *bis.* Each Party shall, by 1 January 2019 or within three months of the date of entry into force of this paragraph for it, whichever is later, establish and implement a system for licensing the import and export of new, used, recycled and reclaimed controlled substances in Annex F. Any Party operating under paragraph 1 of Article 5 that decides it is not in a position to establish and implement such a system by 1 January 2019 may delay taking those actions until 1 January 2021.

3. Each Party shall, within three months of the date of introducing its licensing a system, report to the Secretariat on the establishment and operation of that system.

4. The Secretariat shall periodically prepare and circulate to all Parties a list of the Parties that have reported to it on their licensing systems and shall forward this information to the Implementation Committee for consideration and appropriate recommendations to the Parties.

Article 5: Special situation of developing countries

1. Any Party that is a developing country and whose annual calculated level of consumption of the controlled substances in Annex A is less than 0.3 kilograms per capita on the date of the entry into force of the Protocol for it, or any time thereafter until 1 January 1999, shall, in order to meet its basic domestic needs, be entitled to delay for ten years its compliance with the control measures set out in Articles 2A to 2E, provided that any further amendments to the adjustments or Amendment adopted at the Second Meeting of the Parties in London, 29 June 1990, shall apply to the Parties operating under this paragraph after the review provided for in paragraph 8 of this Article has taken place and shall be based on the conclusions of that review.

1 *bis.* The Parties shall, taking into account the review referred to in paragraph 8 of this Article, the assessments made pursuant to Article 6 and any other relevant information, decide by 1 January 1996, through the procedure set forth in paragraph 9 of Article 2:

(a) With respect to paragraphs 1 to 6 Article 2F, what base year, initial levels, control schedules and phase-out date for consumption of the controlled substances in Group I of Annex C will apply to Parties operating under paragraph 1 of this Article;

(b) With respect to Article 2G, what phase-out date of production and consumption of the controlled substances in Group II of Annex C will apply to Parties operating under paragraph 1 of this Article; and

(c) With respect to Article 2H, what base year, initial levels and control schedules for consumption and production of the controlled substance in Annex E will apply to Parties operating under paragraph 1 of this Article.

2. However, any Party operating under paragraph 1 of this Article shall exceed neither an annual calculated level of consumption of the controlled substances in Annex A of 0.3 kilograms per capita nor an annual calculated level of consumption of controlled substances of Annex B of 0.2 kilograms per capita.

3. When implementing the control measures set out in Article 2A to 2E, any Party operating under paragraph 1 of this Article shall be entitled to use:

(a) For controlled substances under Annex A, either the average of its annual calculated level of consumption for the period 1995 to 1997 inclusive or a calculated level of consumption of 0.3 kilograms per capita, whichever is the lower, as the basis for determining its compliance with the control measures relating to consumption.

(b) For controlled substances under Annex B, the average of its annual calculated level of consumption for the period 1998 to 2000 inclusive or a calculated level of consumption of 0.2 kilograms per capita, whichever is the lower, as the basis for determining its compliance with the control measure relating to consumption.

(c) For controlled substances under Annex A, either the average of its annual calculated level of production for the period 1995 to 1997 inclusive or a calculated level of production of 0.3 kilograms per capita, whichever is the lower, as the basis for determining its compliance with the control measures relating to production.

(d) For controlled substances under Annex B, either the average of its annual calculated level of production for the period 1998 to 2000 inclusive or a calculated level of production of 0.2 kilograms per capita, whichever is the lower, as the basis for determining its compliance with the control measures relating to production.

4. If a Party operating under paragraph 1 of this Article, at any time before the control measures obligations in Article 2A to 2J become applicable to it, finds itself unable to obtain an adequate supply of controlled substances, it may notify this to the Secretariat. The Secretariat shall forthwith transmit a copy of such notification to the Parties, which shall consider the matter at their next Meeting, and decide upon appropriate action to be taken.

5. Developing the capacity to fulfil the obligations of the Parties operating under paragraph 1 of this Article to comply with the control measures set out in Articles 2A to 2E and Articles 2I and 2J, and with any control measures in Articles 2F to 2H that are decided pursuant to paragraph 1 *bis* of this Article, and their implementation by those same Parties will depend upon the effective implementation of the financial co-operation as provided by Article 10 and the transfer of technology as provided by Article 10A.

6. Any Party operating under paragraph 1 of this Article may, at any time, notify the Secretariat in writing that, having taken all practicable steps it is unable to implement any or all of the obligations laid down in Articles 2A to 2E and Articles 2I and 2J, or any or all obligations in Articles 2F to 2H that are decided pursuant to paragraph 1 *bis* of this Article, due to the inadequate implementation of Articles 10 and 10A. The Secretariat shall forthwith transmit a copy of the notification to the Parties, which shall consider the matter at their next Meeting, giving due recognition to paragraph 5 of this Article and shall decide upon appropriate action to be taken.

7. During the period between notification and the Meeting of the Parties at which the appropriate action referred to in paragraph 6 above is to be decided, or for further period if the Meeting of the Parties so decides, the non-compliance procedures referred to in Article 8 shall not be invoked against the notifying Party.

8. A Meeting of the Parties shall review, not later than 1995, the situation of the Parties operating under paragraph 1 of this Article, including the effective implementation of financial co-operation and transfer of technology to them, and adopt such revisions that may be deemed necessary regarding the schedule of control measures applicable to those Parties.

8 *bis*. Based on the conclusions of the review referred to in paragraph 8 above:

(a) With respect to the controlled substances in Annex A, a Party operating under paragraph 1 of this Article shall, in order to meet its basic domestic needs, be entitled to delay for ten years its compliance with the control measures adopted by the Second Meeting of the Parties in London, 29 June 1990, and reference by the Protocol to Articles 2A and 2B shall be read accordingly;

(b) With respect to the controlled substances in Annex B, a Party operating under paragraph 1 of this Article shall, in order to meet its basic domestic needs, be entitled to delay for ten years its compliance with the control measures adopted by the Second Meeting of the Parties in London, 29 June 1990, and reference by the Protocol to Articles 2C to 2E shall be read accordingly.

8 *ter*. Pursuant to paragraph 1 *bis* above:

(a) Each Party operating under paragraph 1 of this Article shall ensure that for the twelve-month period commencing on 1 January 2013, and in each twelve-month period thereafter, its calculated level of consumption of the controlled substances in Group I of Annex C does not exceed, annually, the average of its calculated levels of consumption in 2009 and 2010. Each Party operating under paragraph 1 of this Article shall ensure that for the twelve-month period commencing on 1 January 2013 and in each twelve-month period thereafter, its calculated level of production of the controlled substances in Group I of Annex C does not exceed, annually, the average of its calculated levels of production in 2009 and 2010;

(b) Each Party operating under paragraph 1 of this Article shall ensure that for the twelve-month period commencing on 1 January 2020, and in each twelve-month period thereafter, its calculated level of consumption of the controlled substances in Group I of Annex C does not exceed, annually, ninety per cent of the average of its calculated levels of consumption in 2009 and 2010. Each such Party producing one or more of these substances shall, for the same periods, ensure that its calculated level of production of the controlled substances in Group I of Annex C does not exceed, annually, ninety per cent of the average of its calculated levels of production in 2009 and 2010;

(c) Each Party operating under paragraph 1 of this Article shall ensure that for the twelve-month period commencing on 1 January 2020, and in each twelve-month period thereafter, its calculated level of consumption of the controlled substances in Group I of Annex C does not exceed, annually, sixty-five per cent of the average of its calculated levels of consumption in 2009 and 2010. Each such Party producing one or more of these substances shall, for the same periods, ensure that its calculated level of production of the controlled substances in Group I of Annex C does not exceed, annually, sixty-five per cent of the average of its calculated levels of production in 2009 and 2010;

(d) Each Party operating under paragraph 1 of this Article shall ensure that for the twelve-month period commencing on 1 January 2025, and in each twelve-month period thereafter, its calculated level of consumption of the controlled substances in Group I of Annex C does not exceed, annually, thirty-two point five per cent of the average of its calculated levels of consumption in 2009 and 2010. Each such Party producing one or more of these substances shall, for the same periods, ensure that its calculated level of production of the controlled substances in Group I of Annex C does not exceed, annually, thirty-two point five per cent of the average of its calculated levels of production in 2009 and 2010;

(e) Each Party operating under paragraph 1 of this Article shall ensure that for the twelve-month period commencing on 1 January 2030, and in each twelve-month period thereafter, its calculated level of consumption of the controlled substances in Group I of Annex C does not exceed zero. Each such Party producing one or more of these substances shall, for the same periods, ensure that its calculated level of production of the controlled substances in Group I of Annex C does not exceed zero. However:

(i) Each such party may exceed that limit on consumption in any such twelve-month period so long as the sum of its calculated levels of consumption over the ten-year period from 1 January 2030 to 1 January 2040, divided by ten, does not exceed two point five per cent of the average of its calculated levels of consumption in 2009 and 2010, and provided that such consumption shall be restricted to the servicing of refrigeration and air-conditioning equipment existing on 1 January 2030.

(ii) Each such Party may exceed that limit on production in any such twelve-month period so long as the sum of its calculated levels of production over the ten-year period from 1 January 2030 to 1 January 2040, divided by ten, does not exceed two point five per cent of the average of its calculated levels of production in 2009 and 2010, and provided that such production shall be restricted to the servicing of refrigeration and air-conditioning equipment existing on 1 January 2030.

(f) Each Party operating under paragraph 1 of this Article shall comply with Article 2G;

(g) With regard to the controlled substance contained in Annex E:

(i) As of 1 January 2002 each Party operating under paragraph 1 of this Article shall comply with the control measures set out in paragraph 1 of Article 2H and, as the basis for its compliance with these control measures, it shall use the average of its annual calculated level of consumption and production, respectively, for the period of 1995 to 1998 inclusive;

(ii) Each Party operating under paragraph 1 of this Article shall ensure that for the twelve-month period commencing on 1 January 2005, and in each twelve-month period thereafter, its calculated levels of consumption and production of the controlled substance in Annex

E do not exceed, annually, eighty per cent of the average of its annual calculated levels of consumption and production, respectively, for the period of 1995 to 1998 inclusive;

(iii) Each Party operating under paragraph 1 of this Article shall ensure that for the twelve-month period commencing on 1 January 2015 and in each twelve-month period thereafter, its calculated levels of consumption and production of the controlled substance in Annex E do not exceed zero. This paragraph will apply save to the extent that the Parties decide to permit the level of production or consumption that is necessary to satisfy uses agreed by them to be critical uses;

(iv) The calculated levels of consumption and production under this paragraph shall not include the amount used by the Party for quarantine and pre-shipment applications.

8 *qua.* (a) Each Party operating under paragraph 1 of this Article, subject to any adjustments made to the control measures in Article 2J in accordance with paragraph 9 of Article 2, shall be entitled to delay its compliance with the control measures set out in subparagraphs (a) to (e) of paragraph 1 of Article 2J and subparagraphs (a) to (e) of paragraph 3 of Article 2J and modify those measures as follows:

(i) 2024 to 2028: 100 per cent

(ii) 2029 to 2034: 90 per cent

(iii) 2035 to 2039: 70 per cent

(iv) 2040 to 2044: 50 per cent

(v) 2045 and thereafter: 20 per cent

(b) Notwithstanding subparagraph (a) above, the Parties may decide that a Party operating under paragraph 1 of this Article, subject to any adjustments made to the control measures in Article 2J in accordance with paragraph 9 of Article 2, shall be entitled to delay its compliance with the control measures set out in subparagraphs (a) to (e) of paragraph 1 of Article 2J and subparagraphs (a) to (e) of paragraph 3 of Article 2J and modify those measures as follows:

(i) 2028 to 2031: 100 per cent

(ii) 2032 to 2036: 90 per cent

(iii) 2037 to 2041: 80 per cent

(iv) 2042 to 2046: 70 per cent

(v) 2047 and thereafter: 15 per cent

(c) Each Party operating under paragraph 1 of this Article, for the purpose of calculating its consumption baseline under Article 2J, shall be entitled to use the average of its calculated levels of consumption of Annex F controlled substances for the years 2020, 2021 and 2022, plus sixty-five per cent of its baseline consumption of Annex C, Group I, controlled substances as set out in paragraph 8 *ter* of this Article.

(d) Notwithstanding subparagraph (c) above, the Parties may decide that a Party operating under paragraph 1 of this Article, for the purposes of calculating its consumption baseline under Article 2J, shall be entitled to use the average of its calculated levels of consumption of Annex F controlled substances for the years 2024, 2025 and 2026, plus sixty-five per cent of its baseline consumption of Annex C, Group I, controlled substances as set out in paragraph 8 *ter* of this Article.

(e) Each Party operating under paragraph 1 of this Article and producing the controlled substances in Annex F, for the purpose of calculating its production baseline under Article 2J, shall be entitled to use the average of its calculated levels of production of Annex F controlled substances for the years 2020, 2021 and 2022, plus sixty-five per cent of its baseline production of Annex C, Group I, controlled substances as set out in paragraph 8 *ter* of this Article.

(f) Notwithstanding subparagraph (e) above, the Parties may decide that a Party operating under paragraph 1 of this Article and producing the controlled substances in Annex F, for the purposes of calculating its production baseline under Article 2J, shall be entitled to use the average of its calculated levels of production of Annex F controlled substances for the years 2024, 2025 and 2026, plus sixty-five per cent of its baseline production of Annex C, Group I, controlled substances as set out in paragraph 8 *ter* of this Article.

(g) Subparagraphs (a) to (f) of this paragraph will apply to calculated levels of production and consumption save to the extent that a high-ambient-temperature exemption applies based on criteria decided by the Parties.

9. Decisions of the Parties referred to in paragraph 4,6 and 7 of this Article shall be taken according to the same procedure applied to decision-making under Article 10.

Article 6: Assessment and review of control measures

Beginning in 1990, and at least every four years thereafter, the Parties shall assess the control measures provided for in Article 2 and Articles 2A to 2J on the basis of available scie tific, environmental, technical and economic information. At least one year before each assessment, the Parties shall convene appropriate panels of experts qualified in the fields mentioned and determine the composition and terms of reference of any such panels. Within one year of being convened, the panels will report their conclusions, through the Secretariat, to the Parties.

Article 7: Reporting of data

1. Each Party shall provide to the Secretariat, within three months of becoming a Party, statistical data on its production, imports and exports of each of the controlled substances in Annex A for the year 1986, or the best possible estimates of such data where actual data are not available.

2. Each Party shall provide to the Secretariat statistical data on its production, imports and exports of each of the controlled substances
   - in Annex B and Group I and II of Annex C for the year 1989;
   - in Annex E, for the year 1991,
   - in Annex F, for the years 2011 to 2013, except that Parties operating under paragraph 1 of Article 5 shall provide such data for the years 2020 to 2022, but those Parties operating under paragraph 1 of Article 5 to which subparagraphs (d) and (f) of paragraph 8 *qua* of Article 5 applies shall provide such data for the years 2024 to 2026;

   or the best possible estimates of such data where actual data are not available, not later than three months after the date when the provisions set out in the Protocol with regard to the substances in Annexes B, C, E and F respectively enter into force for that Party.

3. Each Party shall provide to the Secretariat statistical data on its annual production (as defined in paragraph 5 of Article 1) of each of the controlled substances listed in Annexes A, B, C, E and F and, separately, for each substance,
   - Amounts used for feedstocks,
   - Amounts destroyed by technologies approved by the Parties, and
   - Imports from and exports to Parties and non-Parties respectively,

   for the year during which provisions concerning the substances in Annexes A, B, C, E and F respectively entered into force for that Party and for each year thereafter. Each Party shall provide to the Secretariat statistical data on the annual amount of the controlled substance listed in Annex E used for quarantine and pre-shipment applications. Data shall be forwarded not later than nine months after the end of the year to which the data relate.

3 *bis*. Each Party shall provide to the Secretariat separate statistical data of its annual imports and exports of each of the controlled substances listed in Group II of Annex A and Group I of Annex C that have been recycled.

3 *ter.* Each Party shall provide to the Secretariat statistical data on its annual emissions of Annex F, Group II, controlled substances per facility in accordance with paragraph 1 (d) of Article 3 of the Protocol.

4. For Parties operating under the provisions of paragraph 8 (a) of Article 2, the requirements in paragraphs 1,2,3 and 3 *bis* of this Article in respect of statistical data on production, imports and exports shall be satisfied if the regional economic integration organization concerned provides data on production, imports and exports between the organization and States that are not members of that organization.

Article 8: Non-compliance

The Parties, at their first meeting, shall consider and approve procedures and institutional mechanisms for determining non-compliance with the provisions of this Protocol and for treatment of Parties found to be in non-compliance.

Article 9: Research, development, public awareness and exchange of information

1. The Parties shall co-operate, consistent with their national laws, regulations and practices and taking into account in particular the needs of developing countries, in promoting, directly or through competent international bodies, research, development and exchange of information on:

   (a) best technologies for improving the containment, recovery, recycling, or destruction of controlled substances or otherwise reducing their emissions;

   (b) possible alternatives to controlled substances, to products containing such substances, and to products manufactures with them; and

   (c) costs and benefits of relevant control strategies.

2. The Parties, individually, jointly or through competent international bodies, shall co-operate in promoting public awareness of the environmental effects of the emissions of controlled substances and other substances that deplete the ozone layer.

3. Within two years of the entry into force of this Protocol and every two years thereafter, each Party shall submit to the Secretariat a summary of the activities it has conducted pursuant to this Article.

Article 10: Financial mechanism

1. The Parties shall establish a mechanism for the purposes of providing financial and technical co-operation, including the transfer of technologies, to Parties operating under paragraph 1 of Article 5 of this Protocol to enable their compliance with the control measures set out in Articles 2A to 2E, Article 2I and Article 2J, and any control measures in Articles 2F to 2H that are decided pursuant to paragraph 1 *bis* Article 5 of the Protocol. The mechanism, contributions to which shall be additional to other financial transfers to Parties operating under that paragraph, shall meet all agreed incremental costs of such Parties in order to enable their compliance with the control measures of the Protocol. An indicative list of the categories of incremental costs shall be decided by the meeting of the Parties. Where a Party operating under paragraph 1 of Article 5 chooses to avail itself of funding from any other financial mechanism that could result in meeting any part of its agreed incremental costs, that part shall not be met by the financial mechanism under Article 10 of this Protocol.

2. The mechanism established under paragraph 1 shall include a Multilateral Fund. It may also include other means of multilateral, regional and bilateral co-operation.

3. The Multilateral Fund shall:

   (a) Meet, on a grant or concessional basis a appropriate, and according to criteria to be decided upon by the Parties, the agreed incremental costs;

   (b) Finance clearing-house functions to:
      (i) Assist Parties operating under paragraph 1 of Article 5, through country specific studies and other technical co-operation, to identify their needs for co-operation;
      (ii) Facilitate technical co-operation to meet these identified needs;
      (iii) Distribute, as provided for in Article 9, information and relevant materials, and hold workshops, training sessions, and other related activities, for the benefit of Parties that are developing countries; and
      (iv) Facilitate and monitor other multilateral, regional and bilateral co-operation available to Parties that are developing countries;

   (c) Finance the secretariat services of the Multilateral Fund and related support costs.

4. The Multilateral Fund shall operate under the authority of the Parties who shall decide on its overall policies.

5. The Parties shall establish an Executive Committee to develop and monitor the implementation of specific operational policies, guidelines and

administrative arrangements, including the disbursement of resources, for the purpose of achieving the objectives of the Multilateral Fund. The Executive Committee shall discharge its tasks and responsibilities, specified in its terms of reference as agreed by the Parties, with the co-operation and assistance of the International Bank for Reconstruction and Development (World Bank), the United Nations Environment Programme, the United Nations Development Programme or other appropriate agencies depending on their respective areas of expertise. The members of the Executive Committee, which shall be selected on the basis of a balanced representation of the Parties operating under paragraph 1 of Article 5 and the Parties not so operating, shall be endorsed by the Parties.

6. The Multilateral Fund shall be financed by contributions from Parties not operating under paragraph 1 of Article 5 in convertible currency or, in certain circumstances, in kind and/or in national currency, on the basis of the United Nations scale of assessments. Contributions by other Parties shall be encouraged. Bilateral and, in particular cases agreed by a decision of the Parties, regional co-operation may, up to a per centage and consistent with any criteria to be specified by decision of the Parties, be considered as a contribution to the Multilateral Fund, provided that such co-operation, as a minimum:

   (a) Strictly relates to compliance with the provisions of this Protocol;

   (b) Provides additional resources; and

   (c) Meets agreed incremental cost.

7. The Parties shall decide upon the programme budget of the Multilateral Fund for each fiscal period and upon the per centage of contributions of the individual Parties thereto.

8. Resources under the Multilateral Fund shall be disbursed with the concurrence of the beneficiary Party.

9. Decisions by the Parties under this Article shall be taken by consensus whenever possible. If all efforts at consensus have been exhausted and no agreement reached, decisions shall be adopted by a two-thirds majority vote of the Parties present and voting, representing a majority of the Parties operating under paragraph 1 of Article 5 present and voting and a majority of the Parties not so operating present and voting.

10. The financial mechanism set out in this Article is without prejudice to any future arrangements that may be developed with respect to other environmental issues.

Article 10A: Transfer of technology

Each Party shall take every practicable step, consistent with the programmes supported by the financial mechanism, to ensure:

(a) that the best available, environmentally safe substitutes and related technologies are expeditiously transferred to Parties operating under paragraph 1 of Article 5; and

(b) that the transfer referred to in subparagraph (a) occur under fair and most favourable conditions.

Article 11: Meetings of the Parties

1. The Parties shall hold meetings at regular interval intervals. The Secretariat shall convene the first meeting of the Parties not later than one year after the date of the entry into force of this Protocol and in conjunction with a meeting of the Conference of the Parties to the Convention, if a meeting of the latter is scheduled within that period.

2. Subsequent ordinary meetings of the Parties shall be held, unless the Parties otherwise decide, in conjunction with meetings of the Conference of the Parties to the Convention. Extraordinary meetings of the Parties shall be held at such other times as may be deemed necessary by a meeting of the Parties, or at the written request of any Party, provided that within six months of such a request being communicated to them by the Secretariat, it is supported by at least one third of the Parties.

3. The Parties, at their first meeting, shall:

   (a) adopt by consensus rules of procedure for their meetings:

   (b) adopt by consensus the financial rules referred to in paragraph 2 of Article 13;

   (c) establish the panels and determine the terms of reference referred to in Article 6;

   (d) consider and approve the procedures and institutional mechanisms specified in Article 8; and

   (e) begin preparation of workplans pursuant to paragraph 3 to Article 10. [The Article 10 in question is that of the original Protocol adopted in 1987.]

4. The functions of the meetings of the Parties shall be to:

   (a) review the implementation of this Protocol;

   (b) decide on any adjustments or reductions referred to in paragraph 9 of Article 2;

   (c) decide on any addition to, insertion in or removal from any annex of substances and on related control measures in accordance with paragraph 10 of Article 2;

   (d) establish, where necessary, guidelines or procedures for reporting of information as provided for in Article 7 and paragraph 3 of Article 9;

   (e) review requests for technical assistance submitted pursuant to paragraph 2 of Article 10;

   (f) review reports prepared by the secretariat pursuant to subparagraph (c) of Article 12;

   (g) assess, in accordance with Article 6, the control measure;

   (h) consider and adopt, as required, proposals for a mendment of this Protocol or any annex and for any new annex;

   (i) consider and adopt the budget for implementing this Protocol; and

   (j) consider and undertake any additional action that may be required for the achievement of the purposes of this Protocol.

5. The United Nations, its specialized agencies and the International Atomic Energy Agency, as well as any State not party to this Protocol, may be represented at meeting of the Parties as observers. Any body or agency, whether national or international, governmental or non-governmental, qualified in fields relating to the protection of the ozone layer which has informed the secretariat of its wish to be represented at a meeting of the Parties as an observer may be admitted unless at least one third of the Parties present object. The admission and participation of observers shall be subject to the rules of procedure adopted by the Parties.

Article 12: Secretariat

For the purposes of this Protocol, the Secretariat shall:

(a) arrange for and service meetings of the Parties as provided for in Article 11;

(b) receive and make available, upon request by a Party, data provided pursuant to Article 7;

(c) prepare and distribute regularly to the Parties reports based on information received pursuant to Articles 7 and 9;

(d) notify the Parties of any request for technical assistance received pursuant to Article 10 so as to facilitate the provision of such assistance;

(e) encourage non-Parties to attend the meetings of the Parties as observers and to act in accordance with the provisions of this Protocol;

(f) provide, as appropriate, the information and requests referred to in subparagraphs (c) and (d) to such non-party observers; and

(g) perform such other functions for the achievement of the purposes of this Protocol as may be assigned to it by the Parties.

Article 13: Financial provisions

1. The funds required for the operation of this Protocol, including those for the functioning of the Secretariat related to this Protocol, shall be charged exclusively against contributions from the Parties.

2. The Parties, at their first meeting, shall adopt by consensus financial rules for the operation of this Protocol.

Article 14: Relationship of this Protocol to the Convention

Except as otherwise provided in this Protocol, the provisions of the Convention relating to its protocol shall apply to this Protocol.

Article 15: Signature

This Protocol shall be open for signature by States and by regional economic integration organizations in Montreal on 16 September 1987, in Ottawa from 17 September 1987 to 16 January 1988, and at United Nations Headquarters in New York from 17 January 1988 to 15 September 1988.

Article 16: Entry into force

1. This Protocol shall enter into force on 1 January 1989, provided that at least eleven instruments of ratification, acceptance, approval of the Protocol or

accession thereto have been deposited by States or regional economic integration organizations representing at least two-thirds of 1986 estimated global consumption of the controlled substances, and the provisions of paragraph 1 of Article 17 of the Convention have been fulfilled. In the event that these conditions have not been fulfilled by that date, the Protocol shall enter into force on the ninetieth day following the date on which the conditions have been fulfilled.

2. For the purposes of paragraph 1, any such instrument deposited by a regional economic integration organization shall not be counted as additional to those deposited by member States of such organization.

3. After the entry into force of this Protocol, any State or regional economic integration organization shall become a Party to it on the ninetieth day following the date of deposit of its instrument of ratification, acceptance, approval or accession.

Article 17: Parties joining after entry into force

Subject to Article 5, any State or regional economic integration organization which becomes a Party to this Protocol after the date of its entry into force, shall fulfil forthwith the sum of the obligations under Article 2, as well as under Articles 2A to 2J and Article 4, that apply at that date to the States and regional economic integration organizations that became Parties on the date the Protocol entered into force.

Article 18: Reservations

No reservations may be made to this Protocol.

Article 19: Withdrawal

Any Party may withdraw from this Protocol by giving written notification to the Depositary at any time after four years of assuming the obligations specified in paragraph 1 of Article 2A. Any such withdrawal shall take effect upon expiry of one year after the date of its receipt by the Depositary, or on such later date as may be specified in the notification of the withdrawal.

Article 20: Authentic texts

The original of this Protocol, of which the Arabic, Chinese, English, French, Russian and Spanish texts are equally authentic, shall be deposited with the Secretary-General of the United Nations.

IN WITNESS WHEREOF the undersigned, being duly authorized to that effect, have signed this Protocol.

DONE at Montreal this sixteenth day of September, one thousand nine hundred and eighty seven.

Annex A: Controlled substances

Group	Substance	Ozone-Depleting Potential*	100-Year Global Warming Potential
Group I			
$CFCl_3$	(CFC-11)	1.0	4 750
CF_2Cl_2	(CFC-12)	1.0	10 900
$C_2F_3Cl_3$	(CFC-113)	0.8	6 130
$C_2F_4Cl_2$	(CFC-114)	1.0	10 000
C_2F_5Cl	(CFC-115)	0.6	7 370
Group II			
CF_2BrCl	(halon-1211)	3.0	
CF_3Br	(halon-1301)	10.0	
$C_2F_4Br_2$	(halon-2402)	6.0	

*These ozone depleting potentials are estimates based on existing knowledge and will be reviewed and revised periodically.

Annex B: Controlled substances

Group	Substance	Ozone-Depleting Potential
Group I		
CF_3Cl	(CFC-13)	1.0
C_2FC_5	(CFC-111)	1.0
$C_2F_2Cl_4$	(CFC-112)	1.0
C_3FCl_7	(CFC-211)	1.0
$C_3F_2Cl_6$	(CFC-212)	1.0
$C_3F_3Cl_5$	(CFC-213)	1.0
$C_3F_4Cl_4$	(CFC-214)	1.0
$C_3F_5Cl_3$	(CFC-215)	1.0

Group	Substance	Ozone-Depleting Potential
$C_3F_6Cl_2$	(CFC-216)	1.0
C_3F_7Cl	(CFC-217)	1.0
Group II		
CCl_4	carbon tetrachloride	1.1
Group III		
$C_2H_3Cl_3$*	1,1,1-trichloroethane* (methyl chloroform)	0.1

*This formula does not refer to 1,1,2-trichloroethane

Annex C: Controlled substances

Group	Substance	Number of isomers	Ozone-Depleting Potential*	Year Global Warming Potential***
Group I				
$CHFCl_2$	(HCFC-21)**	1	0.44	151
CHF_2Cl	(HCFC-22)**	1	0.055	1 810
CH_2FCl	(HCFC-31)	1	0.02	
C_2HFCl_4	(HCFC-121)	2	0.01–0.04	
$C_2HF_2Cl_3$	(HCFC-122)	3	0.02–0.08	
$C_2HF_3Cl_2$	(HCFC-123)	3	0.02–0.06	77
$CHCl_2CF_3$	(HCFC-123)**	–	0.02	
C_2HF_4Cl	(HCFC-124)	2	0.02–0.04	609
$CHFClCF_3$	(HCFC-124)**	–	0.022	
$C_2H_2FCl_3$	(HCFC-131)	3	0.007–0.05	
$C_2H_2F_2Cl_2$	(HCFC-132)	4	0.008–0.05	
$C_2H_2F_3Cl$	(HCFC-133)	3	0.02–0.06	
$C_2H_3FCl_2$	(HCFC-141)	3	0.005–0.07	
CH_3CFCl_2	(HCFC-141b)**	–	0.11	725
$C_2H_3F_2Cl$	(HCFC-142)	3	0.008–0.07	
CH_3CF_2Cl	(HCFC-142b)**	–	0.065	2 310

Group	Substance	Number of isomers	Ozone-Depleting Potential*	Year Global Warming Potential***
C_2H_4FCl	(HCFC-151)	2	0.003–0.005	
C_3HFCl_6	(HCFC-221)	5	0.015–0.07	
$C_3HF_2Cl_5$	(HCFC-222)	9	0.01–0.09	
$C_3HF_3Cl_4$	(HCFC-223)	12	0.01–0.08	
$C_3HF_4Cl_3$	(HCFC-224)	12	0.01–0.09	
$C_3HF_5Cl_2$	(HCFC-225)	9	0.02–0.07	
$CF_3CF_2CHCl_2$	(HCFC-ca)**	–	0.025	122
CF_2ClCF_2CHClF	(HCFC-225cb)**	–	0.033	595
C_3HF_6Cl	(HCFC-226)	5	0.02–0.10	
$C_3H_2FCl_5$	(HCFC-231)	9	0.05–0.09	
$C_3H_2F_2Cl_4$	(HCFC-232	16	0.008–0.10	
$C_3H_2F_3Cl_3$	(HCFC-233)	18	0.007–0.23	
$C_3H_2F_4Cl_2$	(HCFC-234)	16	0.01–0.28	
$C_3H_2F_5Cl$	(HCFC-235)	9	0.03–0.52	
$C_3H_3FCl_4$	(HCFC-241)	12	0.004–0.09	
$C_3H_3F_2Cl_3$	(HCFC-242)	18	0.005–0.13	
$C_3H_3F_3Cl_2$	(HCFC-243)	18	0.007–0.12	
$C_3H_3F_4Cl$	(HCFC-244)	12	0.009–0.14	
$C_3H_4FCl_3$	(HCFC-251)	12	0.001–0.01	
$C_3H_4F_2Cl_2$	(HCFC-252)	16	0.005–0.04	
$C_3H_4F_3Cl$	(HCFC-253)	12	0.003–0.03	
$C_3H_5FCl_2$	(HCFC-261)	9	0.002–0.02	
$C_3H_5F_2Cl$	(HCFC-262)	9	0.002–0.02	
C_3H_6FCl	(HCFC-271)	5	0.001–0.03	
Group II				
$CHFBr_2$		1	1.00	
CHF_2Br	(HBFC-22B1)	1	0.74	
CH_2FBr		1	0.73	

Group	Substance	Number of isomers	Ozone-Depleting Potential*
C_2HFBr_4		2	0.3–0.8
$C_2HF_2Br_3$		3	0.5–1.8
$C_2HF_3Br_2$		3	0.4–1.6
C_2HF_4Br		2	0.7–1.2
$C_2H_2FBr_3$		3	0.1–1.1
$C_2H_2F_2Br_2$		4	0.2–1.5
$C_2H_2F_3Br$		3	0.7–1.6
$C_2H_3FBr_2$		3	0.1–1.7
$C_2H_3F_2Br$		3	0.2–1.1
C_2H_4FBr		2	0.07–0.1
C_3HFBr_6		5	0.3–1.5
$C_3HF_2Br_5$		9	0.2–1.9
$C_3HF_3Br_4$		12	0.3–1.8
$C_3HF_4Br_3$		12	0.5–2.2
$C_3HF_5Br_2$		9	0.9–2.0
C_3HF_6Br		5	0.7–3.3
$C_3H_2FBr_5$		9	0.1–1.9
$C_3H_2F_2Br_4$		16	0.2–2.1
$C_3H_2F_3Br_3$		18	0.2–5.6
$C_3H_2F_4Br_2$		16	0.3–7.5
$C_3H_2F_5Br$		8	0.9–1.4
$C_3H_3FBr_4$		12	0.08–1.9
$C_3H_3F_2Br_3$		18	0.1–3.1
$C_3H_3F_3Br_2$		18	0.1–2.5
$C_3H_3F_4Br$		12	0.3–4.4
$C_3H_4FBr_3$		12	0.03–0.3
$C_3H_4F_2Br_2$		16	0.1–1.0
$C_3H_4F_3Br$		12	0.07–0.8
$C_3H_5FBr_2$		9	0.04–0.4

Group	Substance	Number of isomers	Ozone-Depleting Potential*
$C_3H_5F_2Br$		9	0.07–0.8
C_3H_6FBr		5	0.02–0.7
Group III			
CH_2BrCl	bromochloromethane	1	0.12

*Where a range of ODPs is indicated, the highest value in that range shall be used for the purposes of the Protocol. The ODPs listed as a single value have been determined from calculations based on laboratory measurements. Those listed as range are based on estimates and are less certain. The range pertains to an isomeric group. The upper value is the estimate of the ODP of the isomer with the highest ODP, and the lower value is the estimate of the ODP of the isomer with the lowest ODP.
**Identifies the most commercially viable substances with ODP values listed against them to be used for the purposes of the Protocol.
***For substances for which no GWP is indicated, the default value 0 applies until a GWP value is included by means of the procedure foreseen in paragraph 9(a) (ii) of Article 2.

Annex D:* A list of products** containing controlled substances specified in Annex A

No.	Products	Customs code number
1	Automobile and truck air conditioning units (whether incorporated in vehicles or not)	....
2	Domestic and commercial refrigeration and air conditioning/ heat pump equipment***	....
	e.q. Refrigerators	....
	Freezers	....
	Dehumidifiers	....
	Water coolers	....
	Ice machines	....
	Air conditioning and heat pump units	....
3	Aerosol products, except medical aerosols	....
4	Portable fire extinguisher	....
5	Insulation boards, panels and pipe covers	....
6	Pre-polymers	....

*This Annex was adopted by the Third Meeting of the Parties in Nairobi, 21 June 1991 as required by paragraph 3 of Article 4 of the Protocol.
**Though not when transported in consignments of personal or household effects or in similar non-commercial situations normally exempted from customs attention.
***When containing controlled substances in Annex A as a refrigerant and/or in insulting material of the product.

Annex E: Controlled substances

Group	Substance	Ozone-Depleting Potential
Group I		
CH_3Br	methyl bromide	0.6

Annex F: Controlled substances

Group	Substance	100-Year Global Warming Potential
Group I		
CHF_2CHF_2	HFC-134	1 100
CH_2FCF_3	HFC-134a	1 430
CH_2FCHF_2	HFC-143	353
$CHF_2CH_2CF_3$	HFC-245fa	1 030
$CF_3CH_2CF_2CH_3$	HFC-365mfc	794
CF_3CHFCF_3	HFC-227ea	3 220
$CH_2FCF_2CF_3$	HFC-236cb	1 340
CHF_2CHFCF_3	HFC-236ea	1 370
$CF_3CH_2CF_3$	HFC-236fa	9 810
$CH_2FCF_2CHF_2$	HFC-245ca	693
$CF_3CHFCHFCF_2CF_3$	HFC-43-10mee	1 640
CH_2F_2	HFC-32	675
CHF_2CF_3	HFC-125	3 500
CH_3CF_3	HFC-143a	4 470
CH_3F	HFC-41	92
CH_2FCH_2F	HFC-152	53
CH_3CHF_2	HFC-152a	124
Group II		
CHF_3	HFC-23	14 800

Section 1.2
Summary of control measures under the Montreal Protocol

This summary of control measures takes into account all the Amendments including the Kigali Amendment. An Article 5 party is a party classified as a developing country accoring to the definition provided in Article 5 paragraph 1 of the Montreal Protocol. All other paties are referred to as non-Article 5 parties.

Annex A – Group I: Chlorofluorocarbons (CFC-11, CFC-12, CFC-113, CFC-114 and CFC-115)

Applicable to production and consumption.

Non-Article 5 parties		Article 5 parties	
Baseline	1986	**Baseline**	**Average of 1995–1997**
Freeze	July 1, 1989	Freeze	July 1, 1999
75 per cent reduction	January 1, 1994	50 per cent reduction	January 1, 2005
100 per cent reduction	January 1, 1996 (with possible essential use exemptions)	85 per cent reduction	January 1, 2007
		100 per cent reduction	January 1, 2010 (with possible essential use exemptions)

CFCs (Annex A/I) production/consumption reduction schedule

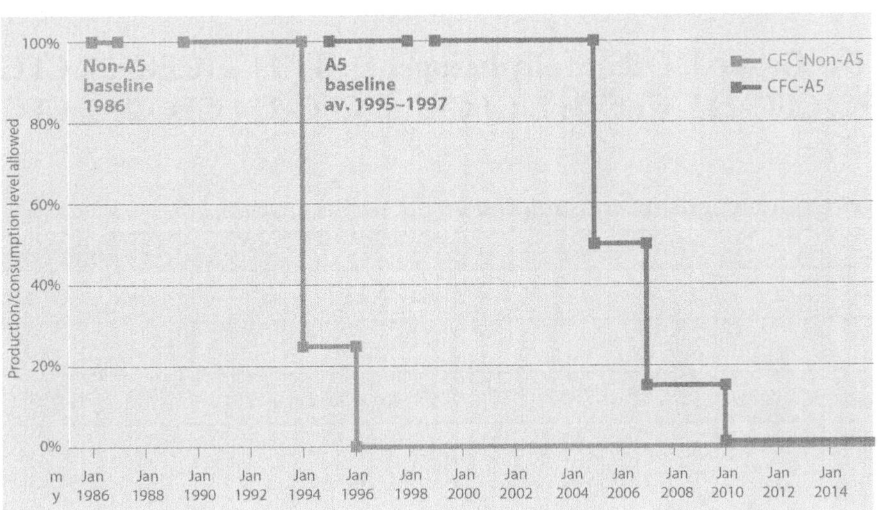

Annex A – Group II: Halons (halon-1211, halon-1301 and halon-2402)

Applicable to production and consumption.

Non-Article 5 parties		Article 5 parties	
Baseline	1986	Baseline	Average of 1995–1997
Freeze	January 1, 1992	Freeze	January 1, 2002
100 per cent reduction	January 1, 1994 (with possible essential use exemptions)	50 per cent reduction	January 1, 2005
		100 per cent reduction	January 1, 2010 (with possible essential use exemptions)

Halon (Annex A/II) production/consumption reduction schedule

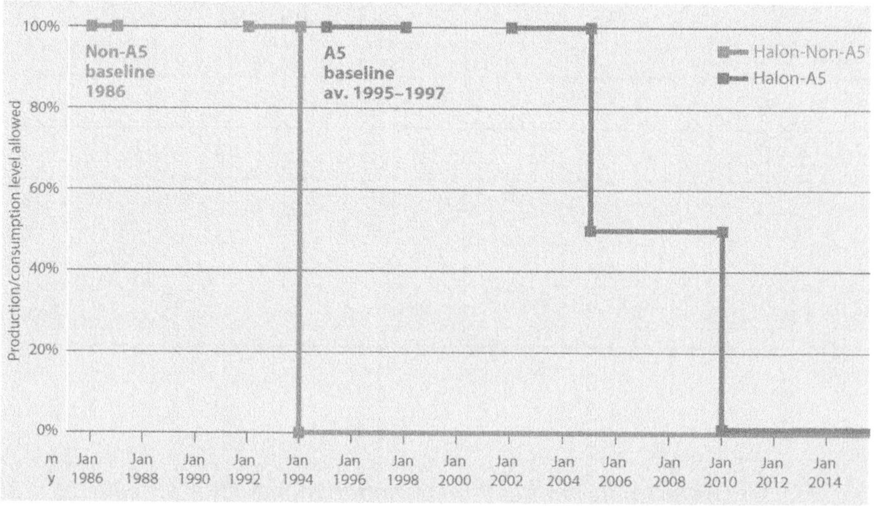

Annex B – Group I: Other fully halogenated CFCs (CFC-13, CFC-111, CFC-112, CFC-211, CFC-212, CFC-213, CFC-214 CFC-215, CFC-216, CFC-217)

Applicable to production and consumption.

Non-Article 5 parties		Article 5 parties	
Baseline	1989	Baseline	Average of 1998–2000
20 per cent reduction	July 1, 1993	20 per cent reduction	January 1, 2003
75 per cent reduction	January 1, 1994	85 per cent reduction	January 1, 2007
100 per cent reduction	January 1, 1996 (with possible essential use exemptions)	100 per cent reduction	January 1, 2010 (with possible essential use exemptions)

Other CFCs (Annex B/I) production/consumption reduction schedule

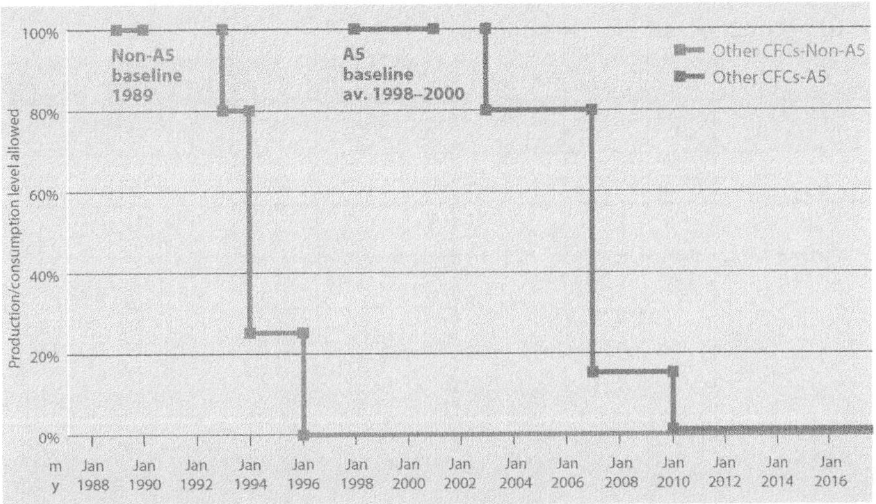

Annex B – Group II: Carbon tetrachloride

Applicable to production and consumption.

Non-Article 5 parties		Article 5 parties	
Baseline	1989	Baseline	Average of 1998–2000
85 per cent reduction	January 1, 1995	85 per cent reduction	January 1, 2005
100 per cent reduction	January 1, 1996 (with possible essential use exemptions)	100 per cent reduction	January 1, 2010 (with possible essential use exemptions)

Carbon tetrachloride (Annex B/II) production/consumption reduction schedule

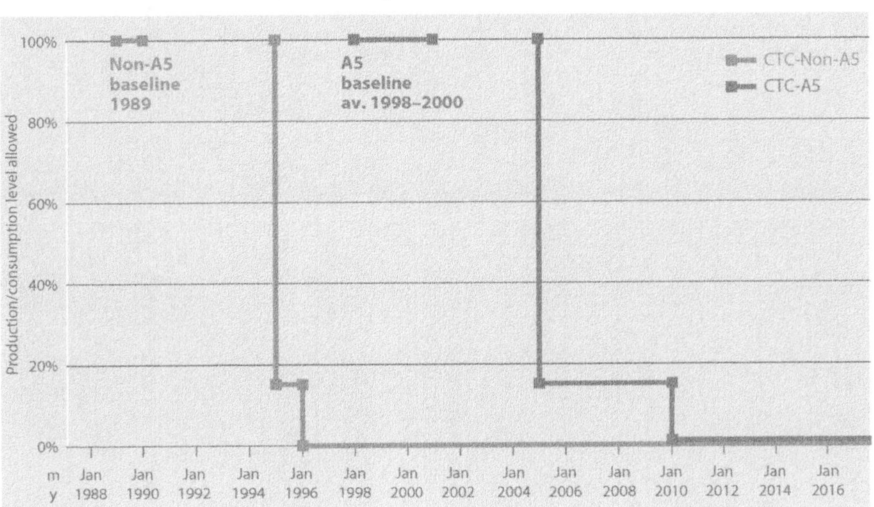

Annex B – Group III: 1,1,1-trichloroethane (methyl chloroform)

Applicable to production and consumption.

Non-Article 5 parties		Article 5 parties	
Baseline	1989	Baseline	Average of 1998–2000
Freeze	January 1, 1993	Freeze	January 1, 2003
85 per cent reduction	January 1, 1995	30 per cent reduction	January 1, 2005
100 per cent reduction	January 1, 1996 (with possible essential use exemptions)	70 per cent reduction	January 1, 2010
		100 per cent reduction	January 1, 2015 (with possible essential use exemptions)

Methyl chloroform (Annex B/III) production/consumption reduction schedule

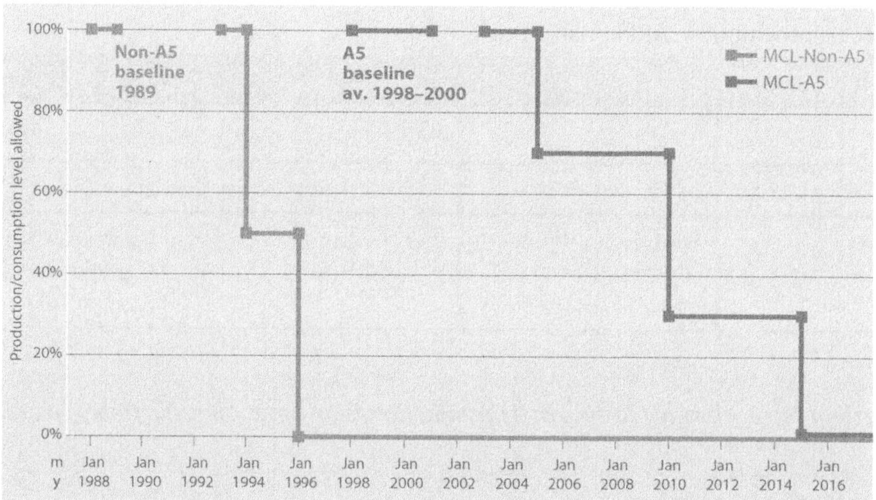

Annex C – Group I: HCFCs

Applicable to consumption.

Non-Article 5 parties		Article 5 parties	
Baseline	1989 HCFC consumption + 2.8 per cent of 1989 CFC consumption	Baseline	Average of 2009–2010
Freeze	1996	Freeze	January 1, 2013
35 per cent reduction	January 1, 2004	10 per cent reduction	January 1, 2015
75 per cent reduction	January 1, 2010	35 per cent reduction	January 1, 2020
90 per cent reduction	January 1, 2015	67.5 per cent reduction	January 1, 2025
100 per cent reduction	January 1, 2020, and thereafter, allowance of 0.5 per cent of baseline consumption until January 1, 2030 for servicing of refrigeration and air-conditioning equipment existing on 1 January 2020	100 per cent reduction	January 1, 2030, and thereafter, allowance of 2.5 per cent of baseline consumption when averaged over ten years 2030-2040 until January 1, 2040 for servicing of refrigeration and air conditioning equipment existing on 1 January 2030

HCFCs (Annex C/I) consumption reduction schedule

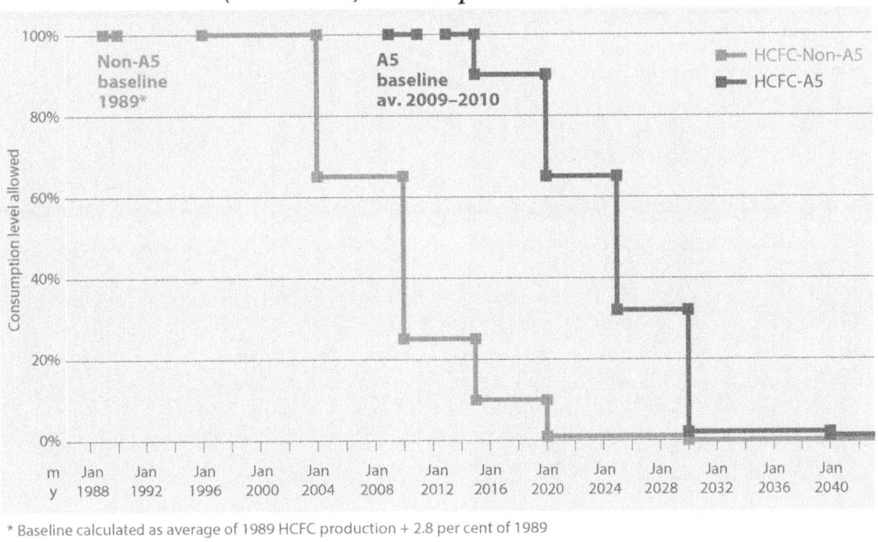

* Baseline calculated as average of 1989 HCFC production + 2.8 per cent of 1989

Annex C – Group I: HCFCs

Applicable to production

Non-Article 5 parties		Article 5 parties	
Baseline	Average of 1989 HCFC production + 2.8 per cent of 1989 CFC production and 1989 HCFC consumption + 2.8 per cent of 1989 CFC consumption	**Baseline**	**Average of 2009–2010**
Freeze	January 1, 2004, at the baseline for production	Freeze	January 1, 2013
		10 per cent reduction	January 1, 2015
75 per cent reduction	January 1, 2010	35 per cent reduction	January 1, 2020
90 per cent reduction	January 1, 2015	67.5 per cent reduction	January 1, 2025
100 per cent reduction	January 1, 2020, and thereafter, allowance of 0.5 per cent of baseline production until January 1, 2030 for servicing of refrigeration and air-conditioning equipment existing on 1 January 2020	100 per cent reduction	January 1, 2030, and thereafter, allowance of 2.5 per cent of baseline production when averaged over ten years 2030-2040 until January 1, 2040 for servicing of refrigeration and air conditioning equipment existing on 1 January 2030

HCFCs (Annex C/I) consumption reduction schedule

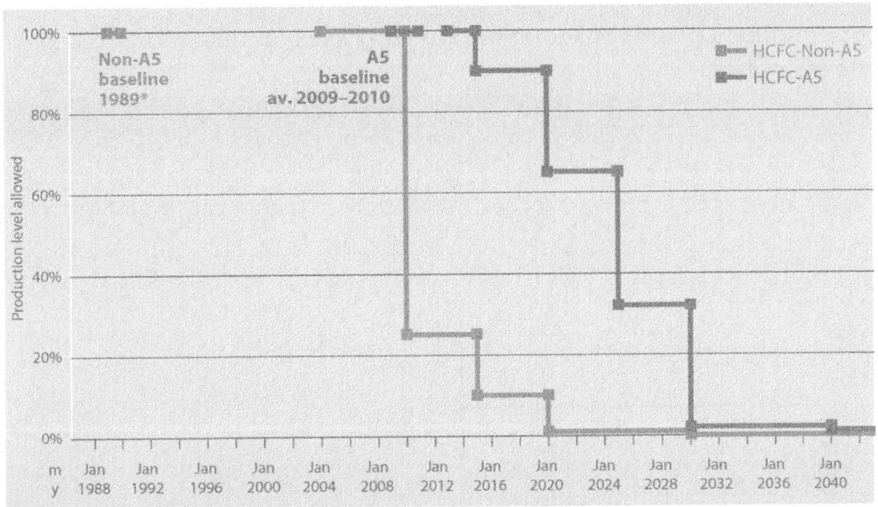

* Baseline calculated as average of 1989 HCFC production + 2.8 per cent of 1989 CFC production and 1989 HCFC consumption + 2.8 per cent of 1989 CFC consumption

Annex C – Group II: HBFCs

Applicable to production and consumption.

Non-Article 5 parties		Article 5 parties	
100 per cent reduction	January 1, 1996 (with possible essential use exemptions)	100 per cent reduction	January 1, 1996 (with possible essential use exemptions)

Annex C – Group III: Bromochloromethane

Applicable to production and consumption.

Non-Article 5 parties		Article 5 parties	
100 per cent reduction	January 1, 2002 (with possible essential use exemptions)	100 per cent reduction	January 1, 2002 (with possible essential use exemptions)

Annex E – Group I: Methyl bromide

Applicable to production and consumption, amounts used for quarantine and preshipment applications exempted.

Non-Article 5 parties		Article 5 parties	
Baseline	1991	Baseline	Average of 1995–1998
Freeze	January 1, 1995	Freeze	January 1, 2002
25 per cent reduction	January 1, 1999	20 per cent reduction	January 1, 2005
50 per cent reduction	January 1, 2001	100 per cent reduction	January 1, 2015 (with possible critical use exemptions)
70 per cent reduction	January 1, 2003		
100 per cent reduction	January 1, 2005 (with possible critical use exemptions)		

Methyl bromide (Annex E) production/consumption reduction schedule

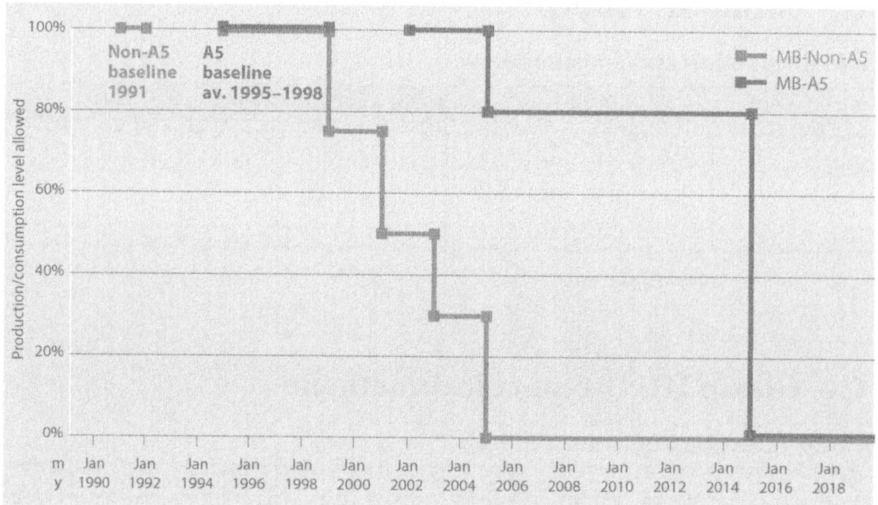

Annex F: Hydrofluorocarbons

Applicable to production and consumption.

Non-Article 5 parties		Article 5 parties – Group 1		Article 5 parties – Group 2	
Baseline	Average HFC for 2011–2013 + 15% of HCFC baseline	Baseline	Average HFC for 2020–2022 + 65% of HCFC baseline	Baseline	Average HFC for 2024–2026 + 65% of HCFC baseline
Freeze	–	Freeze	January 1, 2024	Freeze	January 1, 2028
10* per cent reduction	January 1, 2019	10 per cent reduction	January 1, 2029	10 per cent reduction	January 1, 2032
40* per cent reduction	January 1, 2024	30 per cent reduction	January 1, 2035	20 per cent reduction	January 1, 2037
70 per cent reduction	January 1, 2029	50 per cent reduction	January 1, 2040	30 per cent reduction	January 1, 2042
80 per cent reduction	January 1, 2034	80 per cent reduction	January 1, 2045	85 per cent reduction	January 1, 2047
85 per cent reduction	January 1, 2036				

*For Belarus, Kazakhstan, the Russian Federation, Tajikistan and Uzbekistan, 25% HCFC component of baseline and different initial two steps (1) 5% reduction in 2020 and (2) 35% reduction in 2025

Group 1: Article 5 parties not part of Group 2

Group 2: Bahrain, India, the Islamic Republic of Iran, Iraq, Kuwait, Oman, Pakistan, Qatar, Saudi Arabia and the United Arab Emirates

HFCs (Annex F) production/consumption reduction schedule

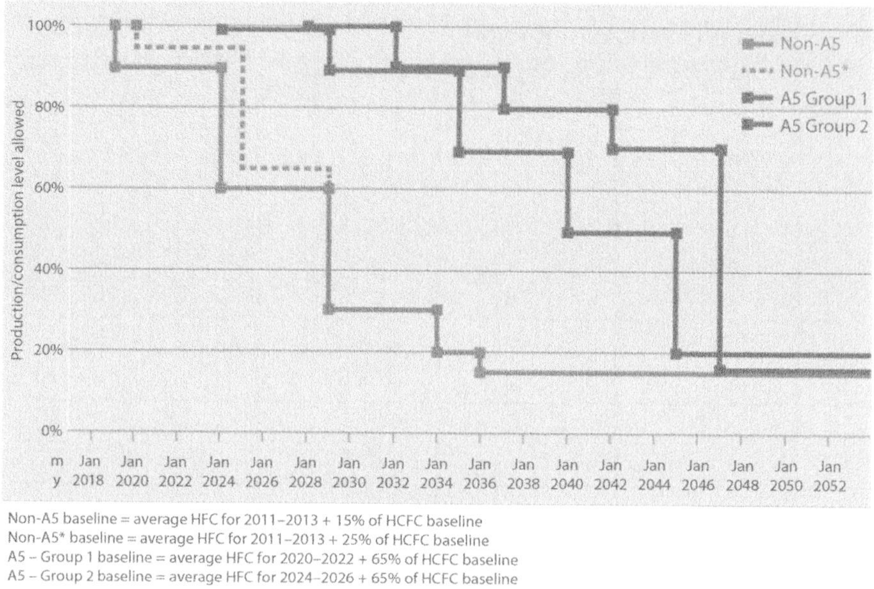

Non-A5 baseline = average HFC for 2011 - 2013 + 15% of HCFC baseline

Non-A5* baseline = average HFC for 2011 - 2013 + 25% of HCFC baseline

A5 - Group 1 baseline = average HFC for 2020 - 2022 + 65% of HCFC baseline

A5 - Group 2 baseline = average HFC for 2024 - 2026 + 65% of HCFC baseline

Section 1.3
Allowance for production to meet the basic domestic needs of Article 5 parties following the Montreal Adjustments in 2007

With regard to summary below, it appears as though the allowance for production to meet the basic domestic needs of Article 5 parties continues indefinitely after the date of the phase out (e.g for Article 5 parties in the case of Annex A substances; for both Article 5 and non-Article 5 parties in the case of Annex B group II and III substances). However, no party can consume controlled substances, except for permitted essential (or critical) uses, after the dates of phase-out for both Article 5 and non-Article 5 parties. Hence, no party can produce controlled substances after such dates, except for essential uses.

Annex A – Group I: CFCs

Non-Article 5 parties		Article 5 parties	
Baseline	Production in 1986	Baseline	Average of production for 1995–1997 or 0.3 kg per capita (whichever is lower)
July 1, 1989	10 per cent of baseline	July 1, 1999	10 per cent of baseline
January 1, 1996	15 per cent of baseline	January 1, 2010	15 per cent of baseline
New baseline for basic domestic needs (effective July 28, 2000)	Annual average production for satisfying basic domestic needs of Article 5 parties for the period 1995–1997 inclusive		
July 28, 2000	100 per cent of new baseline for satisfying basic domestic needs until end of 2002		
January 1, 2003	80 per cent of new baseline		
January 1, 2005	50 per cent of new baseline		
January 1, 2007	15 per cent of new baseline		
January 1, 2010	Zero		

Annex A – Group II: Halons

Non-Article 5 parties		Article 5 parties	
Baseline	Production in 1986	Baseline	Average of production for 1995–1997 or 0.3 kg per capita (whichever is lower)
January 1, 1992	10 per cent of baseline	January 1, 2002	10 per cent of baseline
January 1, 1994	15 per cent of baseline	January 1, 2010	15 per cent of baseline
New baseline for basic domestic needs (effective July 28, 2000)	Annual average production for satisfying basic domestic needs of Article 5 parties for the period 1995–1997 inclusive		
January 1, 2002	100 per cent of new baseline		
January 1, 2005	50 per cent of new baseline		
January 1, 2010	Zero		

Annex B – Group I: Other fully halogenated CFCs

Non-Article 5 parties		Article 5 parties	
Baseline	Production in 1986	Baseline	Average of production for 1998–2000 or 0.2 kg per capita (whichever is lower)
January 1, 1993	10 per cent of baseline	January 1, 2003	10 per cent of baseline
January 1, 1996	15 per cent of baseline	January 1, 2010	15 per cent of baseline
New baseline for basic domestic needs (effective July 28, 2000)	Annual average production for basic domestic needs for the period 1998-2000		
January 1, 2003	80 per cent of the new baseline		
January 1, 2007	15 per cent of the new baseline		
January 1, 2010	Zero		

Annex B – Group II: Carbon tetrachloride

Non-Article 5 parties		Article 5 parties	
Baseline	Production in 1989	Baseline	Average of production for 1998–2000 or 0.2 kg per capita (whichever is lower)
January 1, 1995	10 per cent of baseline	January 1, 2005	10 per cent of baseline
January 1, 1996	15 per cent of baseline	January 1, 2010	15 per cent of baseline

Annex B – Group III: 1,1,1-trichloroethane (methyl chloroform)

Non-Article 5 parties		Article 5 parties	
Baseline	Production in 1989	Baseline	Average of production for 1998–2000 or 0.2 kg per capita (whichever is lower)
January 1, 1993	10 per cent of baseline	January 1, 2003	10 per cent of baseline
January 1, 1996	15 per cent of baseline	January 1, 2015	15 per cent of baseline

Annex C – Group I: HCFCs

Non-Article 5 parties		Article 5 parties
Baseline	Average of (1989 HCFC production + 2.8 per cent of 1989 CFC production) and (1989 HCFC consumption +2.8 per cent of 1989 CFC consumption)	No additional production allowed for basic domestic needs
January 1, 2004	15 per cent of baseline	
January 1, 2010	10 per cent of baseline	
January 1, 2020	Zero	

Annex E – Group I: Methyl bromide

Non-Article 5 parties		Article 5 parties	
Baseline	**Production in 1991**	**Baseline**	**Average of production for 1995–1998**
January 1, 1995	10 per cent of baseline until July 28, 2000	January 1, 2002	10 per cent of baseline
July 28, 2000	15 per cent of baseline until January 1, 2002	January 1, 2005	Zero
New baseline for basic domestic needs (effective July 28, 2000)	Annual average production for basic domestic needs of Article 5 parties for the period 1995 to 1998 inclusive		
January 1, 2002	100 per cent of baseline		
January 1, 2005	80 per cent of new baseline		
January 1, 2015	Zero		

Appendix II

OzonAction Kigali Fact Sheet 6

Next Steps: HFC Phase-Down Strategy

Background:

An important aspect of Kigali Amendment implementation is for each country to consider its optimum phase-down strategy. Numerous questions need to be considered to develop a good strategy. These include:

- How is the current consumption of HFCs[1] and HCFCs split between market sectors?
- How might these markets develop in a "business-as-usual" scenario, taking into account factors such plans to phase-out HCFCs and economic growth?
- What type of actions can be taken to reduce future HFC consumption?
- Which fluids and technologies need to be adopted to achieve the phase-down targets?
- Which market sectors have the greatest potential for cost-effective actions?
- Which industry stakeholders need to make a contribution to strategy development?
- What support do industry stakeholders require for implementation (e.g. improved training)?
- What is the best way of prioritising different actions over the coming years?

This Fact Sheet summarises actions that can be taken to develop a national HFC phase-down strategy.

Action 1: Understanding Consumption:

The first action is to understand the current requirements for HCFCs and HFCs and to build a picture of how consumption might change over the next 10 years if there was no Kigali Amendment i.e. a business-as-usual (BAU) forecast. To do this you need to build a national HCFC and HFC consumption model. The more detail that you can make available through such a model, the easier it is to analyse an appropriate phase-down strategy. A national HCFC and HFC consumption model needs to be based on two distinct types of data:

[1]See Kigali Fact Sheet 14 for a glossary of all acronyms used.

1. **Top-down data on the bulk consumption of HCFCs and HFCs.** For HCFCs this is a simple process – the relevant data is already reported to the Ozone Secretariat on an annual basis. Data should be available for each individual HCFC over a period of many years. Similar data will need to be collected and reported for HFCs under the Kigali Amendment – although at this stage many A5 countries may have little historic top-down data on HFCs.
2. **Bottom-up data on key market sectors and subsectors.** Top-down data is useful, but it does not help understanding in detail the way that consumption is split between different market sectors and sub-sectors. Kigali Fact Sheet 2 provides an overview of the complex mixture of market sectors that make use of HCFCs and HFCs. It shows that the technical options for using low GWP alternatives vary significantly between different sub-sectors of the market. A bottom-up analysis requires the market to be split into appropriate sub-sectors. Each sub-sector is modelled based on the type of equipment used and the typical lifecycle of individual products. See Box 1 for an example.

Box 1: Bottom-up model of car air-conditioning

In 2015 a fictional A5 country had 0.5 million cars with air-conditioning. The market has grown rapidly, from 0.2 million in 2005 and is expected to reach 1 million by 2025. Each car can be represented by a "standard" car air-conditioning system:

- Refrigerant used: 0.7 kg HFC-134a
- Average annual leakage rate: 8%
- Average car life: 10 years

From this information, a bottom-up model can be created to show information such as:

a) The total bank of HFCs in all cars
b) Annual amount of HFCs for new cars
c) Annual amount of HFCs for servicing
d) HFC emissions in use and at end-of-life

Data from bottom-up modelling can be "calibrated" against top-down data to confirm that the input assumptions (e.g. annual leakage rates) are reasonable. When a model of this type has been created, it can act as a powerful tool to support analysis of future options.

Action 2: Understanding "Core Actions":

There are a number of different ways in which future HFC consumption can be reduced. These can be treated as a set of core actions that can be considered for each market sub-sector.

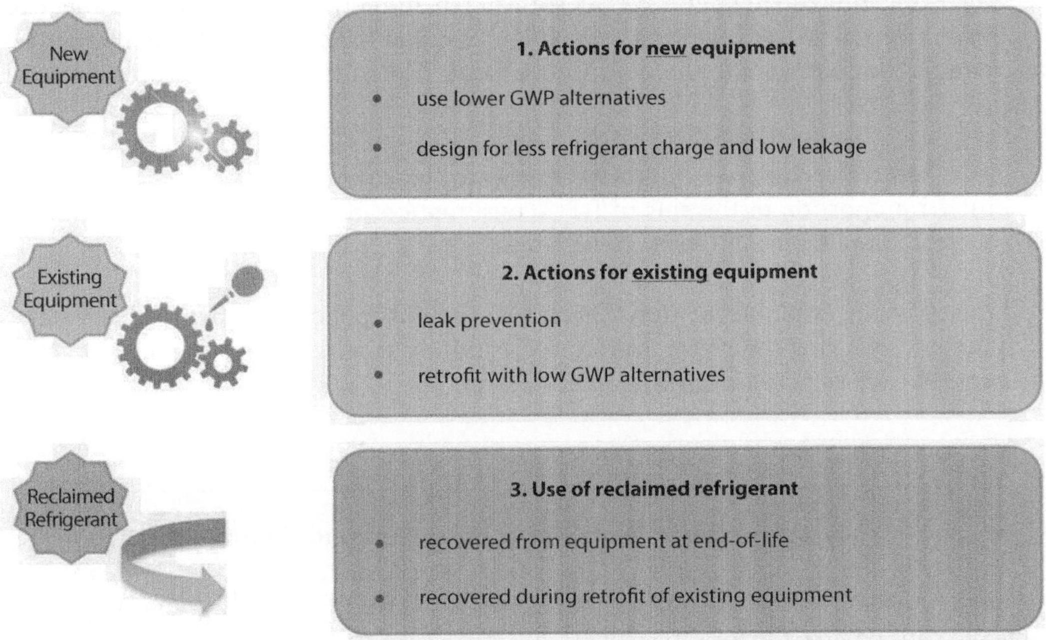

The most important long-term core action is to use low GWP alternatives to HFCs in all new equipment. For example, if new supermarket refrigeration uses an ultra-low GWP refrigerant such as CO_2 (GWP=1) there are big reductions in HFC use compared to using the popular HFC, R-404A (GWP=3 922).

However, other core actions such as leak prevention measures are also worth considering as they may be the most cost-effective ways of reducing HFC usage, especially in the early years of the phase-down process. The most appropriate core actions vary across different parts of the HFC market. For example:

- In market sectors with high leakage rates and long equipment life (e.g. industrial refrigeration or large central supermarket refrigeration) it is important to consider the actions that apply to existing equipment e.g. leak prevention or equipment retrofit.
- In market sectors that use small sealed equipment (e.g. domestic refrigerators, stand-alone retail refrigerators) it is not practical or cost-effective to modify existing equipment. For these sectors the priority is to introduce low GWP refrigerants in new equipment as soon as possible.

Action 3: Development of future consumption scenarios

The national HCFC and HFC consumption model can be used to investigate future scenarios that forecast the national requirements for HFCs. The modelling assumptions for each market sub-sector can be adjusted to predict future HFC demand from that sector. For example, if the car air-conditioning market switches from HFC-134a to an ultra-low GWP alternative, the future demand for HFC-134a will fall, initially in new equipment and, over a period of time, in the car maintenance market. The total HFC forecasts for all market sectors can then be compared to the steps in the Kigali HFC phase-down schedule. For each scenario this shows:

a) Whether the Kigali Amendment targets are being met
b) The proportion of the cuts in HFC consumption being delivered by each market sector and by the different core actions.

This is very powerful information that enables national policy makers to identify different routes to achieving the phase-down targets and to assess which are the most practical and cost-effective ways of prioritising future action.

Box 2: Scenario Modelling for Car Air-conditioning

Using data such as that shown in Box 1, a national HCFC and HFC consumption model can indicate the annual demand for HFC-134a in car air-conditioning, both for the gas required to fill air-conditioning in new vehicles and for the gas required to top-up leaks from all existing cars. The model can then be used to forecast future consumption using different scenarios. In this example, three scenarios have been defined and used to calculate HFC demand. These scenarios are:

Business-as-usual: all new cars continue to use HFC-134a

Phase-down Scenario 1: over a 7-year period from 2026, car air-conditioning in new cars switches from HFC-134a to an ultra-low GWP non-HFC alternative (e.g. HFO-1234yf). Scenario 1 can be considered as a "conservative" phase-down strategy, with a late starting date and a slow switch away from HFC-134a in new cars.

Phase-down Scenario 2: over a 5-year period from 2020, car air-conditioning in new cars switches from HFC-134a to a non-HFC alternative. Scenario 2 is a more aggressive scenario, with an earlier start and a faster switch to a low GWP alternative.

The first graph below illustrates the switch away from HFC-134a in new cars for each of the three scenarios. Based on the scenario assumptions, the second graph shows the results from the model: the annual HFC demand from this market sector, expressed in tonnes CO_2e (see Kigali Fact Sheet 3 for information about GWP and tonnes CCbel).

The benefits of an early switch to a low GWP alternative are clear from the second graph. In the period to 2040, Scenario 2 saves around 3 million tonnes CO_2e whereas Scenario 1 only saves 1.8 million tonnes CO_2e.

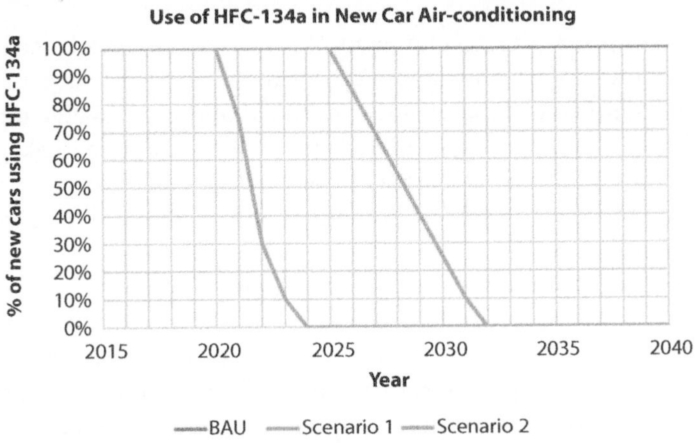

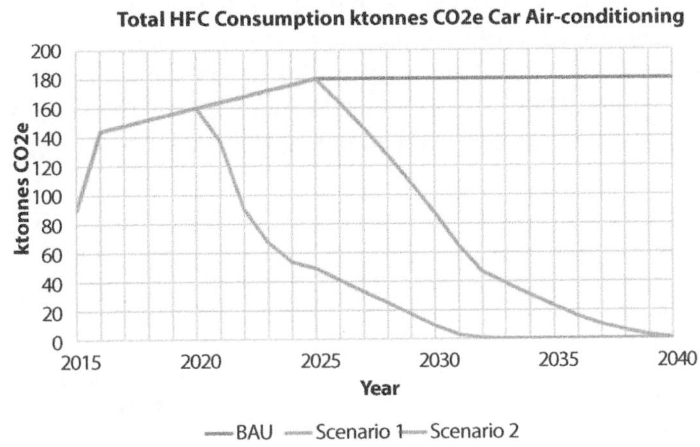

It is interesting to note how the consumption of HFC-134a continues for 10 years after the switch away from HFC-134a in new cars - this is because there are older cars in the fleet that continue to need to be maintained using HFC-134a because of leakage.

Scenario modelling provides very powerful insights into the usage of HFCs and alternatives. Individual assumptions can be made for each market sector or sub-sector, e.g. in terms of the date at which lower GWP gases are introduced. All the different core actions can be modelled, e.g. leak reduction or retrofit of existing equipment as well as switching to low GWP alternatives in new equipment. A range of different scenarios can be tested, helping to identify the actions that will have the greatest impact. The UN OzonAction team can provide National Ozone Units with further guidance on scenario modelling.

Action 4: Consideration of all technical options and timelines

Closely linked to Action 3 (the development of future consumption scenarios) it is necessary to consider the various technical options available in each market sector. Kigali Fact Sheet 4 provides a summary of many of the technical options available for new equipment. Other core actions must also be considered in some sectors (e.g. leakage prevention for industrial refrigeration and supermarket refrigeration). National circumstances must be considered e.g. will the new low GWP alternatives be available and does the workforce have the skills to use them.

The selection of an appropriate timeline is an important consideration as it is likely to influence environmental impact and cost. The analysis in Box 2 shows, in a fictional example, the benefits of an early start to the phase-down process. However, you also need to consider the potential cost impact of starting too early. The adjacent diagram illustrates the typical maturity curve for products and equipment. Early adopters pay more and may not get optimum performance. As a product becomes more mature it usually gets cheaper and provides better performance. Most of the early adopter costs are being absorbed in non-A5 countries. By the early 2020s there will be many products using low GWP alternatives that have reached maturity in terms of low cost and high performance. The car air-conditioning market is a good example. Regulations in the European Union mean that since January 2017 all new cars in the EU must use a low GWP alternative. Costs are still high, but over the next three to five years they are likely to fall rapidly.

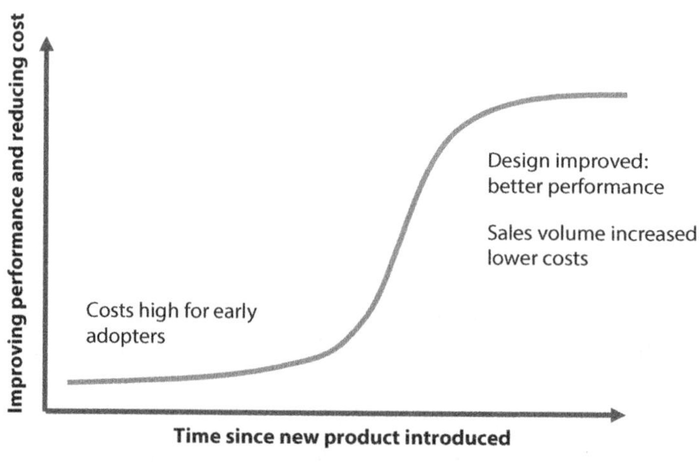

It is worth noting that a "late start" to a phase-down action could create extra costs over an "optimum start". Most innovation by manufacturers will be directed towards products and equipment that use low GWP alternatives. Improvements such as higher energy efficiency will be made to these products, while the older products using high GWP fluids might "stagnate" and create higher running costs and worse environmental impact.

Action 5: Stakeholder input and sort

National Ozone Units should carefully consider their engagement with stakeholders. This is a 2-way process:

- **Support from stakeholders:** some stakeholders can make a big contribution to the strategy development by providing insights into the current markets and opinions on the best technical options and timelines
- **Support to stakeholders:** some stakeholders need lots of support to help them understand the Kigali Amendment and the new products and technologies. There may be a need to support training and to assist in the setup of new infrastructure (e.g. national capability to recover and recycle refrigerants from old equipment reaching end-of-life).

Kigali Fact Sheet 8 provides further details about stakeholder engagement.

Action 6: Strategy reviews

The strategy needs regular review and updating. The availability of low GWP alternatives is changing rapidly. New alternatives are being introduced at a rapid rate by chemical producers and new products using these fluids are being commercialised by equipment manufacturers. The geographical availability of low GWP fluids and equipment is also changing rapidly. Currently the focus is to supply those countries with the most challenging HFC phase-down schedules – in particular, the European Union. However, with the Kigali Amendment in place this is likely to change rapidly and Article 5 countries can expect that access to low GWP technologies will quickly improve. The national strategy should be re-evaluated on an annual basis to assess whether any changes are required.

Appendix III

PARIS AGREEMENT

PARIS AGREEMENT
(Effective November 4, 2016)

The Parties to this Agreement,

Being Parties to the United Nations Framework Convention on Climate Change, hereinafter referred to as "the Convention",

Pursuant to the Durban Platform for Enhanced Action established by decision 1/CP.17 of the Conference of the Parties to the Convention at its seventeenth session,

In pursuit of the objective of the Convention, and being guided by its principles, including the principle of equity and common but differentiated responsibilities and respective capabilities, in the light of different national circumstances,

Recognizing the need for an effective and progressive response to the urgent threat of climate change on the basis of the best available scientific knowledge,

Also recognizing the specific needs and special circumstances of developing country Parties, especially those that are particularly vulnerable to the adverse effects of climate change, as provided for in the Convention,

Taking full account of the specific needs and special situations of the least developed countries with regard to funding and transfer of technology,

Recognizing that Parties may be affected not only by climate change, but also by the impacts of the measures taken in response to it,

Emphasizing the intrinsic relationship that climate change actions, responses and impacts have with equitable access to sustainable development and eradication of poverty,

Recognizing the fundamental priority of safeguarding food security and ending hunger, and the particular vulnerabilities of food production systems to the adverse impacts of climate change,

Taking into account the imperatives of a just transition of the workforce and the creation of decent work and quality jobs in accordance with nationally defined development priorities,

Acknowledging that climate change is a common concern of humankind, Parties should, when taking action to address climate change, respect, promote and consider their respective obligations on human rights, the right to health, the rights of indigenous peoples, local communities, migrants, children, persons with disabilities and people in vulnerable situations and the right to development, as well as gender equality, empowerment of women and intergenerational equity,

Recognizing the importance of the conservation and enhancement, as appropriate, of sinks and reservoirs of the greenhouse gases referred to in the Convention,

Noting the importance of ensuring the integrity of all ecosystems, including oceans, and the protection of biodiversity, recognized by some cultures as Mother Earth, and noting the importance for some of the concept of "climate justice", when taking action to address climate change,

Affirming the importance of education, training, public awareness, public participation, public access to information and cooperation at all levels on the matters addressed in this Agreement,

Recognizing the importance of the engagements of all levels of government and various actors, in accordance with respective national legislations of Parties, in addressing climate change,

Also recognizing that sustainable lifestyles and sustainable patterns of consumption and production, with developed country Parties taking the lead, play an important role in addressing climate change,

Have agreed as follows:

Article 1

For the purpose of this Agreement, the definitions contained in Article 1 of the Convention shall apply. In addition:

(a) "Convention" means the United Nations Framework Convention on Climate Change, adopted in New York on 9 May 1992;

(b) "Conference of the Parties" means the Conference of the Parties to the Convention;

(c) "Party" means a Party to this Agreement.

Article 2

1. This Agreement, in enhancing the implementation of the Convention, including its objective, aims to strengthen the global response to the threat of climate change, in the context of sustainable development and efforts to eradicate poverty, including by:

(a) Holding the increase in the global average temperature to well below 2°C above pre-industrial levels and pursuing efforts to limit the temperature increase to 1.5°C above pre-industrial levels, recognizing that this would significantly reduce the risks and impacts of climate change;

(b) Increasing the ability to adapt to the adverse impacts of climate change and foster climate resilience and low greenhouse gas emissions development, in a manner that does not threaten food production; and

(c) Making finance flows consistent with a pathway towards low greenhouse gas emissions and climate-resilient development.

2. This Agreement will be implemented to reflect equity and the principle of common but differentiated responsibilities and respective capabilities, in the light of different national circumstances.

Article 3

As nationally determined contributions to the global response to climate change, all Parties are to undertake and communicate ambitious efforts as defined in Articles 4, 7, 9, 10, 11 and 13 with the view to achieving the purpose of this Agreement as set out in Article 2. The efforts of all Parties will represent a progression over time, while recognizing the need to support developing country Parties for the effective implementation of this Agreement.

Article 4

1. In order to achieve the long-term temperature goal set out in Article 2, Parties aim to reach global peaking of greenhouse gas emissions as soon as possible, recognizing that peaking will take longer for developing country Parties, and to undertake rapid reductions thereafter in accordance with best available science, so as to achieve a balance between anthropogenic emissions by sources and removals by sinks of greenhouse gases in the second half of this century, on the basis of equity, and in the context of sustainable development and efforts to eradicate poverty.

2. Each Party shall prepare, communicate and maintain successive nationally determined contributions that it intends to achieve. Parties shall pursue domestic mitigation measures, with the aim of achieving the objectives of such contributions.

3. Each Party's successive nationally determined contribution will represent a progression beyond the Party's then current nationally determined contribution and reflect its highest possible ambition, reflecting its common but differentiated responsibilities and respective capabilities, in the light of different national circumstances.

4. Developed country Parties should continue taking the lead by undertaking economy-wide absolute emission reduction targets. Developing country Parties should continue enhancing their mitigation efforts, and are encouraged to move over time towards economy-wide emission reduction or limitation targets in the light of different national circumstances.

5. Support shall be provided to developing country Parties for the implementation of this Article, in accordance with Articles 9, 10 and 11, recognizing that enhanced support for developing country Parties will allow for higher ambition in their actions.

6. The least developed countries and small island developing States may prepare and communicate strategies, plans and actions for low greenhouse gas emissions development reflecting their special circumstances.

7. Mitigation co-benefits resulting from Parties' adaptation actions and/or economic diversification plans can contribute to mitigation outcomes under this Article.

8. In communicating their nationally determined contributions, all Parties shall provide the information necessary for clarity, transparency and understanding in accordance with decision 1/CP.21 and any relevant decisions of the Conference of the Parties serving as the meeting of the Parties to this Agreement.

9. Each Party shall communicate a nationally determined contribution every five years in accordance with decision 1/CP.21 and any relevant decisions of the Conference of the Parties serving as the meeting of the Parties to this Agreement and be informed by the outcomes of the global stocktake referred to in Article 14.

10. The Conference of the Parties serving as the meeting of the Parties to this Agreement shall consider common time frames for nationally determined contributions at its first session.

11. A Party may at any time adjust its existing nationally determined contribution with a view to enhancing its level of ambition, in accordance with guidance adopted by the Conference of the Parties serving as the meeting of the Parties to this Agreement.

12. Nationally determined contributions communicated by Parties shall be recorded in a public registry maintained by the secretariat.

13. Parties shall account for their nationally determined contributions. In accounting for anthropogenic emissions and removals corresponding to their nationally determined contributions, Parties shall promote environmental integrity, transparency, accuracy, completeness, comparability and consistency, and ensure the avoidance of double counting, in accordance with guidance adopted by the Conference of the Parties serving as the meeting of the Parties to this Agreement.

14. In the context of their nationally determined contributions, when recognizing and implementing mitigation actions with respect to anthropogenic emissions and removals, Parties should take into account, as appropriate, existing methods and guidance under the Convention, in the light of the provisions of paragraph 13 of this Article.

15. Parties shall take into consideration in the implementation of this Agreement the concerns of Parties with economies most affected by the impacts of response measures, particularly developing country Parties 16.

Parties, including regional economic integration organizations and their member States, that have reached an agreement to act jointly under paragraph 2 of this Article shall notify the secretariat of the terms of that agreement, including the emission level allocated to each Party within the relevant time period, when they communicate their nationally determined contributions. The secretariat shall in turn inform the Parties and signatories to the Convention of the terms of that agreement.

17. Each party to such an agreement shall be responsible for its emission level as set out in the agreement referred to in paragraph 16 of this Article in accordance with paragraphs 13 and 14 of this Article and Articles 13 and 15.

18. If Parties acting jointly do so in the framework of, and together with, a regional economic integration organization which is itself a Party to this Agreement, each member State of that regional economic integration organization individually, and together with the regional economic integration organization, shall be responsible for its emission level as set out in the agreement communicated under paragraph 16 of this Article in accordance with paragraphs 13 and 14 of this Article and Articles 13 and 15.

19. All Parties should strive to formulate and communicate long-term low greenhouse gas emission development strategies, mindful of Article 2 taking into account their common but differentiated responsibilities and respective capabilities, in the light of different national circumstances.

Article 5

1. Parties should take action to conserve and enhance, as appropriate, sinks and reservoirs of greenhouse gases as referred to in Article 4, paragraph 1 (d), of the Convention, including forests.

2. Parties are encouraged to take action to implement and support, including through results-based payments, the existing framework as set out in related guidance and decisions already agreed under the Convention for: policy approaches and positive incentives for activities relating to reducing emissions from deforestation and forest degradation, and the role of conservation, sustainable management of forests and enhancement of forest carbon stocks in developing countries; and alternative policy approaches, such as joint mitigation and adaptation approaches for the integral and sustainable management of forests, while reaffirming the importance of incentivizing, as appropriate, non-carbon benefits associated with such approaches.

Article 6

1. Parties recognize that some Parties choose to pursue voluntary cooperation in the implementation of their nationally determined contributions to allow for higher ambition in their mitigation and adaptation actions and to promote sustainable development and environmental integrity.

2. Parties shall, where engaging on a voluntary basis in cooperative approaches that involve the use of internationally transferred mitigation outcomes towards nationally determined contributions, promote sustainable development and ensure environmental integrity and transparency, including in governance, and shall apply robust accounting to ensure, inter alia, the avoidance of double counting, consistent with guidance adopted by the Conference of the Parties serving as the meeting of the Parties to this Agreement.

3. The use of internationally transferred mitigation outcomes to achieve nationally determined contributions under this Agreement shall be voluntary and authorized by participating Parties.

4. A mechanism to contribute to the mitigation of greenhouse gas emissions and support sustainable development is hereby established under the authority and guidance of the Conference of the Parties serving as the meeting of the Parties to this Agreement for use by Parties on a voluntary basis. It shall be supervised by a body designated by the Conference of the Parties serving as the meeting of the Parties to this Agreement, and shall aim:

(a) To promote the mitigation of greenhouse gas emissions while fostering sustainable development;

(b) To incentivize and facilitate participation in the mitigation of greenhouse gas emissions by public and private entities authorized by a Party;

(c) To contribute to the reduction of emission levels in the host Party, which will benefit from mitigation activities resulting in emission reductions that can also be used by another Party to fulfil its nationally determined contribution; and

(d) To deliver an overall mitigation in global emissions.

5. Emission reductions resulting from the mechanism referred to in paragraph 4 of this Article shall not be used to demonstrate achievement of the host Party's nationally determined contribution if used by another Party to demonstrate achievement of its nationally determined contribution.

6. The Conference of the Parties serving as the meeting of the Parties to this Agreement shall ensure that a share of the proceeds from activities under the mechanism referred to in paragraph 4 of this Article is used to cover administrative expenses as well as to assist developing country Parties that are particularly vulnerable to the adverse effects of climate change to meet the costs of adaptation.

7. The Conference of the Parties serving as the meeting of the Parties to this Agreement shall adopt rules, modalities and procedures for the mechanism referred to in paragraph 4 of this Article at its first session.

8. Parties recognize the importance of integrated, holistic and balanced non-market approaches being available to Parties to assist in the implementation of their nationally

determined contributions, in the context of sustainable development and poverty eradication, in a coordinated and effective manner, including through, inter alia, mitigation, adaptation, finance, technology transfer and capacity-building, as appropriate. These approaches shall aim to:

(a) Promote mitigation and adaptation ambition;

(b) Enhance public and private sector participation in the implementation of nationally determined contributions; and

(c) Enable opportunities for coordination across instruments and relevant institutional arrangements.

9. A framework for non-market approaches to sustainable development is hereby defined to promote the non-market approaches referred to in paragraph 8 of this Article.

Article 7

1. Parties hereby establish the global goal on adaptation of enhancing adaptive capacity, strengthening resilience and reducing vulnerability to climate change, with a view to contributing to sustainable development and ensuring an adequate adaptation response in the context of the temperature goal referred to in Article 2.

2. Parties recognize that adaptation is a global challenge faced by all with local, subnational, national, regional and international dimensions, and that it is a key component of and makes a contribution to the long-term global response to climate change to protect people, livelihoods and ecosystems, taking into account the urgent and immediate needs of those developing country Parties that are particularly vulnerable to the adverse effects of climate change.

3. The adaptation efforts of developing country Parties shall be recognized, in accordance with the modalities to be adopted by the Conference of the Parties serving as the meeting of the Parties to this Agreement at its first session.

4. Parties recognize that the current need for adaptation is significant and that greater levels of mitigation can reduce the need for additional adaptation efforts, and that greater adaptation needs can involve greater adaptation costs.

5. Parties acknowledge that adaptation action should follow a country-driven, gender-responsive, participatory and fully transparent approach, taking into consideration vulnerable groups, communities and ecosystems, and should be based on and guided by the best available science and, as appropriate, traditional knowledge, knowledge of indigenous peoples and local knowledge systems, with a view to integrating adaptation into relevant socioeconomic and environmental policies and actions, where appropriate.

6. Parties recognize the importance of support for and international cooperation on adaptation efforts and the importance of taking into account the needs of developing country Parties, especially those that are particularly vulnerable to the adverse effects of climate change.

7. Parties should strengthen their cooperation on enhancing action on adaptation, taking into account the Cancun Adaptation Framework, including with regard to:

(a) Sharing information, good practices, experiences and lessons learned, including, as appropriate, as these relate to science, planning, policies and implementation in relation to adaptation actions;

(b) Strengthening institutional arrangements, including those under the Convention that serve this Agreement, to support the synthesis of relevant information and knowledge, and the provision of technical support and guidance to Parties;

(c) Strengthening scientific knowledge on climate, including research, systematic observation of the climate system and early warning systems, in a manner that informs climate services and supports decision-making;

(d) Assisting developing country Parties in identifying effective adaptation practices, adaptation needs, priorities, support provided and received for adaptation actions and efforts, and challenges and gaps, in a manner consistent with encouraging good practices; and

(e) Improving the effectiveness and durability of adaptation actions.

8. United Nations specialized organizations and agencies are encouraged to support the efforts of Parties to implement the actions referred to in paragraph 7 of this Article, taking into account the provisions of paragraph 5 of this Article.

9. Each Party shall, as appropriate, engage in adaptation planning processes and the implementation of actions, including the development or enhancement of relevant plans, policies and/or contributions, which may include:

(a) The implementation of adaptation actions, undertakings and/or efforts;

(b) The process to formulate and implement national adaptation plans;

(c) The assessment of climate change impacts and vulnerability, with a view to formulating nationally determined prioritized actions, taking into account vulnerable people, places and ecosystems;

(d) Monitoring and evaluating and learning from adaptation plans, policies, programmes and actions; and

(e) Building the resilience of socioeconomic and ecological systems, including through economic diversification and sustainable management of natural resources.

10. Each Party should, as appropriate, submit and update periodically an adaptation communication, which may include its priorities, implementation and support needs, plans and actions, without creating any additional burden for developing country Parties.

11. The adaptation communication referred to in paragraph 10 of this Article shall be, as appropriate, submitted and updated periodically, as a component of or in conjunction with other communications or documents, including a national adaptation plan, a nationally determined contribution as referred to in Article 4, paragraph 2, and/or a national communication.

12. The adaptation communications referred to in paragraph 10 of this Article shall be recorded in a public registry maintained by the secretariat.

13. Continuous and enhanced international support shall be provided to developing country Parties for the implementation of paragraphs 7, 9, 10 and 11 of this Article, in accordance with the provisions of Articles 9, 10 and 11.

14. The global stocktake referred to in Article 14 shall, inter alia:

(a) Recognize adaptation efforts of developing country Parties;

(b) Enhance the implementation of adaptation action taking into account the adaptation communication referred to in paragraph 10 of this Article;

(c) Review the adequacy and effectiveness of adaptation and support provided for adaptation; and

(d) Review the overall progress made in achieving the global goal on adaptation referred to in paragraph 1 of this Article.

Article 8

1. Parties recognize the importance of averting, minimizing and addressing loss and damage associated with the adverse effects of climate change, including extreme weather events and slow onset events, and the role of sustainable development in reducing the risk of loss and damage.

2. The Warsaw International Mechanism for Loss and Damage associated with Climate Change Impacts shall be subject to the authority and guidance of the Conference of the Parties serving as the meeting of the Parties to this Agreement and may be enhanced and strengthened, as determined by the Conference of the Parties serving as the meeting of the Parties to this Agreement.

3. Parties should enhance understanding, action and support, including through the Warsaw International Mechanism, as appropriate, on a cooperative and facilitative basis with respect to loss and damage associated with the adverse effects of climate change.

4. Accordingly, areas of cooperation and facilitation to enhance understanding, action and support may include:

(a) Early warning systems;

(b) Emergency preparedness;

(c) Slow onset events;

(d) Events that may involve irreversible and permanent loss and damage;

(e) Comprehensive risk assessment and management;

(f) Risk insurance facilities, climate risk pooling and other insurance solutions;

(g) Non-economic losses; and

(h) Resilience of communities, livelihoods and ecosystems.

5. The Warsaw International Mechanism shall collaborate with existing bodies and expert groups under the Agreement, as well as relevant organizations and expert bodies outside the Agreement.

Article 9

1. Developed country Parties shall provide financial resources to assist developing country Parties with respect to both mitigation and adaptation in continuation of their existing obligations under the Convention.

2. Other Parties are encouraged to provide or continue to provide such support voluntarily.

3. As part of a global effort, developed country Parties should continue to take the lead in mobilizing climate finance from a wide variety of sources, instruments and channels, noting the significant role of public funds, through a variety of actions, including supporting country-driven strategies, and taking into account the needs and priorities of developing country Parties. Such mobilization of climate finance should represent a progression beyond previous efforts.

4. The provision of scaled-up financial resources should aim to achieve a balance between adaptation and mitigation, taking into account country-driven strategies, and the priorities and needs of developing country Parties, especially those that are particularly

vulnerable to the adverse effects of climate change and have significant capacity constraints, such as the least developed countries and small island developing States, considering the need for public and grant-based resources for adaptation.

5. Developed country Parties shall biennially communicate indicative quantitative and qualitative information related to paragraphs 1 and 3 of this Article, as applicable, including, as available, projected levels of public financial resources to be provided to developing country Parties. Other Parties providing resources are encouraged to communicate biennially such information on a voluntary basis.

6. The global stocktake referred to in Article 14 shall take into account the relevant information provided by developed country Parties and/or Agreement bodies on efforts related to climate finance.

7. Developed country Parties shall provide transparent and consistent information on support for developing country Parties provided and mobilized through public interventions biennially in accordance with the modalities, procedures and guidelines to be adopted by the Conference of the Parties serving as the meeting of the Parties to this Agreement, at its first session, as stipulated in Article 13, paragraph 13. Other Parties are encouraged to do so.

8. The Financial Mechanism of the Convention, including its operating entities, shall serve as the financial mechanism of this Agreement.

9. The institutions serving this Agreement, including the operating entities of the Financial Mechanism of the Convention, shall aim to ensure efficient access to financial resources through simplified approval procedures and enhanced readiness support for developing country Parties, in particular for the least developed countries and small island developing States, in the context of their national climate strategies and plans.

Article 10

1. Parties share a long-term vision on the importance of fully realizing technology development and transfer in order to improve resilience to climate change and to reduce greenhouse gas emissions.

2. Parties, noting the importance of technology for the implementation of mitigation and adaptation actions under this Agreement and recognizing existing technology deployment and dissemination efforts, shall strengthen cooperative action on technology development and transfer.

3. The Technology Mechanism established under the Convention shall serve this Agreement.

4. A technology framework is hereby established to provide overarching guidance to the work of the Technology Mechanism in promoting and facilitating enhanced action

on technology development and transfer in order to support the implementation of this Agreement, in pursuit of the long-term vision referred to in paragraph 1 of this Article.

5. Accelerating, encouraging and enabling innovation is critical for an effective, long-term global response to climate change and promoting economic growth and sustainable development. Such effort shall be, as appropriate, supported, including by the Technology Mechanism and, through financial means, by the Financial Mechanism of the Convention, for collaborative approaches to research and development, and facilitating access to technology, in particular for early stages of the technology cycle, to developing country Parties.

6. Support, including financial support, shall be provided to developing country Parties for the implementation of this Article, including for strengthening cooperative action on technology development and transfer at different stages of the technology cycle, with a view to achieving a balance between support for mitigation and adaptation. The global stocktake referred to in Article 14 shall take into account available information on efforts related to support on technology development and transfer for developing country Parties.

Article 11

1. Capacity-building under this Agreement should enhance the capacity and ability of developing country Parties, in particular countries with the least capacity, such as the least developed countries, and those that are particularly vulnerable to the adverse effects of climate change, such as small island developing States, to take effective climate change action, including, inter alia, to implement adaptation and mitigation actions, and should facilitate technology development, dissemination and deployment, access to climate finance, relevant aspects of education, training and public awareness, and the transparent, timely and accurate communication of information.

2. Capacity-building should be country-driven, based on and responsive to national needs, and foster country ownership of Parties, in particular, for developing country Parties, including at the national, subnational and local levels. Capacity-building should be guided by lessons learned, including those from capacity-building activities under the Convention, and should be an effective, iterative process that is participatory, cross-cutting and gender-responsive.

3. All Parties should cooperate to enhance the capacity of developing country Parties to implement this Agreement. Developed country Parties should enhance support for capacity-building actions in developing country Parties.

4. All Parties enhancing the capacity of developing country Parties to implement this Agreement, including through regional, bilateral and multilateral approaches, shall regularly communicate on these actions or measures on capacity-building. Developing country Parties should regularly communicate progress made on implementing capacity-building plans, policies, actions or measures to implement this Agreement.

5. Capacity-building activities shall be enhanced through appropriate institutional arrangements to support the implementation of this Agreement, including the appropriate institutional arrangements established under the Convention that serve this Agreement. The Conference of the Parties serving as the meeting of the Parties to this Agreement shall, at its first session, consider and adopt a decision on the initial institutional arrangements for capacity-building.

Article 12

Parties shall cooperate in taking measures, as appropriate, to enhance climate change education, training, public awareness, public participation and public access to information, recognizing the importance of these steps with respect to enhancing actions under this Agreement.

Article 13

1. In order to build mutual trust and confidence and to promote effective implementation, an enhanced transparency framework for action and support, with built-in flexibility which takes into account Parties' different capacities and builds upon collective experience is hereby established.

2. The transparency framework shall provide flexibility in the implementation of the provisions of this Article to those developing country Parties that need it in the light of their capacities. The modalities, procedures and guidelines referred to in paragraph 13 of this Article shall reflect such flexibility.

3. The transparency framework shall build on and enhance the transparency arrangements under the Convention, recognizing the special circumstances of the least developed countries and small island developing States, and be implemented in a facilitative, non-intrusive, non-punitive manner, respectful of national sovereignty, and avoid placing undue burden on Parties.

4. The transparency arrangements under the Convention, including national communications, biennial reports and biennial update reports, international assessment and review and international consultation and analysis, shall form part of the experience drawn upon for the development of the modalities, procedures and guidelines under paragraph 13 of this Article.

5. The purpose of the framework for transparency of action is to provide a clear understanding of climate change action in the light of the objective of the Convention as set out in its Article 2, including clarity and tracking of progress towards achieving Parties' individual nationally determined contributions under Article 4, and Parties' adaptation actions under Article 7, including good practices, priorities, needs and gaps, to inform the global stocktake under Article 14.

6. The purpose of the framework for transparency of support is to provide clarity on support provided and received by relevant individual Parties in the context of climate change actions under Articles 4, 7, 9, 10 and 11, and, to the extent possible, to provide a full overview of aggregate financial support provided, to inform the global stocktake under Article 14.

7. Each Party shall regularly provide the following information:

(a) A national inventory report of anthropogenic emissions by sources and removals by sinks of greenhouse gases, prepared using good practice methodologies accepted by the Intergovernmental Panel on Climate Change and agreed upon by the Conference of the Parties serving as the meeting of the Parties to this Agreement; and

(b) Information necessary to track progress made in implementing and achieving its nationally determined contribution under Article 4.

8. Each Party should also provide information related to climate change impacts and adaptation under Article 7, as appropriate.

9. Developed country Parties shall, and other Parties that provide support should, provide information on financial, technology transfer and capacity-building support provided to developing country Parties under Articles 9, 10 and 11.

10. Developing country Parties should provide information on financial, technology transfer and capacity-building support needed and received under Articles 9, 10 and 11.

11. Information submitted by each Party under paragraphs 7 and 9 of this Article shall undergo a technical expert review, in accordance with decision 1/CP.21. For those developing country Parties that need it in the light of their capacities, the review process shall include assistance in identifying capacity-building needs. In addition, each Party shall participate in a facilitative, multilateral consideration of progress with respect to efforts under Article 9, and its respective implementation and achievement of its nationally determined contribution.

12. The technical expert review under this paragraph shall consist of a consideration of the Party's support provided, as relevant, and its implementation and achievement of its nationally determined contribution. The review shall also identify areas of improvement for the Party, and include a review of the consistency of the information with the modalities, procedures and guidelines referred to in paragraph 13 of this Article, taking into account the flexibility accorded to the Party under paragraph 2 of this Article. The review shall pay particular attention to the respective national capabilities and circumstances of developing country Parties.

13. The Conference of the Parties serving as the meeting of the Parties to this Agreement shall, at its first session, building on experience from the arrangements related

to transparency under the Convention, and elaborating on the provisions in this Article, adopt common modalities, procedures and guidelines, as appropriate, for the transparency of action and support.

14. Support shall be provided to developing countries for the implementation of this Article.

15. Support shall also be provided for the building of transparency-related capacity of developing country Parties on a continuous basis.

Article 14

1. The Conference of the Parties serving as the meeting of the Parties to this Agreement shall periodically take stock of the implementation of this Agreement to assess the collective progress towards achieving the purpose of this Agreement and its long-term goals (referred to as the "global stocktake"). It shall do so in a comprehensive and facilitative manner, considering mitigation, adaptation and the means of implementation and support, and in the light of equity and the best available science.

2. The Conference of the Parties serving as the meeting of the Parties to this Agreement shall undertake its first global stocktake in 2023 and every five years thereafter unless otherwise decided by the Conference of the Parties serving as the meeting of the Parties to this Agreement.

3. The outcome of the global stocktake shall inform Parties in updating and enhancing, in a nationally determined manner, their actions and support in accordance with the relevant provisions of this Agreement, as well as in enhancing international cooperation for climate action.

Article 15

1. A mechanism to facilitate implementation of and promote compliance with the provisions of this Agreement is hereby established.

2. The mechanism referred to in paragraph 1 of this Article shall consist of a committee that shall be expert-based and facilitative in nature and function in a manner that is transparent, non-adversarial and non-punitive. The committee shall pay particular attention to the respective national capabilities and circumstances of Parties.

3. The committee shall operate under the modalities and procedures adopted by the Conference of the Parties serving as the meeting of the Parties to this Agreement at its first session and report annually to the Conference of the Parties serving as the meeting of the Parties to this Agreement.

Article 16

1. The Conference of the Parties, the supreme body of the Convention, shall serve as the meeting of the Parties to this Agreement.

2. Parties to the Convention that are not Parties to this Agreement may participate as observers in the proceedings of any session of the Conference of the Parties serving as the meeting of the Parties to this Agreement. When the Conference of the Parties serves as the meeting of the Parties to this Agreement, decisions under this Agreement shall be taken only by those that are Parties to this Agreement.

3. When the Conference of the Parties serves as the meeting of the Parties to this Agreement, any member of the Bureau of the Conference of the Parties representing a Party to the Convention but, at that time, not a Party to this Agreement, shall be replaced by an additional member to be elected by and from amongst the Parties to this Agreement.

4. The Conference of the Parties serving as the meeting of the Parties to this Agreement shall keep under regular review the implementation of this Agreement and shall make, within its mandate, the decisions necessary to promote its effective implementation. It shall perform the functions assigned to it by this Agreement and shall:

(a) Establish such subsidiary bodies as deemed necessary for the implementation of this Agreement; and

(b) Exercise such other functions as may be required for the implementation of this Agreement.

5. The rules of procedure of the Conference of the Parties and the financial procedures applied under the Convention shall be applied *mutatis mutandis* under this Agreement, except as may be otherwise decided by consensus by the Conference of the Parties serving as the meeting of the Parties to this Agreement.

6. The first session of the Conference of the Parties serving as the meeting of the Parties to this Agreement shall be convened by the secretariat in conjunction with the first session of the Conference of the Parties that is scheduled after the date of entry into force of this Agreement. Subsequent ordinary sessions of the Conference of the Parties serving as the meeting of the Parties to this Agreement shall be held in conjunction with ordinary sessions of the Conference of the Parties, unless otherwise decided by the Conference of the Parties serving as the meeting of the Parties to this Agreement.

7. Extraordinary sessions of the Conference of the Parties serving as the meeting of the Parties to this Agreement shall be held at such other times as may be deemed necessary by the Conference of the Parties serving as the meeting of the Parties to this Agreement or at the written request of any Party, provided that, within six months of the request being communicated to the Parties by the secretariat, it is supported by at least one third of the Parties.

8. The United Nations and its specialized agencies and the International Atomic Energy Agency, as well as any State member thereof or observers thereto not party to the Convention, may be represented at sessions of the Conference of the Parties serving as the meeting of the Parties to this Agreement as observers. Any body or agency, whether national or international, governmental or non-governmental, which is qualified in matters covered by this Agreement and which has informed the secretariat of its wish to be represented at a session of the Conference of the Parties serving as the meeting of the Parties to this Agreement as an observer, may be so admitted unless at least one third of the Parties present object. The admission and participation of observers shall be subject to the rules of procedure referred to in paragraph 5 of this Article.

Article 17

1. The secretariat established by Article 8 of the Convention shall serve as the secretariat of this Agreement.

2. Article 8, paragraph 2, of the Convention on the functions of the secretariat, and Article 8, paragraph 3, of the Convention, on the arrangements made for the functioning of the secretariat, shall apply *mutatis mutandis* to this Agreement. The secretariat shall, in addition, exercise the functions assigned to it under this Agreement and by the Conference of the Parties serving as the meeting of the Parties to this Agreement.

Article 18

1. The Subsidiary Body for Scientific and Technological Advice and the Subsidiary Body for Implementation established by Articles 9 and 10 of the Convention shall serve, respectively, as the Subsidiary Body for Scientific and Technological Advice and the Subsidiary Body for Implementation of this Agreement. The provisions of the Convention relating to the functioning of these two bodies shall apply *mutatis mutandis* to this Agreement. Sessions of the meetings of the Subsidiary Body for Scientific and Technological Advice and the Subsidiary Body for Implementation of this Agreement shall be held in conjunction with the meetings of, respectively, the Subsidiary Body for Scientific and Technological Advice and the Subsidiary Body for Implementation of the Convention.

2. Parties to the Convention that are not Parties to this Agreement may participate as observers in the proceedings of any session of the subsidiary bodies. When the subsidiary bodies serve as the subsidiary bodies of this Agreement, decisions under this Agreement shall be taken only by those that are Parties to this Agreement.

3. When the subsidiary bodies established by Articles 9 and 10 of the Convention exercise their functions with regard to matters concerning this Agreement, any member of the bureaux of those subsidiary bodies representing a Party to the Convention but, at that time, not a Party to this Agreement, shall be replaced by an additional member to be elected by and from amongst the Parties to this Agreement.

Article 19

1. Subsidiary bodies or other institutional arrangements established by or under the Convention, other than those referred to in this Agreement, shall serve this Agreement upon a decision of the Conference of the Parties serving as the meeting of the Parties to this Agreement. The Conference of the Parties serving as the meeting of the Parties to this Agreement shall specify the functions to be exercised by such subsidiary bodies or arrangements.

2. The Conference of the Parties serving as the meeting of the Parties to this Agreement may provide further guidance to such subsidiary bodies and institutional arrangements.

Article 20

1. This Agreement shall be open for signature and subject to ratification, acceptance or approval by States and regional economic integration organizations that are Parties to the Convention. It shall be open for signature at the United Nations Headquarters in New York from 22 April 2016 to 21 April 2017. Thereafter, this Agreement shall be open for accession from the day following the date on which it is closed for signature. Instruments of ratification, acceptance, approval or accession shall be deposited with the Depositary.

2. Any regional economic integration organization that becomes a Party to this Agreement without any of its member States being a Party shall be bound by all the obligations under this Agreement. In the case of regional economic integration organizations with one or more member States that are Parties to this Agreement, the organization and its member States shall decide on their respective responsibilities for the performance of their obligations under this Agreement. In such cases, the organization and the member States shall not be entitled to exercise rights under this Agreement concurrently.

3. In their instruments of ratification, acceptance, approval or accession, regional economic integration organizations shall declare the extent of their competence with respect to the matters governed by this Agreement. These organizations shall also inform the Depositary, who shall in turn inform the Parties, of any substantial modification in the extent of their competence.

Article 21

1. This Agreement shall enter into force on the thirtieth day after the date on which at least 55 Parties to the Convention accounting in total for at least an estimated 55 per cent of the total global greenhouse gas emissions have deposited their instruments of ratification, acceptance, approval or accession.

2. Solely for the limited purpose of paragraph 1 of this Article, "total global greenhouse gas emissions" means the most up-to-date amount communicated on or before the date of adoption of this Agreement by the Parties to the Convention.

3. For each State or regional economic integration organization that ratifies, accepts or approves this Agreement or accedes thereto after the conditions set out in paragraph 1 of this Article for entry into force have been fulfilled, this Agreement shall enter into force on the thirtieth day after the date of deposit by such State or regional economic integration organization of its instrument of ratification, acceptance, approval or accession.

4. For the purposes of paragraph 1 of this Article, any instrument deposited by a regional economic integration organization shall not be counted as additional to those deposited by its member States.

Article 22

The provisions of Article 15 of the Convention on the adoption of amendments to the Convention shall apply *mutatis mutandis* to this Agreement.

Article 23

1. The provisions of Article 16 of the Convention on the adoption and amendment of annexes to the Convention shall apply *mutatis mutandis* to this Agreement.

2. Annexes to this Agreement shall form an integral part thereof and, unless otherwise expressly provided for, a reference to this Agreement constitutes at the same time a reference to any annexes thereto. Such annexes shall be restricted to lists, forms and any other material of a descriptive nature that is of a scientific, technical, procedural or administrative character.

Article 24

The provisions of Article 14 of the Convention on settlement of disputes shall apply *mutatis mutandis* to this Agreement.

Article 25

1. Each Party shall have one vote, except as provided for in paragraph 2 of this Article.

2. Regional economic integration organizations, in matters within their competence, shall exercise their right to vote with a number of votes equal to the number of their member States that are Parties to this Agreement. Such an organization shall not exercise its right to vote if any of its member States exercises its right, and vice versa.

Article 26

The Secretary-General of the United Nations shall be the Depositary of this Agreement.

Article 27

No reservations may be made to this Agreement.

Article 28

1. At any time after three years from the date on which this Agreement has entered into force for a Party, that Party may withdraw from this Agreement by giving written notification to the Depositary.

2. Any such withdrawal shall take effect upon expiry of one year from the date of receipt by the Depositary of the notification of withdrawal, or on such later date as may be specified in the notification of withdrawal.

3. Any Party that withdraws from the Convention shall be considered as also having withdrawn from this Agreement.

Article 29

The original of this Agreement, of which the Arabic, Chinese, English, French, Russian and Spanish texts are equally authentic, shall be deposited with the Secretary-General of the United Nations.

DONE at Paris this twelfth day of December two thousand and fifteen.

IN WITNESS WHEREOF, the undersigned, being duly authorized to that effect, have signed this Agreement.

Index

Printed and bound by CPI Group (UK) Ltd, Croydon, CR0 4YY